Introductory
Matter
Physics

Introductory
Matter
Physics

Francesco Simoni

Università Politecnica delle Marche, Italy

World Scientific

NEW JERSEY · LONDON · SINGAPORE · BEIJING · SHANGHAI · HONG KONG · TAIPEI · CHENNAI · TOKYO

Published by

World Scientific Publishing Co. Pte. Ltd.

5 Toh Tuck Link, Singapore 596224

USA office: 27 Warren Street, Suite 401-402, Hackensack, NJ 07601

UK office: 57 Shelton Street, Covent Garden, London WC2H 9HE

Library of Congress Cataloging-in-Publication Data

Names: Simoni, Francesco, author.

Title: Introductory matter physics / Francesco Simoni, Università Politecnica delle Marche, Italy.

Description: Singapore ; Hackensack, NJ : World Scientific Publishing Co. Pte. Ltd., [2018]

Identifiers: LCCN 2018018464| ISBN 9789813235717 (hardcover ; alk. paper) |
    ISBN 9813235713 (hardcover ; alk. paper)

Subjects: LCSH: Matter--Properties.

Classification: LCC QC173.3 .S56 2018 | DDC 530.4--dc23

LC record available at https://lccn.loc.gov/2018018464

British Library Cataloguing-in-Publication Data

A catalogue record for this book is available from the British Library.

For any available supplementary material, please visit
http://www.worldscientific.com/worldscibooks/10.1142/10869#t=suppl

Typeset by Stallion Press
Email: enquiries@stallionpress.com

To Elena

and

To my grandchildren:

Nicolò
Arianna
Sveva
Emilia

Preface

This book introduces the reader to the basic concepts of matter physics, describing how the fundamental properties of atoms, molecules and condensed matter are affected by the properties of electrons and by their interaction with electromagnetic waves. It is based on lectures given for courses on matter physics and solid state physics over more than a decade.

While teaching undergraduate students in electrical engineering, I encountered the problem of presenting those different subjects in a way suitable for an audience with only a basic background in general physics and differential calculus, without losing mathematical accuracy. An additional problem had been providing the source of information to the students since generally the subjects can only be found spread among several different books, each of them devoted to a single topic. For this reason, I had to collect my lectures into notes that I made available to my students to pass the final examination.

After many years, I considered that this approach to the subject could be useful to a student community wider than my own, and therefore I decided to revise my lecture notes and write this book. It innovates by presenting, in a single volume, the basic concepts of wave optics, quantum mechanics, atomic physics, solid state physics and quantum electronics. In this way, the reader can find in a single source all the information necessary for understanding this diverse set of topics.

The book is organized into six chapters. The first chapter is a short review of the basic properties of electromagnetic waves, giving the basic concepts related to wave propagation to assist in easily understanding the subsequent topics. The next chapter, on quantum mechanics, helps to understand the quantum properties of matter using the simplest formalizations. Chapter 3 enters into the core of the book by using quantum mechanics to describe the electronic properties of the atom. Then, after an introduction to atomic bonding, molecules and condensed matter are discussed, before approaching the structural properties of crystals and soft matter. Chapters 4 and 5 are devoted to electrical and optical properties and address the main topics related to solid state and semiconductor physics, as well as light–matter interaction. Finally, Chapter 6 deals with the basic properties, of lasers, owing to the relevance of these light sources in everyday life, and their widespread use in all branches of engineering.

Despite the great effort spent to present these topics with the necessary accuracy without giving unnecessary details, I may have failed to achieve this goal in some pages of the book; therefore I would appreciate receiving from colleagues any comments and advice that may help me in improving the material in the future.

I have to thank many people who helped me in completing this work. First of all, my friends and colleagues who revised parts of this book, suggesting changes and improvements: first, my very good friend Cesare Umeton for his extremely careful reading of chapter 6, then Gabriella Cipparrone, Luigino Criante, Daniele Lucchetta, Liana Lucchetti, Francesco Vita, Riccardo Castagna; Jack Dyson and Paolo Spegni who gave me considerable help in getting the figures suitable for printing. I also appreciate the contribution that Oriano Francescangeli made through the discussions we had during several years on specific issues related to basic concepts addressed in this book. I acknowledge the support of the whole staff at World Scientific headquarters whom I had the opportunity to meet in Singapore; among them I like to mention Lakshmi Narayanan for assisting me in the final stage of the publication process.

Finally I want to thank my wife Elena for supporting me and all my family for providing me the good feeling to carry out this work, with special thanks to my daughter-in-law Lauren for revising the presentation of this book.

Contents

Chapter 1

Wave Optics

1.1. Wave propagation in isotropic media

The Maxwell equations describe the properties of electric and magnetic fields by connecting the vector quantities $\mathbf{E}$ (electric field), $\mathbf{H}$ (magnetic field), $\mathbf{B}$ (magnetic induction), $\mathbf{D}$ (electric displacement) and $\mathbf{J}$ (electric current density) and the density of free charges ρ. In MKSA units they are

$$\nabla \cdot \mathbf{D} = \rho, \tag{1.1}$$

$$\nabla \cdot \mathbf{B} = 0, \tag{1.2}$$

$$\nabla \times \mathbf{E} = -\frac{\partial \mathbf{B}}{\partial t}, \tag{1.3}$$

$$\nabla \times \mathbf{H} = \mathbf{J} + \frac{\partial \mathbf{D}}{\partial t}. \tag{1.4}$$

Since $\mathbf{B}$, $\mathbf{D}$ and $\mathbf{J}$ take into account the material response to the fields, the Maxwell equations have to be accompanied by the constitutive relations linking $\mathbf{B}$, $\mathbf{D}$ and $\mathbf{J}$ to the fields:

$$\mathbf{D} = \underline{\epsilon}\mathbf{E} = \epsilon_0 \mathbf{E} + \mathbf{P}, \tag{1.5}$$

$$\mathbf{B} = \underline{\mu}\mathbf{H} = \mu_0(\mathbf{H} + \mathbf{M}), \tag{1.6}$$

$$\mathbf{J} = \underline{\sigma}\mathbf{E}, \tag{1.7}$$

where, as usual, $\mathbf{P}$ is the electric polarization, $\mathbf{M}$ the magnetization, $\underline{\epsilon}$ the dielectric permittivity, $\underline{\mu}$ the magnetic permeability and $\underline{\sigma}$ the electric conductivity; ϵ_0 and μ_0 are the corresponding values in vacuum, $\underline{\epsilon} = \epsilon_0 \underline{\epsilon}_r$ and $\underline{\mu} = \mu_0 \underline{\mu}_r$, with $\underline{\epsilon}_r$ and $\underline{\mu}_r$ being the corresponding values relative to vacuum.

1

When one is dealing with the Maxwell equations, the Gauss cgs units are often used. In this case, they read

$$\nabla \cdot \mathbf{D} = 4\pi\rho,$$

$$\nabla \cdot \mathbf{B} = 0,$$

$$\nabla \times \mathbf{E} = -\frac{1}{c}\frac{\partial \mathbf{B}}{\partial t},$$

$$\nabla \times \mathbf{H} = \frac{4\pi}{c}\mathbf{J} + \frac{1}{c}\frac{\partial \mathbf{D}}{\partial t}.$$

The quantities $\underline{\epsilon}$, $\underline{\mu}$ and $\underline{\sigma}$ have been underlined to point out that in the more general case they are tensors, i.e. at a given frequency they depend on the vibration direction of the field (polarization). The main consequence is that $\mathbf{D}$ is not parallel to $\mathbf{E}$, $\mathbf{B}$ is not parallel to $\mathbf{H}$, and $\mathbf{J}$ is not parallel to $\mathbf{E}$. Now we limit our discussion to isotropic materials where these functions reduce to scalar parameters, while we will deal with anisotropic media in the last section of Chapter 3. We also consider the material parameters not dependent on the field amplitude, thus regarding the constitutive relations as *linear equations*. In doing that, we restrict ourselves to the frame of *linear optics*. This approximation drops when fields are high enough to modify the electronic distribution or molecular orientation in the medium, leading to a wide landscape of phenomena that is the subject of *nonlinear optics*.

We additionally consider nonmagnetic materials ($\mu_r = 1$) and absence of free charges ($\rho = 0$). Then the Maxwell equations become

$$\nabla \cdot \mathbf{E} = 0, \tag{1.8}$$

$$\nabla \cdot \mathbf{H} = 0, \tag{1.9}$$

$$\nabla \times \mathbf{E} = -\mu_0 \frac{\partial \mathbf{H}}{\partial t}, \tag{1.10}$$

$$\nabla \times \mathbf{H} = \mathbf{J} + \epsilon \frac{\partial \mathbf{E}}{\partial t}. \tag{1.11}$$

From Eqs. (1.10) and (1.11) we get

$$\nabla \times (\nabla \times \mathbf{E}) = -\mu_0 \frac{\partial(\nabla \times \mathbf{H})}{\partial t} = -\mu_0 \frac{\partial \mathbf{J}}{\partial t} - \mu_0\epsilon \frac{\partial^2 \mathbf{E}}{\partial t^2}. \tag{1.12}$$

Using the identity

$$\nabla \times (\nabla \times \mathbf{E}) = -\nabla^2 \mathbf{E} + \nabla(\nabla \cdot \mathbf{E}) = -\nabla^2 \mathbf{E}, \tag{1.13}$$

where we have used Eq. (1.8), we finally get

$$\nabla^2 \mathbf{E} - \mu_0\sigma \frac{\partial \mathbf{E}}{\partial t} - \mu_0\epsilon \frac{\partial^2 \mathbf{E}}{\partial t^2} = 0. \tag{1.14}$$

Propagation in lossless media

When $\sigma = 0$, we have

$$\nabla^2 \mathbf{E} - \mu_0\epsilon \frac{\partial^2 \mathbf{E}}{\partial t^2} = 0. \tag{1.15}$$

This equation describes a wave traveling at a speed

$$v = \sqrt{\frac{1}{\mu_0\epsilon}} = \sqrt{\frac{1}{\mu_0\epsilon_0\epsilon_r}}. \tag{1.16}$$

Then

$$\nabla^2 \mathbf{E} - \frac{1}{v^2} \frac{\partial^2 \mathbf{E}}{\partial t^2} = 0. \tag{1.17}$$

Since in vacuum we have $\epsilon_r = 1$, from Eq. (1.16) we get the speed of the wave in vacuum,

$$c = \sqrt{\frac{1}{\mu_0\epsilon_0}}, \tag{1.18}$$

and

$$v = \frac{c}{\sqrt{\epsilon_r}}, \tag{1.19}$$

which defines the *refractive index* of the medium as

$$n \equiv \sqrt{\epsilon_r} = \frac{c}{v}. \tag{1.20}$$

Therefore, Eq. (1.17) can also be written as

$$\nabla^2 \mathbf{E} - \frac{n^2}{c^2} \frac{\partial^2 \mathbf{E}}{\partial t^2} = 0. \tag{1.21}$$

Starting from Eq. (1.11), we can obtain the same result for the magnetic field $\mathbf{H}$, taking into account Eqs. (1.7) and (1.10):

$$\nabla \times (\nabla \times \mathbf{H}) = \nabla \times \mathbf{J} + \epsilon \frac{\partial(\nabla \times \mathbf{E})}{\partial t} = -\sigma\mu_0 \frac{\partial \mathbf{H}}{\partial t} - \epsilon\mu_0 \frac{\partial^2 \mathbf{H}}{\partial t^2}.$$

Then, using the identity (1.13) applied to $\mathbf{H}$ and Eq. (1.9),

$$\nabla^2 \mathbf{H} - \sigma\mu_0 \frac{\partial \mathbf{H}}{\partial t} - \epsilon\mu_0 \frac{\partial^2 \mathbf{H}}{\partial t^2} = 0,$$

and if $\sigma = 0$

$$\nabla^2 \mathbf{H} - \epsilon\mu_0 \frac{\partial^2 \mathbf{H}}{\partial t^2} = 0$$

or

$$\nabla^2 \mathbf{H} - \frac{n^2}{c^2} \frac{\partial^2 \mathbf{H}}{\partial t^2} = 0.$$

Then both the electric field $\mathbf{E}(\mathbf{r}, t)$ and the magnetic field $\mathbf{H}(\mathbf{r}, t)$ propagate in space and time as a wave that is called an *electromagnetic wave*.

The *plane wave* solution to Eq. (1.21) can be written as

$$\mathbf{E} = \mathbf{E}_0 e^{i(\mathbf{k}\cdot\mathbf{r}-\omega t)}, \tag{1.22}$$

with the meaning of considering the real part as the physical quantity describing the electric field. In Eq. (1.22), $\mathbf{k}$ is the wave vector indicating the propagation direction and ω the angular frequency of the wave. The orientation of the amplitude $\mathbf{E}_0$ defines the instantaneous polarization of the field. In the following, we choose the propagation direction as the $\mathbf{z}$ axis in order to write

$$\mathbf{E} = \mathbf{E}_0 e^{i(kz-\omega t)}. \tag{1.23}$$

$\mathbf{E}(\mathbf{r}, t)$, given by Eq. (1.22) or Eq. (1.23), is a periodic function in both space and time. Periodicity in space is given by the *wavelength* λ, i.e. the distance between two points of the wave with the same phase:

$$\mathbf{E}_0 e^{i(kz-\omega t)} = \mathbf{E}_0 e^{i(k[z+\lambda]-\omega t)}. \tag{1.24}$$

It implies $e^{ik\lambda} = 1$. Then

$$\lambda = \frac{2\pi}{k}. \tag{1.25}$$

We can apply the same concept of periodicity to time and define the period T as the time delay between two points of equal phase:

$$\mathbf{E}_0 e^{i(kz-\omega t)} = \mathbf{E}_0 e^{i(kz-\omega[t+T])}. \tag{1.26}$$

Then $e^{ikT} = 1$, and

$$T = \frac{2\pi}{\omega}, \tag{1.27}$$

and the frequency is defined as

$$\nu = \frac{1}{T} = \frac{\omega}{2\pi}. \tag{1.28}$$

By introducing the plane wave solution in Eqs. (1.10) and (1.11) with $\sigma = 0$ (i.e. $\mathbf{J} = 0$), we have

$$\nabla \times [\mathbf{E}_0 e^{i(\mathbf{k} \cdot \mathbf{r} - \omega t)}] = -\mu_0 \frac{\partial}{\partial t} [\mathbf{H}_0 e^{i(\mathbf{k} \cdot \mathbf{r} - \omega t)}], \tag{1.29}$$

$$\nabla \times [\mathbf{H}_0 e^{i(\mathbf{k} \cdot \mathbf{r} - \omega t)}] = \epsilon \frac{\partial}{\partial t} [\mathbf{E}_0 e^{i(\mathbf{k} \cdot \mathbf{r} - \omega t)}]. \tag{1.30}$$

Then

$$i\mathbf{k} \times [\mathbf{E}_0 e^{i(\mathbf{k} \cdot \mathbf{r} - \omega t)}] = i\omega \mu_0 [\mathbf{H}_0 e^{i(\mathbf{k} \cdot \mathbf{r} - \omega t)}] \rightarrow \mathbf{k} \times \mathbf{E} = \omega \mu_0 \mathbf{H}, \tag{1.31}$$

$$i\mathbf{k} \times [\mathbf{H}_0 e^{i(\mathbf{k} \cdot \mathbf{r} - \omega t)}] = -i\omega \epsilon [\mathbf{E}_0 e^{i(\mathbf{k} \cdot \mathbf{r} - \omega t)}] \rightarrow \mathbf{k} \times \mathbf{H} = -\omega \epsilon \mathbf{E}. \tag{1.32}$$

These relations show that $\mathbf{E}$, $\mathbf{H}$ and $\mathbf{k}$ are orthogonal to each other, and thus the electromagnetic wave is a transverse wave where the fields oscillate in a plane orthogonal to the propagation direction. If $\mathbf{E}$ is oscillating along the $\mathbf{x}$ axis, $\mathbf{H}$ will be oscillating along the $\mathbf{y}$ direction, $\mathbf{k}$ being directed along $\mathbf{z}$. Such a transverse linearly polarized plane wave is depicted in Fig. 1.1. The *wave fronts* are the equal phase planes (i.e. planes with a uniform field) and are parallel to the $\mathbf{xy}$ plane.

Additionally, using the plane wave solution (1.23) in the wave equation (1.21), we find the link between the space and time periodicity — in other words, the dispersion relation

$$\nabla^2 [\mathbf{E}_0 e^{i(kz-\omega t)}] - \frac{n^2}{c^2} \frac{\partial^2 [\mathbf{E}_0 e^{i(kz-\omega t)}]}{\partial t^2} = 0, \tag{1.33}$$

i.e.

$$-k^2 \mathbf{E} + \frac{n^2}{c^2} \omega^2 \mathbf{E} = 0. \tag{1.34}$$

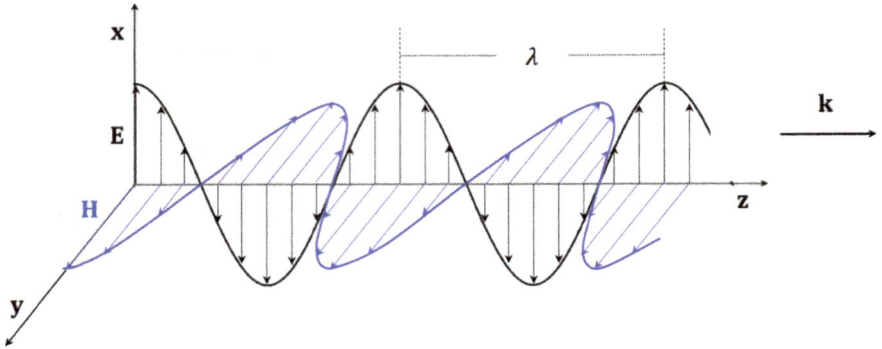

Fig. 1.1.   Linearly polarized electromagnetic wave.

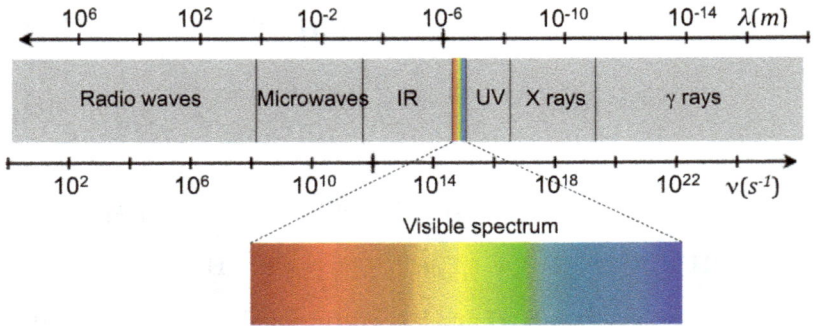

Fig. 1.2.   The spectrum of the electromagnetic waves and the visible part of it.

Therefore,

$$k = n\frac{\omega}{c} = nk_0, \tag{1.35}$$

defining $k_0 = \omega/c$, the vacuum wave vector.

From Eq. (1.35) we can also write

$$k = n\frac{2\pi\nu}{c}, \quad \lambda = \frac{c}{n} \cdot \frac{1}{\nu}. \tag{1.36}$$

In Fig. 1.2 is sketched the wide spectrum of the known electromagnetic waves spanning over 24 orders of magnitude in frequency or wavelength, showing that our eyes are sensitive to a very small part of it (approximately the range with wavelength between 350 nm and 700 nm).

Propagation in media with losses

1. *Conductive materials*

Let us rewrite Eq. (1.14) by employing the refractive index definition (1.20):

$$\nabla^2 \mathbf{E} - \mu_0 \sigma \frac{\partial \mathbf{E}}{\partial t} - \frac{n^2}{c^2} \frac{\partial^2 \mathbf{E}}{\partial t^2} = 0. \tag{1.37}$$

Using a plane wave solution (1.23) we get

$$-k^2 \mathbf{E} - i\omega \mu_0 \sigma \mathbf{E} + \frac{n^2}{c^2} \omega^2 \mathbf{E} = 0. \tag{1.38}$$

This means that

$$k^2 = \frac{n^2}{c^2} \omega^2 + i\omega \mu_0 \sigma = \frac{n^2}{c^2} \omega^2 \left[1 + i \frac{\mu_0 \sigma c^2}{n^2 \omega} \right] \tag{1.39}$$

or

$$k = n k_0 \sqrt{\left[1 + i \frac{\mu_0 \sigma c^2}{n^2 \omega} \right]}. \tag{1.40}$$

That is to say, Eq. (1.35) is still fulfilled just by considering the refractive index as a complex quantity:

$$k = \tilde{n} k_0, \tag{1.41}$$

where

$$\tilde{n} = n \sqrt{\left[1 + i \frac{\mu_0 \sigma c^2}{n^2 \omega} \right]} \equiv n' + in''. \tag{1.42}$$

Here it is necessary to underline that $n = \sqrt{\epsilon_r}$ has been considered a real quantity. In the approximation $\mu_0 \sigma c^2 / n^2 \omega \ll 1$, Eq. (1.42) can be written as

$$\tilde{n} \approx n \left[1 + i \frac{\mu_0 \sigma c^2}{2n^2 \omega} \right], \tag{1.43}$$

i.e.

$$n' = n, \quad n'' = \frac{\mu_0 \sigma c^2}{2n^2 \omega}. \tag{1.44}$$

By considering a complex number,

$$z^2 = a + ib, \text{ i.e. } z = \sqrt{a + ib},$$

it can be shown that the square root can be written as

$$z = \sqrt{\frac{\sqrt{a^2 + b^2} + a}{2}} + i\frac{b}{|b|}\sqrt{\frac{\sqrt{a^2 + b^2} - a}{2}}.$$

Then, from Eq. (1.42), we have

$$z = \sqrt{\left[1 + i\frac{\mu_0 \sigma c^2}{n^2 \omega}\right]},$$

where we can define

$$a = 1, \quad b = \frac{\mu_0 \sigma c^2}{n^2 \omega}.$$

By taking $b \ll 1$, we get

$$\sqrt{1 + b^2} \approx 1 + \frac{b^2}{2}; \quad \text{then } z \approx \sqrt{\frac{1 + b^2/2 + 1}{2}} + i\sqrt{\frac{1 + b^2/2 - 1}{2}}$$

$$= \sqrt{1 + \frac{b^2}{4}} + i\frac{b}{2} \approx 1 + i\frac{b}{2}.$$

Then

$$z \approx 1 + i\frac{\mu_0 \sigma c^2}{2n^2 \omega},$$

giving Eq. (1.43).

It is also useful to check the fulfillment of the condition $\mu_0 \sigma c^2 / n^2 \omega \ll 1$ taking into account the actual values of these quantities:

$$c = 3 \cdot 10^8 \frac{m}{s}, \mu_0 = 1.26 \cdot 10^{-6} \frac{H}{m} \text{ and } \omega = 4 \cdot 10^{15} \text{ s}^{-1} \text{ for the center of}$$
the visible spectrum.

Since $n \geq 1$ we can neglect it, because it always plays positively for the above approximation.

1. *Semiconductors*

Taking into account, for instance, the pure silicon conductivity $\sigma = 4 \cdot 10^{-4}$ S/m (S = Siemens), we get

$$\frac{\mu_0 \sigma c^2}{\omega} = \frac{1.26 \cdot 4 \cdot 9 \cdot 10^6}{4 \cdot 10^{15}} = 1.1 \cdot 10^{-8} \ll 1.$$

Therefore, the approximation is fulfilled even for doped semiconductors where the electrical conductivity is higher, being of the order of $\sigma \approx 10^3$ S/m.

2. Metals

On the contrary, by considering an average value of metal conductivity at optical frequencies as $\sigma = 5 \cdot 10^7$ S/m, we get

$$\frac{\mu_0 \sigma c^2}{\omega} = \frac{1.26 \cdot 5 \cdot 9 \cdot 10^{17}}{4 \cdot 10^{15}} = 1.4 \cdot 10^3 \gg 1,$$

not fulfilling the used approximation.

In this case, the imaginary part of the refractive index at optical frequencies is not negligible with respect to the real part, and therefore by defining again

$$z = \sqrt{\left[1 + i \frac{\mu_0 \sigma c^2}{n^2 \omega} \right]}, \quad \text{with } a = 1 \quad \text{and} \quad b = \frac{\mu_0 \sigma c^2}{n^2 \omega},$$

we now have $a \ll b$. Then

$$z = \sqrt{a + ib} = \sqrt{\frac{\sqrt{a^2 + b^2} + a}{2}} + i \frac{b}{|b|} \sqrt{\frac{\sqrt{a^2 + b^2} - a}{2}} \approx \sqrt{\frac{b}{2}} + i \sqrt{\frac{b}{2}}.$$

From Eq. (1.42),

$$\tilde{n} \approx \sqrt{\frac{\mu_0 \sigma c^2}{2\omega}} + i \sqrt{\frac{\mu_0 \sigma c^2}{2\omega}}$$

and the field amplitude decays as

$$\mathbf{E}_0 e^{-\sqrt{(\mu_0 \sigma c^2 / 2\omega)} k_0 z},$$

thus reducing to $\mathbf{E}_0/e$ after traveling for a distance

$$d = \frac{1}{\sqrt{(\mu_0 \sigma c^2 / 2\omega)} k_0} = \sqrt{\frac{2}{\mu_0 \sigma \omega}}.$$

Using the same values as above,

$$d = \sqrt{\frac{2}{1.26 \cdot 4 \cdot 5 \cdot 10^{16}}} \approx 0.3 \cdot 10^{-8} \text{ m},$$

which shows the very small *penetration depth* of the optical wave in a metal.

The meaning of the real and imaginary parts of the refractive index is easily understood by considering a plane wave traveling along the **z** axis:

$$\mathbf{E}_0 e^{i(kz-\omega t)} = \mathbf{E}_0 e^{i(\tilde{n}k_0 z-\omega t)} = \mathbf{E}_0 e^{i([n'+in'']k_0 z-\omega t)}$$

$$= \mathbf{E}_0 e^{-n''k_0 z} e^{i(n'k_0 z-\omega t)} . \tag{1.45}$$

Then we have a plane wave traveling with a phase velocity determined by the real part n' (the usual refractive index far from resonance) with an amplitude $\mathbf{E}_0 e^{-n''k_0 z}$ suffering an exponential decay due to the energy dissipation consequence of the conductivity σ of the medium, leading the electric field of the wave to produce work on the charge carriers.

2. *Dielectric materials*

When electric conduction is negligible, $\sigma \approx 0$ losses can be induced by phenomena leading to an electrical polarization out of phase with respect to the incoming field. An example of such phenomena is molecular reorientation consequent to alignment of microscopic dipoles in the medium.

In this case we can again start from the wave equation (1.21), but we will consider the refractive index as a complex quantity:

$$\tilde{n} = \sqrt{\epsilon_r' + i\epsilon_r''} \equiv n' + in'', \tag{1.46}$$

where the dielectric permittivity is

$$\epsilon = \epsilon' + i\epsilon'' = \epsilon_0(\epsilon_r' + i\epsilon_r''). \tag{1.47}$$

Then Eq. (1.21) is modified to

$$\nabla^2 \mathbf{E} - \frac{\tilde{n}^2}{c^2} \frac{\partial^2 \mathbf{E}}{\partial t^2} = 0, \tag{1.48}$$

giving again a plane wave solution with an exponential decay of the field amplitude as described above.

It is interesting to note that Eq. (1.5) can be written as

$$\mathbf{P} = (\epsilon - \epsilon_0)\mathbf{E} = \epsilon_0(\epsilon_r - 1)\mathbf{E},$$

which in the case of a complex dielectric permittivity becomes

$$\mathbf{P} = \epsilon_0(\epsilon_r' - 1)\mathbf{E} + i\epsilon_0\epsilon_r''\mathbf{E}.$$

By considering the incoming plane wave $\mathbf{E}_0 e^{i(\mathrm{k}_0 z - \omega t)}$,

$$\mathbf{P} = \epsilon_0(\epsilon'_r - 1)\mathbf{E}_0 e^{i(\mathrm{k}_0 z - \omega t)} + i\epsilon_0\epsilon''_r\mathbf{E}_0 e^{i(\mathrm{k}_0 z - \omega t)}.$$

The second term is a polarization *out of phase* by $\pi/2$, accounting for energy dissipation. This is easily shown by recalling the identity $ie^{i\alpha} = e^{i[\alpha + \frac{\pi}{2}]}$, thus reducing the second term to $\epsilon_0\epsilon''_r\mathbf{E}_0 e^{i(\mathrm{k}_0 z - \omega t + \frac{\pi}{2})}$. The first term gives an *in-phase* polarization accounting for the wave propagation; in fact, from the wave equation in the form

$$\nabla^2\mathbf{E} - \mu_0\epsilon\frac{\partial^2\mathbf{E}}{\partial t^2} = 0$$

and using $\mathbf{P} = \epsilon_0(\epsilon'_r - 1)\mathbf{E}$, we get

$$\nabla^2\mathbf{E} - \mu_0\epsilon_0\frac{\partial^2\mathbf{E}}{\partial t^2} = \mu_0\frac{\partial^2\mathbf{P}}{\partial t^2},$$

which differs from the vacuum wave equation for the term on the right-hand side, where the polarization $\mathbf{P}$ acts as the source of the field.

1.2. Energy, intensity and momentum of the electromagnetic wave

Energy and intensity

The wave carries energy associated with the electric and magnetic fields. The variation of electromagnetic energy U in a finite volume V is due to the energy crossing the limiting surfaces of this volume:

$$\frac{\partial U}{\partial t} = -\int_A \mathbf{S} \cdot d\mathbf{A}, \tag{1.49}$$

where the right-hand side represents the flux of the vector $\mathbf{S}$ out of the surface A limiting the volume V, $\mathbf{S}$ being the energy crossing the unit surface in the unit time.

The Gauss theorem allows writing the integral over the surface A as a volume integral over the volume V:

$$\int_A \mathbf{S} \cdot d\mathbf{A} = \int_V \nabla \cdot \mathbf{S}\, dV. \tag{1.50}$$

Defining the volume energy density u, we have $U = \int_V u\,dV$. Then, using Eqs. (1.49) and (1.50), we obtain the *continuity equation*:

$$\frac{\partial u}{\partial t} = -\nabla \cdot \mathbf{S}. \tag{1.51}$$

The energy density u of the electromagnetic field in nonmagnetic materials is usually written as (SI unit system)

$$u = \frac{1}{2}\epsilon \mathbf{E} \cdot \mathbf{E} + \frac{1}{2}\mu_0 \mathbf{H} \cdot \mathbf{H}. \tag{1.52}$$

Then

$$\frac{\partial u}{\partial t} = \epsilon \mathbf{E} \cdot \frac{\partial \mathbf{E}}{\partial t} + \mu_0 \mathbf{H} \cdot \frac{\partial \mathbf{H}}{\partial t}. \tag{1.53}$$

Taking into account the Maxwell equations (1.10) and (1.11) with $\mathbf{J} = 0$ to obtain the time derivatives of the fields, we have

$$\frac{\partial u}{\partial t} = \mathbf{E} \cdot (\nabla \times \mathbf{H}) - \mathbf{H} \cdot (\nabla \times \mathbf{E}) = -\nabla \cdot (\mathbf{E} \times \mathbf{H}). \tag{1.54}$$

Therefore, from (1.51) we identify

$$\mathbf{S} = \mathbf{E} \times \mathbf{H}, \tag{1.55}$$

which is called the *Poynting vector*. It is orthogonal to both fields; then it is parallel to the wave vector $\mathbf{k}$. According to Eq. (1.49), the flux of $\mathbf{S}$ through a surface A is equal to the electromagnetic energy of the wave crossing the surface in the unit time.

Equation (1.54) is easily demonstrated by developing the corresponding calculation:

$$\mathbf{E} \cdot (\nabla \times \mathbf{H}) = \left(\frac{\partial H_z}{\partial y} - \frac{\partial H_y}{\partial z}\right) E_x + \left(\frac{\partial H_x}{\partial z} - \frac{\partial H_z}{\partial x}\right) E_y$$

$$+ \left(\frac{\partial H_y}{\partial x} - \frac{\partial H_x}{\partial y}\right) E_z,$$

$$\mathbf{H} \cdot (\nabla \times \mathbf{E}) = \left(\frac{\partial E_z}{\partial y} - \frac{\partial E_y}{\partial z}\right) H_x + \left(\frac{\partial E_x}{\partial z} - \frac{\partial E_z}{\partial x}\right) H_y$$

$$+ \left(\frac{\partial E_y}{\partial x} - \frac{\partial E_x}{\partial y}\right) H_z$$

and

$$\boldsymbol{\nabla} \cdot (\mathbf{E} \times \mathbf{H}) = \frac{\partial}{\partial x} (E_y H_z - E_z H_y) + \frac{\partial}{\partial y} (E_z H_x - E_x H_z)$$

$$+ \frac{\partial}{\partial z} (E_x H_y - E_y H_x)$$

$$= \frac{\partial E_y}{\partial x} H_z + \frac{\partial H_z}{\partial x} E_y - \frac{\partial E_z}{\partial x} H_y - \frac{\partial H_y}{\partial x} E_z + \frac{\partial E_z}{\partial y} H_x$$

$$+ \frac{\partial H_x}{\partial y} E_z - \frac{\partial E_x}{\partial y} H_z - \frac{\partial H_z}{\partial y} E_x$$

$$\frac{\partial E_x}{\partial z} H_y + \frac{\partial H_y}{\partial z} E_x - \frac{\partial E_y}{\partial z} H_x - \frac{\partial H_x}{\partial z} E_y.$$

By collecting and comparing the corresponding terms, one gets Eq. (1.54):

$$-\boldsymbol{\nabla} \cdot (\mathbf{E} \times \mathbf{H}) = \left(\frac{\partial H_z}{\partial y} - \frac{\partial H_y}{\partial z} \right) E_x + \left(\frac{\partial H_x}{\partial z} - \frac{\partial H_z}{\partial x} \right) E_y$$

$$+ \left(\frac{\partial H_y}{\partial x} - \frac{\partial H_x}{\partial y} \right) E_z - \left(\frac{\partial E_z}{\partial y} - \frac{\partial E_y}{\partial z} \right) H_x$$

$$- \left(\frac{\partial E_x}{\partial z} - \frac{\partial E_z}{\partial x} \right) H_y - \left(\frac{\partial E_y}{\partial x} - \frac{\partial E_x}{\partial y} \right) H_z.$$

Combining Eq. (1.31) with Eq. (1.32), we have

$$kE = \omega\mu_0 H, \quad kH = \omega\epsilon E \rightarrow \frac{\omega\mu_0 H}{E} = \frac{\omega\epsilon E}{H}. \tag{1.56}$$

Then

$$E = \sqrt{\frac{\mu_0}{\epsilon}} H = \frac{c\mu_0}{n} H \tag{1.57}$$

and we get

$$S = \frac{c\mu_0}{n} H^2 = \frac{n}{c\mu_0} E^2 = \sqrt{\frac{\epsilon}{\mu_0}} E^2. \tag{1.58}$$

Since for a plane wave traveling in the direction of the **z** axis $\mathbf{S} = S\mathbf{z}$, in this case the continuity equation (1.51) can be written as

$$\frac{\partial u}{\partial t} = -\frac{\partial S}{\partial z}. \tag{1.59}$$

The amplitude of the Poynting vector represents the energy crossing the unit surface per unit time, as can be easily checked through the dimensions of the two terms of Eq. (1.59). On the other hand, the *intensity* of the electromagnetic wave impinging on a surface is defined as the time average of the energy falling on the unit surface $[\mathrm{J}/(\mathrm{m}^2 \cdot \mathrm{s})]$. As a consequence the intensity is the time average of S:

$$I = \langle S \rangle. \tag{1.60}$$

Using the plane wave (1.23), the real part of the field is given by

$$\Re[\mathrm{E}_0 e^{i(kz-\omega t)}] = \mathrm{E}_0 \cos(kz - \omega t). \tag{1.61}$$

In order to get the intensity I, we must consider the time average of Eq. (1.58), where the only time-varying function is the electric field. Therefore, we have to evaluate the time average of the square of Eq. (1.61). Since the average value of $\cos^2(x)$ is $1/2$, we simply have $\langle E^2 \rangle = (1/2)E_0^2$, and finally

$$I = \frac{1}{2}\sqrt{\frac{\epsilon}{\mu_0}}E_0^2 = \frac{1}{2}\epsilon_0 n c E_0^2. \tag{1.62}$$

It is useful to show that this represents the average power impinging on the unit surface.

From Eqs. (1.52) and (1.57) we can write

$$u = \frac{1}{2}\epsilon_0\epsilon_r E^2 + \frac{1}{2}\mu_0\sqrt{\frac{\epsilon}{\mu_0}}E^2 = \frac{1}{2}\epsilon_0\epsilon_r E^2 + \frac{1}{2}\epsilon_0\epsilon_r E^2 = \epsilon_0\epsilon_r E^2, \tag{1.63}$$

where it becomes evident that the contribution of the electric field and of the magnetic field to energy density are the same. Since the energy in this case travels with the speed of the wave, in the time interval dt it propagates for a distance $dz = (c/n)dt$. In the same time the energy crossing the surface A orthogonal to the propagation direction is $d\mathcal{E} = uA(c/n)dt$, with $A(c/n)dt$ the crossed volume. Therefore, the power impinging on A, taking into account Eq. (1.63), is

$$\frac{d\mathcal{E}/dt}{A} = u\left(\frac{c}{n}\right) = \epsilon_0 n c E^2, \tag{1.64}$$

which after time averaging becomes

$$\left\langle \frac{d\mathcal{E}/dt}{A} \right\rangle = \frac{1}{2}\epsilon_0 n c E_0^2 = I. \tag{1.65}$$

Energy quantization

It is necessary to anticipate here a concept that will be discussed in the next chapter: the energy of the electromagnetic wave cannot be considered as a continuous quantity, but as the sum of minimum energy packets, each of them called a *photon* and given by

$$\varepsilon = h\nu, \tag{1.66}$$

with ν being the wave frequency and

$$h = 6.6 \cdot 10^{-34} \text{ J} \cdot \text{s} \tag{1.67}$$

the Planck constant.

In this way, the total energy carried by the wave is

$$\mathcal{E} = \mathcal{N}h\nu, \tag{1.68}$$

where $\mathcal{N}$ is the total number of photons, usually a very large number. This is actually the only limitation on the wave description of light as far as many photons are involved in the process and we will take it into account in the discussion of the light–matter interaction.

Linear momentum

A medium absorbing energy from an electromagnetic wave gets a pulse in the direction of the wave vector $\mathbf{k}$. In order to show it, let us consider a linearly polarized plane wave (see Section 1.3) interacting with a positive charge q located in the origin of the coordinate system. The energy transferred from the wave during a period T is due to the work done on the charge q,

$$\mathcal{E} = \int_0^T \mathbf{F} \cdot \mathbf{ds} = \int_0^T \mathbf{F} \cdot \mathbf{v}dt = \int_0^T q(\mathbf{E} + \mathbf{v} \times \mathbf{B}) \cdot \mathbf{v}dt$$

$$= q \int_0^T \mathbf{E} \cdot \mathbf{v}dt, \tag{1.69}$$

in fact $\mathbf{v} \times \mathbf{B} \cdot \mathbf{v}dt = 0$ because $\mathbf{v} \times \mathbf{B}$ is orthogonal to $\mathbf{v}$. Therefore, only the electric field is responsible for the provided energy.

On the other hand, the momentum transfer is

$$\mathbf{Q} = \int_0^T \mathbf{F}dt = \int_0^T q(\mathbf{E} + \mathbf{v} \times \mathbf{B})dt = q \int_0^T (\mathbf{v} \times \mathbf{B})dt. \tag{1.70}$$

Here we have used the expression (1.23) for the electric field which makes[1] $\int_0^T q\mathbf{E}dt = 0$. We see that only the magnetic field is responsible for momentum transfer. From Eqs. (1.31) and (1.35) we have

$$\mu_0\mathbf{H} = \mathbf{B} = \mathbf{i}_z \times \frac{n}{c}\mathbf{E}, \tag{1.71}$$

with $\mathbf{i}_z$ being the unit vector along $\mathbf{z}$. Then

$$\mathbf{Q} = q\frac{n}{c}\int_0^T (\mathbf{v} \times \mathbf{i}_z \times \mathbf{E})dt = \mathbf{i}_z q\frac{n}{c}\int_0^T (\mathbf{v} \cdot \mathbf{E})dt, \tag{1.72}$$

where we have used the vector identity $\mathbf{v} \times \mathbf{i}_z \times \mathbf{E} = (\mathbf{v} \cdot \mathbf{E})\mathbf{i}_z - (\mathbf{v} \cdot \mathbf{i}_z)\mathbf{E}$ and the condition $\mathbf{v} \cdot \mathbf{i}_z = 0$ since the charge q moves on the plane of the field $(\mathbf{xy})$.

By comparing Eq. (1.72) to Eq. (1.69), we find that

$$\mathbf{Q} = \mathbf{i}_z\frac{n}{c}\mathcal{E} \quad \text{or} \quad \mathcal{E} = \frac{c}{n}|\mathbf{Q}|. \tag{1.73}$$

This means that a change of linear momentum of the charge is associated with energy absorption and it occurs in the same direction and orientation of the wave propagation $(\mathbf{i}_z)$.

Of course, if a wave is able to transfer momentum to a particle, it is necessary to associate with the wave a linear momentum. This can be done according to the result (1.73). In other words, the additional momentum transferred to the charge is that of the wave, and as a consequence it can be written as

$$\mathbf{p} = \frac{n}{c}\mathcal{E}\hat{k} = \frac{\mathcal{E}}{\omega}\mathbf{k}. \tag{1.74}$$

For a wave traveling in a homogeneous medium (constant n and $\mathbf{k}$), changes of momentum can be due only to changes of $\mathcal{E}$, and therefore the average force due to the momentum variation is

$$\mathbf{F}_{av} = \frac{\Delta\mathbf{p}}{\Delta t} = \frac{n}{c}\frac{\Delta E}{\Delta t}\hat{k} = \frac{n}{c}P\hat{k}.$$

That is a force originated by the absorbed radiation proportional to the optical power P pushing the object in the propagation direction $\hat{k}$. Such a force is the basic origin of the well-known *radiation pressure*.

---

[1]In fact, the integral of $\cos\omega t$ and $\sin\omega t$ over an oscillation period T is zero.

On the other hand, if the wave changes **k** owing to refraction without changing energy, we have

$$\mathbf{F}_{\text{av}} = \frac{\Delta \mathbf{p}}{\Delta t} = \frac{\mathcal{E}}{\omega} \frac{\Delta \mathbf{k}}{\Delta t}.$$

That is to say, Eq. (1.74) suggests that also in this case a force induced by light on matter is possible and it acts along a direction nearly orthogonal to the initial **k**. It may become relevant on small particles in the case of strongly focused beams.

Taking into account Eq. (1.68), we can also write

$$\mathbf{p} = \frac{\mathcal{N}h\nu}{\omega}\mathbf{k} \equiv \frac{\mathcal{N}h}{2\pi}\mathbf{k} = \mathcal{N}\hbar\mathbf{k}. \tag{1.75}$$

Therefore, the linear momentum associated with a single photon is

$$\boldsymbol{p} = \hbar\mathbf{k}. \tag{1.76}$$

Angular momentum

We can also show that an angular momentum can be associated with an electromagnetic wave which becomes able to transfer a torque **M** to matter. Since $\mathbf{M} = \mathbf{r} \times \mathbf{F} = d\mathbf{L}/dt$, the angular momentum **L** is

$$\mathbf{L} = \int_0^T (\mathbf{r} \times \mathbf{F})dt = \int_0^T q\mathbf{r} \times (\mathbf{E} + \mathbf{v} \times \mathbf{B})dt = q \int_0^T (\mathbf{r} \times \mathbf{E})dt, \tag{1.77}$$

with $\int_0^T (\mathbf{r} \times \mathbf{v} \times \mathbf{B})dt = 0$, because $\mathbf{r}, \mathbf{v}, \mathbf{B}$ oscillate at the same frequency, and therefore the time average of their product vanishes. From Eq. (1.77) we notice that when **E** is linearly polarized the displacement **r** is parallel to **E** and $\mathbf{L} = 0$.

On the contrary, for a circularly polarized wave traveling along **z**, the electric field rotates with angular frequency ω (see Section 1.3):

$$\frac{\partial \mathbf{E}}{\partial t} = \pm \omega \mathbf{i}_z \times \mathbf{E}. \tag{1.78}$$

On the other hand, from Eq. (1.69) we have

$$\mathcal{E} = q \int_0^T \mathbf{E} \cdot \mathbf{v}dt = q \int_0^T \mathbf{E} \cdot \frac{d\mathbf{r}}{dt}dt = q \left[(\mathbf{E} \cdot \mathbf{r}) - \int_0^T \mathbf{r} \cdot \frac{d\mathbf{E}}{dt}dt \right],$$

$$\tag{1.79}$$

and using Eqs. (1.78) and (1.77) taking into account displacements normal to light polarization ($\mathbf{E} \cdot \mathbf{r} = 0$),

$$\mathcal{E} = -q \int_0^T \mathbf{r} \cdot \frac{d\mathbf{E}}{dt} dt = \mp q\omega \int_0^T \mathbf{r} \cdot \mathbf{i}_z \times \mathbf{E} \, dt = \pm \omega \mathbf{i}_z \cdot q \int_0^T (\mathbf{r} \times \mathbf{E}) dt$$

$$= \pm \omega \, \mathbf{i}_z \cdot \mathbf{L}. \tag{1.80}$$

Therefore, a material system absorbing the energy $\mathcal{E}$ from a circularly polarized wave of energy $\mathcal{E}$ undergoes a change of angular momentum,

$$\mathbf{L} = \pm \mathbf{i}_z \cdot \frac{\mathcal{E}}{\omega}, \tag{1.81}$$

allowing a rotation clockwise or counterclockwise, depending on the sign of circular polarization of light. For this reason, the expression (1.81) represents the angular momentum of an electromagnetic wave of angular frequency ω and energy $\mathcal{E}$.

Taking into account the energy quantization (1.68), we have

$$\mathbf{L} = \pm \mathcal{N} \hbar \mathbf{i}_z. \tag{1.82}$$

That is to say, each photon of a circularly polarized wave carries an angular momentum oriented along the propagation direction:

$$\boldsymbol{\ell} = \pm \hbar \mathbf{i}_z. \tag{1.83}$$

Both the linear and the angular momentum discussed above depend on the vector properties of the electromagnetic wave, namely the propagation direction given by the wave vector $\mathbf{k}$ and the polarization of electric field $\mathbf{E}$. These mechanical-like properties of the electromagnetic wave were pointed out in the early stage of the analysis of Maxwell equations, even if the experimental demonstration occurred later. More recently, it has been demonstrated that linear and angular momentum transfer is also related to the nonplanar wave front. Linear momentum can be efficiently transferred in the case of a strong field gradient in the transversal plane obtained by strong focusing, while a strong angular momentum exists when regular phase variation is present in the wave front (i.e. the *optical vortex*). These effects are exploited in the *optical tweezers*, which allow manipulation of microparticles or cells using only light.

1.3. Wave polarization

Polarization is one of the basic properties of a wave, being the direction and orientation of the electric (or the magnetic) field. In Eq. (1.23) it is given by the vector $\mathbf{E}_0$. When its orientation is constant in time, we speak of *linear polarization*. Usually the wave equation is discussed for this case and then an extension is made to any polarization, by exploiting the *superposition principle*, which is fulfilled by the wave equation: if $\mathbf{E}_1$ and $\mathbf{E}_2$ are separately solutions to the wave equation, the sum $\mathbf{E}_1 + \mathbf{E}_2$ is also a solution. Let us consider in the same region of space two linearly polarized fields at the same frequency and both traveling along $\mathbf{z}$, but with polarization orthogonal to each other:

$$
\begin{aligned}
\mathbf{E}_x &= \hat{\mathbf{x}} E_1 e^{i(\mathrm{k}z - \omega t + \phi_1)}, \\
\mathbf{E}_y &= \hat{\mathbf{y}} E_2 e^{i(\mathrm{k}z - \omega t + \phi_2)},
\end{aligned}
\tag{1.84}
$$

ϕ_1 and ϕ_2 being the specific phase terms. According to the superposition principle, the total field is given by the vector sum $\mathbf{E} = \mathbf{E}_x + \mathbf{E}_y$, whose components are

$$
\begin{aligned}
E_x &= E_1 e^{i(kz - \omega t + \phi_1)}, \\
E_y &= E_2 e^{i(kz - \omega t + \phi_2)}.
\end{aligned}
\tag{1.85}
$$

Following Born and Wolf,[2] we write the real part of the field as

$$
\frac{E_x}{E_1} = \cos(kz - \omega t + \phi_1) = \cos(\tau + \phi_1) = \cos(\tau)\cos(\phi_1) - \sin(\tau)\sin(\phi_1),
$$

$$
\frac{E_y}{E_2} = \cos(kz - \omega t + \phi_2) = \cos(\tau + \phi_2) = \cos(\tau)\cos(\phi_2) - \sin(\tau)\sin(\phi_2),
$$

where $\tau = kz - \omega t$. Then we multiply the first line by $\sin(\phi_2)$ and the second line by $\sin(\phi_1)$, and get the difference,

$$
\begin{aligned}
\frac{E_x}{E_1}\sin(\phi_2) - \frac{E_y}{E_2}\sin(\phi_1) &= \cos(\tau)\cos(\phi_1)\sin(\phi_2) - \sin(\tau)\sin(\phi_1)\sin(\phi_2) \\
&\quad - \cos(\tau)\cos(\phi_2)\sin(\phi_1) \\
&\quad + \sin(\tau)\sin(\phi_2)\sin(\phi_1) = \cos(\tau)\sin(\phi_2 - \phi_1),
\end{aligned}
$$

---

[2]M. Born and E. Wolf, *Principles of Optics*, sixth edition (Pergamon, 1980).

and in a similar way,

$$\frac{E_x}{E_1}\cos(\phi_2) - \frac{E_y}{E_2}\cos(\phi_1) = \sin(\tau)\sin(\phi_2 - \phi_1).$$

Taking the square of each equation and summing, we obtain

$$\left(\frac{E_x}{E_1}\right)^2 + \left(\frac{E_y}{E_2}\right)^2 - 2\left(\frac{E_x}{E_1}\right)\left(\frac{E_y}{E_2}\right)[\cos(\phi_1)\cos(\phi_2) + \sin(\phi_1)\sin(\phi_2)]$$

$$= \sin^2(\phi_2 - \phi_1).$$

Then we get

$$\left(\frac{E_x}{E_1}\right)^2 + \left(\frac{E_y}{E_2}\right)^2 - 2\left(\frac{E_x}{E_1}\right)\left(\frac{E_y}{E_2}\right)\cos\phi = \sin^2\phi, \qquad (1.86)$$

where $\phi = \phi_2 - \phi_1$ is the phase difference between the two components.

In the $\mathbf{E_x E_y}$ plane, Eq. (1.86) is an ellipse equation describing the amplitude $E = \sqrt{E_x^2 + E_y^2}$ of the field vector whose vertex follows the elliptical trajectory with angular frequency ω, and the corresponding wave is said to be elliptically polarized. In the most general case, the ellipse axes do not correspond to the directions $\mathbf{x}$ and $\mathbf{y}$; an easy calculation leads to the results

$$a^2 + b^2 = E_1^2 + E_2^2, \quad \mathrm{tg}\,2\psi = (\mathrm{tg}\,2\alpha)\cos\phi, \quad \mathrm{tg}\,2\chi = (\sin 2\alpha)\sin\phi, \qquad (1.87)$$

where

$$\mathrm{tg}\,\chi = \mp\frac{b}{a}, \quad \mathrm{tg}\,\alpha = \frac{E_2}{E_1}, \qquad (1.88)$$

and the other relevant quantities are shown in the Fig. 1.3.

Therefore, all the ellipse parameters are determined by the amplitude and phase of the two components of the field. It is easy to understand that the field rotates describing the ellipse in the $\mathbf{xy}$ plane since each component has the periodicity given by the phase factor $\tau = kz - \omega t$, and hence their sum keeps the same periodicity in space (λ) and time (T). Then the field rotates in a plane perpendicular to the propagation direction with amplitude values oscillating between the minimum and maximum values that represent the half-axes of the ellipse.

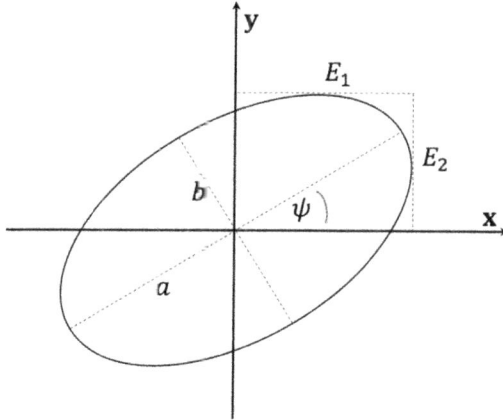

Fig. 1.3. The ellipse describing the field of an elliptically polarized wave.

Linear and circular polarization can be found as particular cases of Eq. (1.86).

Any time the phase difference $\phi = m\pi$ $(m = 0, \pm1, \pm2, \ldots)$ from Eq. (1.87) we get $\operatorname{tg} 2\chi = (\sin 2\alpha) \sin \phi = 0$ and $b = a \operatorname{tg} \chi = 0$, i.e. the ellipse reduces to a line, and therefore we have linear polarization.

On the other hand,

$$\left(\frac{E_x}{E_1}\right)^2 + \left(\frac{E_y}{E_2}\right)^2 - 2(-1)^m \left(\frac{E_x}{E_1}\right)\left(\frac{E_y}{E_2}\right) = 0,$$

i.e.

$$\left[\left(\frac{E_x}{E_1}\right) - (-1)^m \left(\frac{E_y}{E_2}\right)\right]^2 = 0 \rightarrow \left(\frac{E_y}{E_x}\right) = (-1)^m \left(\frac{E_2}{E_1}\right).$$

The component ratio corresponds to the amplitude ratio, and therefore we have linear polarization along a direction at an angle ψ with respect to the **x** axis, being

$$\operatorname{tg} 2\psi = (-1)^m (\operatorname{tg} 2\alpha) \rightarrow \psi = (-1)^m \alpha = (-1)^m \operatorname{arctg}\left(\frac{E_2}{E_1}\right).$$

When $E_2 = E_1$ the angle between the linear polarization and the **x** axis is either $\pi/4$ or $3\pi/4$.

When the phase difference $\phi = (\pi/2 + 2m\pi)$ $(m = 0, \pm 1, \pm 2, \ldots)$ and $E_2 = E_1 = E_0$, Eq. (1.86) becomes

$$\left(\frac{E_x}{E_0}\right)^2 + \left(\frac{E_y}{E_0}\right)^2 = 1,$$

i.e. a circumference with radius E_0, describing a *circularly polarized* wave, which is *right-handed* if $\sin\phi > 0$ and *left-handed* if $\sin\phi < 0$.

From Eq. (1.85) we can write

$$\frac{E_y}{E_x} = \frac{E_2}{E_1} e^{i\phi}. \tag{1.89}$$

Therefore, we can summarize the particular cases:

(1) *Linear polarization*, $\phi = m\pi$ $(m = 0, \pm 1, \pm 2, \ldots)$:

$$\frac{E_y}{E_x} = (-1)^m \frac{E_2}{E_1}. \tag{1.90}$$

(2) *Circular polarization right-handed*, $\phi = (\pi/2 + 2m\pi)$ $(m = 0, \pm 1, \pm 2, \ldots)$, $E_2 = E_1$:

$$\frac{E_y}{E_x} = i. \tag{1.91}$$

(3) *Circular polarization left-handed*, $\phi = (-\pi/2 + 2m\pi)$ $(m = 0, \pm 1, \pm 2, \ldots)$, $E_2 = E_1$:

$$\frac{E_y}{E_x} = -i. \tag{1.92}$$

Given Eqs. (1.91) and (1.92), the circularly polarized field can be written as

$$\mathbf{E} = E_0 e^{i(kz - \omega t)} (\mathbf{x} + i\mathbf{y}) \quad \text{(right-handed)},$$

$$\mathbf{E} = E_0 e^{i(kz - \omega t)} (\mathbf{x} - i\mathbf{y}) \quad \text{(left-handed)}. \tag{1.93}$$

The different polarization states dependent on the phase difference between the electric field components are shown in Fig. 1.4. The figures correspond to a general condition $E_2 \neq E_1$, giving rise to elliptical polarization becoming linear for $\phi = m\pi$. In the particular condition

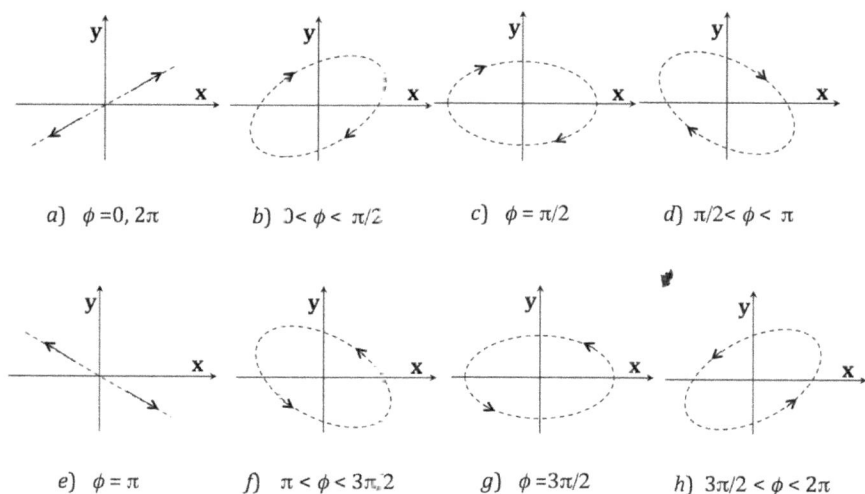

a) $\phi = 0, 2\pi$ *b)* $0 < \phi < \pi/2$ *c)* $\phi = \pi/2$ *d)* $\pi/2 < \phi < \pi$

e) $\phi = \pi$ *f)* $\pi < \phi < 3\pi/2$ *g)* $\phi = 3\pi/2$ *h)* $3\pi/2 < \phi < 2\pi$

Fig. 1.4. Polarization states dependent on the phase difference between the electric field components of the electromagnetic wave. The dashed lines of the ellipses represent the apex of the field vector.

$E_2 = E_1$, the cases shown in Figs. 1.4(c) and 1.4(g) transform into circular polarization states respectively right-handed and left-handed.

The light polarization states are also described through the *Stokes parameters*, which are achievable from simple experiments. They are defined as

$$S_0 = E_1^2 + E_2^2,$$
$$S_1 = E_1^2 - E_2^2,$$
$$S_2 = 2E_1 E_2 \cos \phi,$$
$$S_3 = 2E_1 E_2 \sin \phi,$$

(1.94)

and the following relationships hold with the parameters of the polarization ellipse

$$S_0 = S_1^2 + S_2^2 + S_3^2,$$
$$S_1 = S_0 \cos 2\chi \cos 2\psi,$$
$$S_2 = S_0 \cos 2\chi \sin 2\psi,$$
$$S_3 = S_0 \sin 2\chi.$$

(1.95)

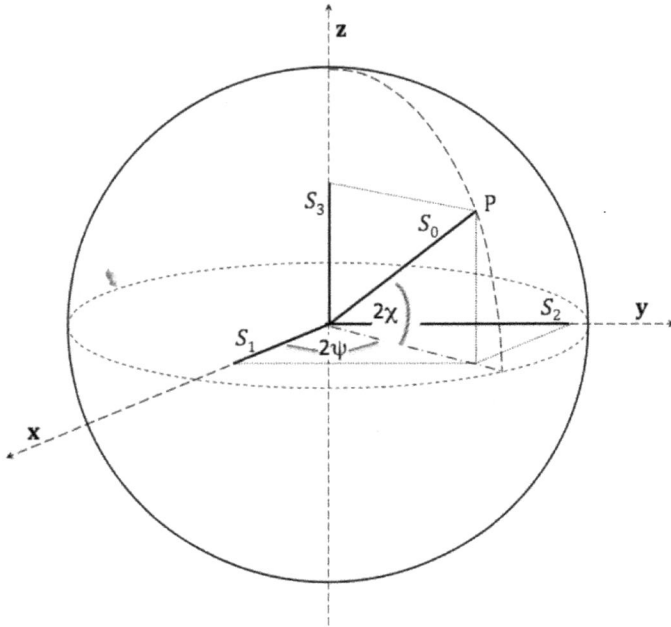

Fig. 1.5.   The Poincaré sphere where each point of the surface represents a possible state of polarization of a wave.

It is clear that Eqs. (1.95) can be considered a transformation from the Cartesian coordinates S_1, S_2 and S_3 and the spherical coordinates $S_0, 2\chi$ and 2ψ. In this way, each point of the sphere whose radius is S_0 represents a specific state of polarization, as shown in Fig. 1.5.

In order to better understand the meaning of the Poincaré sphere, let us make a few examples. For linear polarization $2\chi = 0$, and therefore $S_3 = 0$, it means that the points of the circumference at $z = 0$ represent linear polarization states. On the other hand, in the case of circular polarization $S_1 = S_2 = 0$, being $\cos\phi = 0$, and moreover $\sin 2\chi = \pm 1$ and $|S_3| = S_0$. Therefore, the poles on the **z** axis represent the right-handed circular polarization ($S_3 > 0$) and the left-handed circular polarization ($S_3 < 0$).

Many experimental techniques based on ellipsometry analyze light polarization through the Stokes parameters.

Finally, note that we speak of a polarized wave if the electric field **E** (and as a consequence the magnetic field **H**) follows a regular periodic variation in time and space (elliptical, circular or linear), while the wave is not polarized when the field has a random variation in time and space.

1.4. Waves at the media interface

The phenomena occurring when an optical wave meets the interface between two *optically different* media are very important in our everyday life. In fact, the consequent light reflection allows seeing any object around us and together with refraction followed by light transmission and/or absorption determines how the objects appear to us. The key point is the determination of what "optical difference" is, since only in this case do reflection and refraction take place. We speak of optical difference between two media when they have different refractive indices. In fact, two media may be different in structure and composition but can have the same refractive index; in this case they are not different for an electromagnetic wave, which is not affected at all by crossing their interface.[3] On the contrary, when $n_1 \neq n_2$ we get both reflection and refraction.

The starting point to investigate the properties of a wave at the interface between two media when $n_1 \neq n_2$ is the settlement of the continuity conditions for the field components parallel to the surface:

$$E_{1\parallel} = E_{2\parallel}, \quad H_{1\parallel} = H_{2\parallel}. \tag{1.96}$$

On one side of the interface we have the incident and the reflected waves whose fields are $\mathbf{E}_i$ and $\mathbf{E}_r$; on the other side we have the transmitted wave with field $\mathbf{E}_t$:

$$\mathbf{E}_i = \mathbf{E}_{0i} e^{i(\mathbf{k_i} \cdot \mathbf{r} - \omega t)},$$

$$\mathbf{E}_r = \mathbf{E}_{0r} e^{i(\mathbf{k_r} \cdot \mathbf{r} - \omega t)}, \tag{1.97}$$

$$\mathbf{E}_t = \mathbf{E}_{0t} e^{i(\mathbf{k_t} \cdot \mathbf{r} - \omega t)}.$$

The following hold:

$$|\mathbf{k}_i| = |\mathbf{k}_r| = n_1 k_0, \quad |\mathbf{k}_t| = n_2 k_0, \tag{1.98}$$

with k_0 being the vacuum wave number defined in Eq. (1.35).

---

[3]In the case of waves falling in the visible part of the electromagnetic spectrum, the interface would not be visible.

The condition (1.96) becomes

$$E_{i\parallel} + E_{r\parallel} = E_{t\parallel}, \quad H_{i\parallel} + H_{r\parallel} = H_{t\parallel}. \tag{1.99}$$

Phase conditions

The application of Eqs. (1.99) taking into account Eqs. (1.97) leads to the following conditions for the phases of the three waves:

$$\mathbf{k}_i \cdot \mathbf{r} = \mathbf{k}_r \cdot \mathbf{r} = \mathbf{k}_t \cdot \mathbf{r}, \tag{1.100}$$

since they must have the same phase on the interface. This means that for any $\mathbf{r}$ on this surface

$$\begin{aligned}
(\mathbf{k}_i - \mathbf{k}_r) \cdot \mathbf{r} &= 0, \\
(\mathbf{k}_i - \mathbf{k}_t) \cdot \mathbf{r} &= 0,
\end{aligned} \tag{1.101}$$

and therefore $(\mathbf{k}_i - \mathbf{k}_r)$ and $(\mathbf{k}_i - \mathbf{k}_t)$ are perpendicular to $\mathbf{r}$. If the interface is chosen as the **xy** plane ($z = 0$), this means $(\mathbf{k}_i - \mathbf{k}_r)$ and $(\mathbf{k}_i - \mathbf{k}_t)$ are along the **z** direction. Then

$$\begin{aligned}
\mathbf{k}_i - \mathbf{k}_r &= a\,\hat{\mathbf{z}} \rightarrow \mathbf{k}_r = \mathbf{k}_i - a\,\hat{\mathbf{z}}, \\
\mathbf{k}_i - \mathbf{k}_t &= b\,\hat{\mathbf{z}} \rightarrow \mathbf{k}_t = \mathbf{k}_i - b\,\hat{\mathbf{z}},
\end{aligned} \tag{1.102}$$

with a and b being scalar factors.

Equations (1.102) tell us that $\mathbf{k}_r$ and $\mathbf{k}_t$ belong to the $\mathbf{k}_i\hat{z}$ plane denominated as the *incidence plane*, given by the propagation direction of the incident wave and the line perpendicular to the interface. As a consequence, all of the vectors $\mathbf{k}_r$, $\mathbf{k}_t$ and $\mathbf{k}_i$ belong to the same plane (*the first law of reflection and refraction*) or (in other words typical of geometrical optics) "the incident, reflected and refracted rays and the normal to the interface passing through the incidence point belong to the same plane." The incidence point is the one where $\mathbf{k}_i$ crosses the interface.

Let us choose the incidence plane as the **yz** plane. Then, by looking at Fig. 1.6, it is easy to verify that

$$\begin{aligned}
\mathbf{k}_i &= n_1 k_0 (\sin\theta_i \hat{\mathbf{y}} - \cos\theta_i \hat{\mathbf{z}}), \\
\mathbf{k}_t &= n_2 k_0 (\sin\theta_t \hat{\mathbf{y}} - \cos\theta_t \hat{\mathbf{z}}), \\
\mathbf{k}_r &= n_1 k_0 (\sin\theta_r \hat{\mathbf{y}} + \cos\theta_r \hat{\mathbf{z}}),
\end{aligned} \tag{1.103}$$

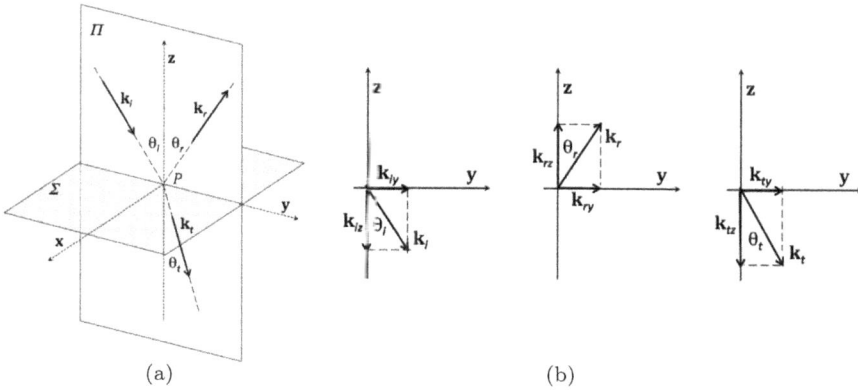

Fig. 1.6. (a) The wave vectors $\mathbf{k}_i$, $\mathbf{k}_t$ and $\mathbf{k}_r$ in the incidence plane (Π) on the interface (Σ) between two media with different refractive indices. The incidence angle θ_i, the reflection angle θ_r and the refraction angle θ_t are shown. (b) Projections of the wave vectors on the axes $\mathbf{y}$ and $\mathbf{z}$.

i.e.

$$\mathbf{k}_i - \mathbf{k}_r = n_1 k_0 (\sin \theta_i - \sin \theta_r) \hat{\mathbf{y}} - n_1 k_0 (\cos \theta_i - \cos \theta_r) \hat{\mathbf{z}},$$

$$\mathbf{k}_i - \mathbf{k}_t = k_0 (n_1 \sin \theta_i - n_2 \sin \theta_t) \hat{\mathbf{y}} - k_0 (n_1 \cos \theta_i - n_2 \cos \theta_t) \hat{\mathbf{z}}.$$

$$(1.104)$$

Since, according to Eqs. (1.102), the y component must be zero, we have

$$\sin \theta_i - \sin \theta_r = 0 \rightarrow \theta_i = \theta_r \quad \text{(law of reflection)},$$

$$n_1 \sin \theta_i - n_2 \sin \theta_t = 0 \rightarrow n_1 \sin \theta_i = n_2 \sin \theta_t \quad \text{(Snell's law of refraction)}.$$

$$(1.105)$$

The law of refraction gives information about the change of propagation direction of the wave after crossing the interface; since $\theta_i \leq \pi/2$ and $\theta_t \leq \pi/2$, we have

$$\theta_t < \theta_i \quad \text{if } n_2 > n_1,$$
$$\theta_t > \theta_i \quad \text{if } n_2 < n_1,$$

$$(1.106)$$

and if $\theta_i = 0$ also $\theta_t = 0$ as well as for the reflected wave $\theta_r = 0$.

It is important to note that when $n_2 < n_1$ a *critical angle* θ_c of incidence exists over which no refraction is possible and we get the condition of *total reflection*. The critical angle is defined as the incidence angle corresponding to $\theta_t = \pi/2$:

$$n_1 \sin \theta_c = n_2 \rightarrow \theta_c = \arcsin\left(\frac{n_2}{n_1}\right). \qquad (1.107)$$

Let us consider the glass–air interface. For common glass we have $n_2 = 1.52$, while we take $n_1 \approx 1.00$ for air. Then

$$\theta_c = \arcsin\left(\frac{n_2}{n_1}\right) = \arcsin(0.66) \rightarrow \theta_c \approx 41.3°,$$

i.e. a plane wave incident on the glass–air interface with $\theta_i > 41.3°$ will be totally reflected.

The working mechanism of *waveguides* and *optical fibers* is based on this effect which makes a glass cylinder in air an actual pipe for light that can be constrained to travel inside the glass for large distances, being always reflected at the internal interface of the cylinder. Of course, the optimization of the system will take into account the wave propagation properties and the limited wave front. For this reason, minimization of losses requires the fabrication of optical fibers made by an internal section (*core*) of high refractive index surrounded by a material (*cladding*) of lower index necessary to get total internal reflection.

A very common device based on this effect is the *total reflection prism*, used in several optical instruments (cameras, microscopes, etc.) in order to reduce losses, since, in principle, total reflection leads to 100% reflection of the impinging light, higher than for any mirror based on metallic layer properties.

Amplitude conditions

Since the wave vector $\mathbf{k}_i$ belongs to the incidence plane and is orthogonal to the wave front, which is the plane of field oscillation, we can write

$$\mathbf{E} = E_s \hat{\mathbf{s}} + E_p \hat{\mathbf{p}}, \qquad (1.108)$$

with $\hat{\mathbf{s}}$ being perpendicular and $\hat{\mathbf{p}}$ being parallel to the incidence plane. Therefore, according to Eqs. (1.103), $\hat{\mathbf{s}} \equiv \hat{\mathbf{x}}$, and looking at Fig. 1.7 we have

$$\mathbf{E}_i = E_{si}\hat{\mathbf{x}} + E_{pi}(\cos\theta_i\hat{\mathbf{y}} + \sin\theta_i\hat{\mathbf{z}}),$$

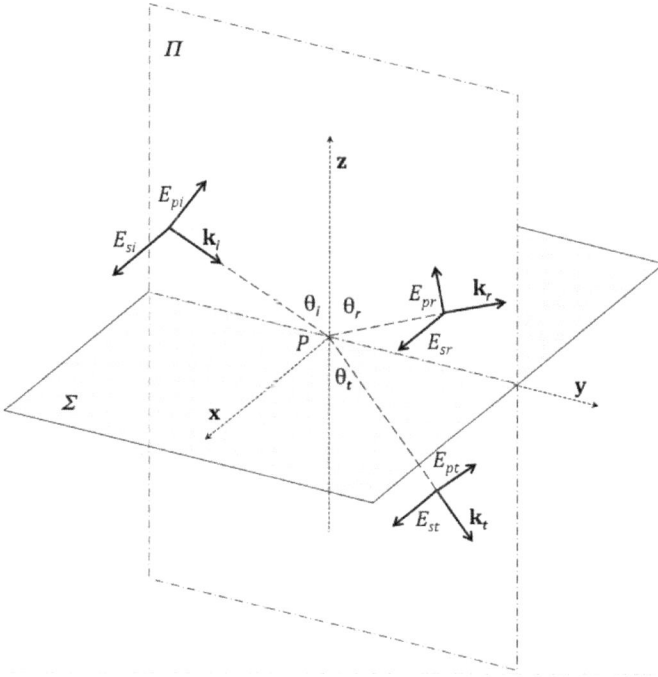

Fig. 1.7. The s and p components of the incident, transmitted and reflected fields at an optical interface.

$$\mathbf{E}_t = E_{st}\hat{\mathbf{x}} + E_{pt}(\cos\theta_t\hat{\mathbf{y}} + \sin\theta_t\hat{\mathbf{z}}),$$

$$\mathbf{E}_r = E_{sr}\hat{\mathbf{x}} - E_{pr}(-\cos\theta_r\hat{\mathbf{y}} + \sin\theta_r\hat{\mathbf{z}}). \tag{1.109}$$

From (1.31),

$$\mathbf{H} = \frac{1}{\omega\mu_0}\mathbf{k}\times\mathbf{E} = \frac{n}{c\mu_0}\hat{\mathbf{k}}\times\mathbf{E} = n\sqrt{\frac{\epsilon_0}{\mu_0}}\hat{\mathbf{k}}\times\mathbf{E}. \tag{1.110}$$

Then

$$\mathbf{H}_i = n_1\sqrt{\frac{\epsilon_0}{\mu_0}}[E_{pi}\hat{\mathbf{x}} - E_{si}(\cos\theta_i\hat{\mathbf{y}} + \sin\theta_i\hat{\mathbf{z}})],$$

$$\mathbf{H}_t = n_2\sqrt{\frac{\epsilon_0}{\mu_0}}[E_{pt}\hat{\mathbf{x}} - E_{st}(\cos\theta_t\hat{\mathbf{y}} + \sin\theta_t\hat{\mathbf{z}})], \tag{1.111}$$

$$\mathbf{H}_r = n_1\sqrt{\frac{\epsilon_0}{\mu_0}}[E_{pr}\hat{\mathbf{x}} + E_{sr}(\cos\theta_r\hat{\mathbf{y}} - \sin\theta_r\hat{\mathbf{z}})].$$

By applying the continuity conditions (1.96) to Eqs. (1.109) and (1.111), taking into account that $\theta_i = \theta_r$, we have

$$E_{si} + E_{sr} = E_{st},$$

$$(E_{pi} - E_{pr})\cos\theta_i = E_{pt}\cos\theta_t,$$

$$n_1(E_{pi} + E_{pr}) = n_2 E_{pt}, \tag{1.112}$$

$$-n_1(E_{si} - E_{sr})\cos\theta_i = -n_2 E_{st}\cos\theta_t.$$

It is remarkable that Eqs. (1.112) are two couples of independent equations, with one including only the s components and the other only the p components. This makes it easy to find the reflection and transmission ratio for the two cases:

$$r_s = \frac{E_{sr}}{E_{si}} = \frac{n_1\cos\theta_i - n_2\cos\theta_t}{n_1\cos\theta_i + n_2\cos\theta_t},$$

$$t_s = \frac{E_{st}}{E_{si}} = \frac{2n_1\cos\theta_i}{n_1\cos\theta_i + n_2\cos\theta_t}, \tag{1.113}$$

$$r_p = \frac{E_{pr}}{E_{pi}} = \frac{n_2\cos\theta_i - n_1\cos\theta_t}{n_2\cos\theta_i + n_1\cos\theta_t},$$

$$t_p = \frac{E_{pt}}{E_{pi}} = \frac{2n_1\cos\theta_i}{n_2\cos\theta_i + n_1\cos\theta_t}. \tag{1.114}$$

These are the *Fresnel formulae*. By using Snell's law they can be written as

$$r_s = -\frac{\sin(\theta_i - \theta_t)}{\sin(\theta_i + \theta_t)},$$

$$t_s = \frac{2\sin\theta_t\cos\theta_i}{\sin(\theta_i + \theta_t)}, \tag{1.115}$$

$$r_p = \frac{\text{tg}(\theta_i - \theta_t)}{\text{tg}(\theta_i + \theta_t)},$$

$$t_p = \frac{2\sin\theta_t\cos\theta_i}{\sin(\theta_i + \theta_t)\cos(\theta_i - \theta_t)}. \tag{1.116}$$

In the case of lossless media (real refractive indices) and not considering total reflection, all the reported quantities are real numbers and therefore the phase of the reflected and transmitted field is the same as that of the incident field or differs from it by π. We note that t_p and t_s are always positive, which means that the phase of the transmitted wave is the same as that of the incident wave. On the contrary, r_p and r_s may change sign depending on the value $\theta_i - \theta_t$. For instance, when $n_2 > n_1$, with $\theta_i > \theta_t$, we have $r_p > 0$ and $r_s < 0$; as a consequence, the s component of the reflected wave undergoes a π phase change, while the p component does not change phase.

The most important consequence is the energy distribution between the transmitted and reflected waves.

The Poynting vector of the incident wave has the same direction of $\mathbf{k}_i$. Then

$$\mathbf{S} = S_i \sin\theta_i \hat{\mathbf{y}} - S_i \cos\theta_i \hat{\mathbf{z}}. \tag{1.117}$$

Therefore, the incident light intensity is

$$I_i = \langle S_i \rangle \cos\theta_i = \frac{\epsilon_0 c n_1}{2} |E_i|^2 \cos\theta_i \tag{1.118}$$

and the intensity of the reflected and of the transmitted wave is

$$I_r = \langle S_r \rangle \cos\theta_i = \frac{\epsilon_0 c n_1}{2} |E_r|^2 \cos\theta_i, \tag{1.119}$$

$$I_t = \langle S_i \rangle \cos\theta_t = \frac{\epsilon_0 c n_2}{2} |E_t|^2 \cos\theta_t, \tag{1.120}$$

and the reflection coefficient R and transmission coefficient T are easily found to be

$$R = \frac{I_r}{I_i} = \frac{|E_r|^2}{|E_i|^2}, \tag{1.121}$$

$$T = \frac{I_t}{I_i} = \frac{|E_t|^2}{|E_i|^2} \left(\frac{n_2 \cos\theta_t}{n_1 \cos\theta_i} \right). \tag{1.122}$$

Obviously, in lossless media $R + T = 1$, as can be verified by these expressions.

Let us apply these results to a linearly polarized wave at an angle α with respect to the incidence plane:

$$E_p = E_i \cos\alpha, \quad E_s = E_i \sin\alpha.$$

Then

$$I_p = I_i \cos^2\alpha, \quad I_s = I_i \sin^2\alpha.$$

And, according to Eqs. (1.119) and (1.120), we can define $I_{pr}, I_{pt}, I_{sr}, I_{st}$. Therefore,

$$R = \frac{I_r}{I_i} = \frac{I_{pr}}{I_i} + \frac{I_{sr}}{I_i} = \frac{I_{pr}}{I_p} \cos^2 \alpha + \frac{I_{sr}}{I_s} \sin^2 \alpha \equiv R_p \cos^2 \alpha + R_s \sin^2 \alpha,$$

where

$$R_p = |r_p|^2 = \left| \frac{n_2 \cos \theta_i - n_1 \cos \theta_t}{n_2 \cos \theta_i + n_1 \cos \theta_t} \right|^2,$$

$$R_s = |r_s|^2 = \left| \frac{n_1 \cos \theta_i - n_2 \cos \theta_t}{n_1 \cos \theta_i + n_2 \cos \theta_t} \right|^2.$$

In a similar way,

$$T = \frac{I_t}{I_i} \left(\frac{n_2 \cos \theta_t}{n_1 \cos \theta_i} \right) = \left[\frac{I_{pt}}{I_p} \cos^2 \alpha + \frac{I_{st}}{I_s} \sin^2 \alpha \right] \left(\frac{n_2 \cos \theta_t}{n_1 \cos \theta_i} \right)$$

$$= T_p \cos^2 \alpha + T_s \sin^2 \alpha,$$

where

$$T_p = |t_p|^2 \left(\frac{n_2 \cos \theta_t}{n_1 \cos \theta_i} \right) = \left| \frac{2n_1 \cos \theta_i}{n_2 \cos \theta_i + n_1 \cos \theta_t} \right|^2 \left(\frac{n_2 \cos \theta_t}{n_1 \cos \theta_i} \right),$$

$$T_s = |t_s|^2 \left(\frac{n_2 \cos \theta_t}{n_1 \cos \theta_i} \right) = \left| \frac{2n_1 \cos \theta_i}{n_1 \cos \theta_i + n_2 \cos \theta_t} \right|^2 \left(\frac{n_2 \cos \theta_t}{n_1 \cos \theta_i} \right).$$

These expressions confirm the concept of optically different media. In fact, if two media do not show optical difference $n_2 = n_1$, then $\theta_i = \theta_t$, and thus $R = 0$, $T = 1$ (there is neither reflection nor deviation of the transmitted wave).

Then, for the linearly polarized wave, using Snell's law we get the reflection and transmission coefficient for the components s and p of the wave:

$$R_p = \frac{\text{tg}^2(\theta_i - \theta_t)}{\text{tg}^2(\theta_i + \theta_t)},$$

$$R_s = \frac{\sin^2(\theta_i - \theta_t)}{\sin^2(\theta_i + \theta_t)},$$

(1.123)

$$T_p = \frac{\sin 2\theta_t \sin 2\theta_i}{\sin^2(\theta_i + \theta_t)\cos^2(\theta_i - \theta_t)},$$

$$T_s = \frac{\sin 2\theta_t \sin 2\theta_i}{\sin^2(\theta_i + \theta_t)}.$$

(1.124)

It is remarkable that if $\theta_i + \theta_t = \pi/2$, then $R_p \to 0$ and the reflected wave is linearly polarized perpendicular to the incidence plane. The incidence angle θ_B which fulfills this condition $[n_1 \sin \theta_B = n_2 \sin(\pi/2 - \theta_B)]$ is called *Brewster's angle*, θ_B, and can easily be found:

$$\theta_B = \text{arctg}\left(\frac{n_2}{n_1}\right).$$

(1.125)

This property is very important and is widely used in several optical devices for different purposes. It becomes nonnegligible even with non-polarized light, since for a wide range of incidence angles $R_s \gg R_p$, i.e. reflection pushes light polarization toward a direction perpendicular to the incidence plane. In other words, any light polarization is generally modified by reflection, with the exception of a pure s or p incident wave.

This is clearly shown in Fig. 1.8, where the reflectivities R_s and R_p are reported for the air–glass interface ($n_1 = 1.00$ and $n_2 = 1.52$).

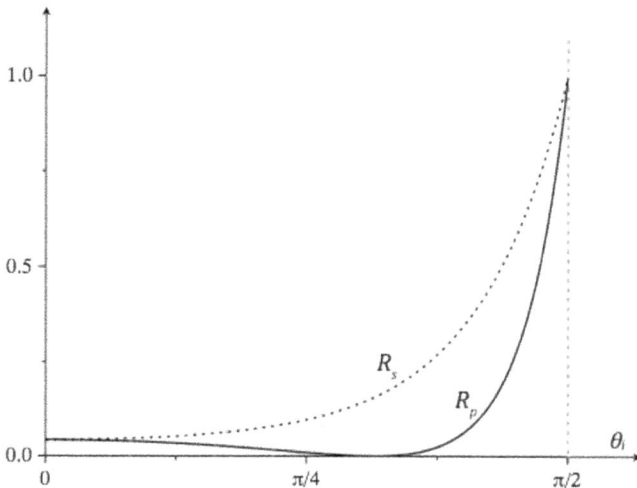

Fig. 1.8. Plot of R_s and R_p versus the incidence angle θ_i for the air–glass interface ($n_1 = 1.00$ and $n_2 = 1.52$).

An estimation of the effect of the interface on the energy of the traveling wave is easily done by considering the previous calculation and setting $\theta_i = \theta_t = 0$ (normal incidence):

$$R = R_p = R_s = \left| \frac{n_2 - n_1}{n_1 + n_2} \right|^2,$$

and

$$T = T_p = T_s = \left| \frac{2n_1}{n_2 + n_1} \right|^2 \left(\frac{n_2}{n_1} \right) = \frac{4n_1 n_2}{(n_2 + n_1)^2}$$

since at normal incidence there is no difference between the s and p directions.

By considering the air–glass interface ($n_1 = 1.00$ and $n_2 = 1.52$), we get $R = 0.04$ and $T = 0.96$. This is a very important result, since it tells us that each time the light meets such an interface the wave traveling in the forward direction loses at least 4% of its energy (because at a higher angle the reflectivity becomes higher). It is an issue concerning the fabrication of optical instruments where several of these interfaces are present owing to lenses and glass plates that strongly affect the performances in terms of the actually available light.

We must note that the Fresnel formulae hold even if the refractive index is a complex quantity, i.e. in the presence of losses. One very important case concerns the interface between air and a conducting material like a metal where we have to take into account that $n_2 = n_2' + in_2''$. When $n_2'' \gg n_2'$ in the visible range, we approximate $n_2 \approx in_2''$. As a consequence, at normal incidence we have

$$R \approx \left| \frac{in_2'' - 1}{1 + in_2''} \right|^2 = 1,$$

showing the expected high reflectivity of a metal surface.

Evanescent wave at the interface

When the condition of total internal reflection is fulfilled, $\theta_i > \theta_c$, the transmitted wave does not exist since $\sin \theta_t = (n_1/n_2) \sin \theta_i > 1$ and then θ_t is not a real quantity. We can write

$$\cos \theta_t = \pm \sqrt{1 - \sin^2 \theta_t} = \pm i \sqrt{\sin^2 \theta_t - 1}. \tag{1.126}$$

The transmitted field becomes

$$\mathbf{E}_t = \mathbf{E}_{0t}e^{i(\mathbf{k_t}\cdot\mathbf{r}-\omega t)} = \mathbf{E}_{0t}e^{i(n_2 k_c(\sin\theta_t\hat{\mathbf{y}}-\cos\theta_t\hat{\mathbf{z}})\cdot\mathbf{r}-\omega t)}$$

$$= \mathbf{E}_{0t}e^{i(n_2 k_0(\sin\theta_t y-\cos\theta_t z)-\omega t)} = \mathbf{E}_{0t}e^{[in_2 k_0\sin\theta_t y\mp n_2 k_0\left(\sqrt{\sin^2\theta_t-1}\right)z-i\omega t]}$$

$$= \mathbf{E}_{0t}e^{i(n_2 k_0\sin\theta_t y-\omega t)}e^{-\left(n_2 k_0\sqrt{\sin^2\theta_i(n_1/n_2)^2-1}\right)z}, \qquad (1.127)$$

where the negative sign has been chosen, being the only one with a physical meaning. We see that along **z** the field has an exponential decay given by the last term.

We can define a penetration length δ setting

$$e^{-\left(n_2 k_0\sqrt{\sin^2\theta_i(n_1/n_2)^2-1}\right)z} = e^{-\frac{z}{\delta}}. \qquad (1.128)$$

That is to say,

$$\delta = \frac{1}{\left(n_2 k_0\sqrt{\sin^2\theta_i(n_1/n_2)^2-1}\right)} = \frac{\lambda_0}{2\pi}\frac{1}{\left(n_2\sqrt{\sin^2\theta_i(n_1/n_2)^2-1}\right)}. \qquad (1.129)$$

Equation (1.129) shows that the penetration depth is of the order of a fraction of the light wavelength. For instance, if we consider the glass–air interface ($n_1 = 1.52; n_2 = 1$) and an incidence angle $\theta_i = 45°$, we get $\delta = 0.40\lambda_0$. This means that we have an *evanescent wave* traveling in **z** for a fraction of the wavelength. On the other hand, Eq. (1.127) points out the existence of a wave traveling on the surface along **y**.

When one is considering a beam of finite cross section, this wave penetration through the interface produces the Goos–Hanken effect, a small shift of the symmetry axis of the reflected beam, which is neglected under the usual condition.

The excitation of surface waves at the interface between a metal and a dielectric material under specific boundary conditions gives rise to *surface plasmons*, whose characteristics have caused the basis of the field of *plasmonics* to grow very quickly in the past years. The excitation of surface plasmons is strongly dependent on the light wavelength and incidence angle as well as on the dielectric permittivity of the material at the interface. For this reason, devices based on the excitation of surface plasmons are studied and developed for several different applications, such as optical biosensing.

Light scattering

Light scattering refers to a wide spectrum of phenomena that lead to irregular light diffusion in directions different from that of the incoming light beam. The features of light scattering are dependent on the specific conditions under which it occurs, and therefore it is investigated independently of reflection and refraction, sometimes appearing as a new phenomenon occurring when light meets interfaces between different media. On the contrary, the physical origin of scattering is reflection and refraction through many interfaces, but it needs a specific approach owing to the large number of the involved interfaces usually not uniformly oriented in space.

Initially, light scattering by small particles was fully investigated, and later on it was discovered also in pure and transparent liquid and gases. Generally, if the particle sizes are comparable to the light wavelength the amount of scattered light increases at a lower wavelength, thus explaining why the blue component is stronger in it. The different ranges of the ratio between particle size a and wavelength λ allow treating light scattering under different approximations (geometrical optics, $a \gg \lambda$; Rayleigh approximation, $a \ll \lambda$; Lorenz–Mie's regime, $a \sim \lambda$), leading to the description of phenomena very important in our everyday life.

For instance, we recall that light scattering is responsible for the blue color of the sky, since the scattering of air molecules is stronger at short wavelengths of the visible range. Another example is the observation of a laser light beam realized by introducing smoke in the propagation direction: scattering by the smoke particles deviates a portion of the energy in other directions, making visible the light beam.

1.5. Interference and coherence

The wave characteristics of the fields **E** and **H** point out that superposition of different waves in the same region may lead to a distribution of the electromagnetic energy not uniform in space, with features which sometimes are unexpected if compared to everyday experience. For the sake of clarity, we will consider two linearly polarized beams E_1 and E_2 at the same frequency and both traveling along the **z** direction superimposed in the same region. The superposition principle gives us the total amplitude of the field as

$$E = E_1 e^{i(kz-\omega t)} + E_2 e^{i(kz-\omega t+\phi)}. \tag{1.130}$$

Our detectors being sensitive to the electromagnetic energy proportional to $|E|^2$, we are interested in

$$|E|^2 = E_1^2 + E_2^2 + E_1 E_2 e^{-i\phi} + E_1 E_2 e^{i\phi} = E_1^2 + E_2^2 + 2E_1 E_2 \cos(\phi).$$
(1.131)

And using Eq. (1.62) we get the total intensity as

$$I = I_1 + I_2 + 2\sqrt{I_1 I_2}\langle \cos \phi \rangle,$$
(1.132)

where ϕ is the phase difference between the two waves and $\langle \rangle$ stands for the time average performed on $\cos \phi$. Therefore, the total intensity may have different values in the superposition area varying from a maximum,

$$I_{\max} = I_1 + I_2 + 2\sqrt{I_1 I_2},$$
(1.133)

and a minimum,

$$I_{\min} = I_1 + I_2 - 2\sqrt{I_1 I_2},$$
(1.134)

corresponding respectively to *constructive interference*,

$$\langle \cos \phi \rangle = 1, \quad \text{i.e. } \phi = 2m\pi, \quad \text{with } m = 0, 1, 2, 3, \ldots,$$
(1.135)

and *destructive interference*,

$$\langle \cos \phi \rangle = -1, \quad \text{i.e. } \phi = (2m + 1)\pi, \quad \text{with } m = 0, 1, 2, 3, \ldots$$
(1.136)

When the intensity modulation given by the last term of Eq. (1.132) can be detected, we speak of *interference* between the waves, appearing as high intensity areas $(I > I_1 + I_2)$ alternate to low intensity ones $(I < I_1 + I_2)$. Therefore, the possibility of observing this effect depends strongly on the time behavior of ϕ. If it is constant in time at each point where the two waves are crossing, the intensity has a definite value between $I_{\max}$ and $I_{\min}$. On the contrary, when ϕ changes quickly with respect to the time of observation, one gets[4] $\langle \cos \phi \rangle = 0$ and the total intensity is just the sum of the intensities of each wave. The intensity modulation is more evident for equal intensity waves $(I_1 = I_2 \equiv I_0)$, since in this case

$$I_{\min} = 0 \quad \text{and} \quad I_{\max} = 4I_0.$$
(1.137)

---

[4]The average of the function $\cos(x)$ over a range L equal to the period Λ is zero: $\int_0^L \cos(x)dx = 0$.

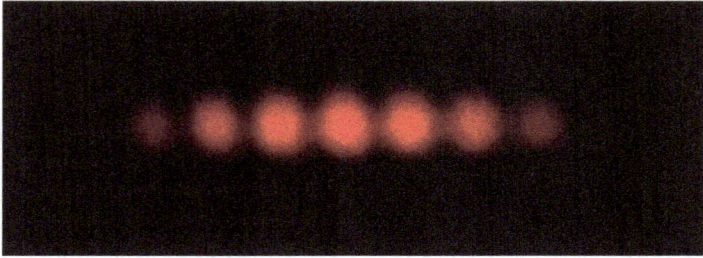

Fig. 1.9.   Interference fringes originated by a light beam crossing two vertical apertures. The waves coming from the apertures give rise to the interference pattern. The difference in intensity of maxima is due to the superimposed effect of diffraction (see Section 1.7).

In the visible it gives rise to a pattern of dark and bright fringes. An example of such a pattern is given in Fig. 1.9, where a picture of the interference of two waves coming from vertical apertures is shown.

In the case where the phase difference is due only to a different path d traveled by the two waves, with the corresponding phase difference[5] $\phi = k_0 n d = (2\pi/\lambda_0)nd$ and defining the optical path as $\delta = nd$, the conditions (1.135) and (1.136) become

$$k_0\delta = 2m\pi \rightarrow \delta = 2m\frac{\pi}{k_0} = 2m\frac{\lambda_0}{2}, \quad m = 0,1,2,3,\ldots, \tag{1.138}$$

$$k_0\delta = (2m+1)\pi \rightarrow \delta = (2m+1)\frac{\pi}{k_0} = (2m+1)\frac{\lambda_0}{2}, \quad m = 0,1,2,3,\ldots, \tag{1.139}$$

These are the most-used relationships to figure out when one gets constructive interference (optical path difference integer multiple of wavelength) or destructive interference (optical path difference odd multiple of half wavelength).

The two cases are depicted in Fig. 1.10, where the fields of two waves of equal amplitude are plotted versus time. In the first case the condition (1.138) is fulfilled, while in the second case the condition (1.139) is fulfilled.

The corresponding intensity distribution versus the optical path difference is shown in Fig. 1.11.

---

[5]In fact, for a plane wave with a spatial phase factor e^{ikz}, an additional path d would modify the phase factor as $e^{ik(z+d)} = e^{ikz}e^{ikd} \equiv e^{i(kz+\phi)}$.

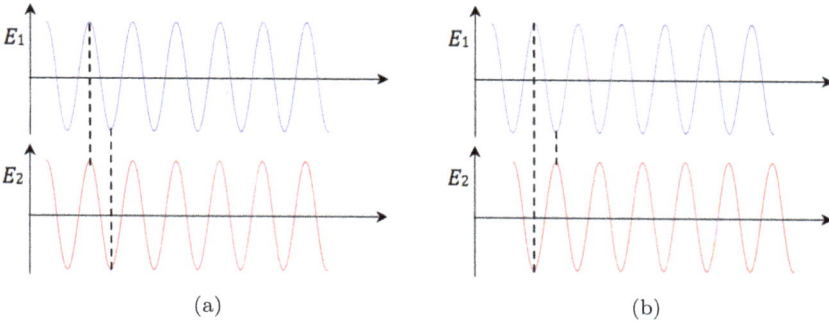

Fig. 1.10. Superposition of equal amplitude waves with the same phase (a) and with opposite phases (b).

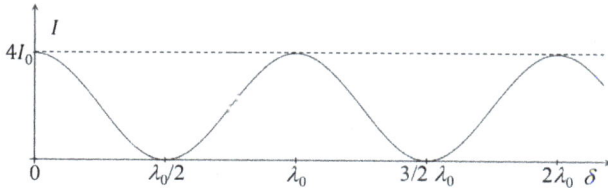

Fig. 1.11. Intensity distribution versus optical path difference for two interfering waves of equal intensity I_0. Here λ_0 is the wavelength in vacuum.

Usually the contrast of the interference pattern is measured by the *fringe visibility*, defined as

$$V = \frac{I_{max} - I_{min}}{I_{max} + I_{min}} = \frac{2\sqrt{I_1 I_2}}{I_1 + I_2}, \tag{1.140}$$

which falls in the range between 0 ($I_{max} = I_{min}$; no fringes) and 1 ($I_{min} = 0$; maximum visibility of the fringes). The latter condition can be realized in the case of equal intensities ($I_1 = I_2 \equiv I_0$):

$$V = \frac{2\sqrt{I_0 I_0}}{I_0 + I_0} = 1. \tag{1.141}$$

We have discussed here the simplest situation of interference between two waves of equal linear polarization, with the same frequency and wave vector. When the two waves have different polarization and a limited frequency bandwidth and wave vector distribution, we can apply the above treatment to each component of the waves (in frequency and wave vector) and, obviously, we will get a more complex result. However, the basic assumption that has been made is that of perfect sinusoidal waves, i.e.

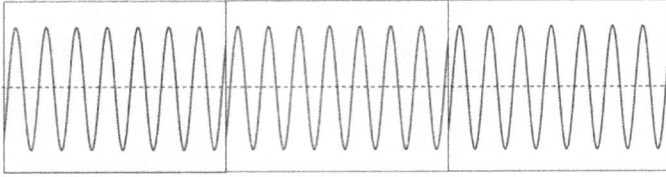

Fig. 1.12.   A sequence of wave packets with sudden changes of phase.

perfect plane waves. Since this condition is quite far from reality we must drop it, introducing the concept of *coherence*.

Let us consider the light emission of a generic source made up of a continuous sequence of wave packets, which are sinusoidal waves of definite length/duration. This sequence is characterized by sudden changes of phase from one wave packet to the next, as depicted in Fig. 1.12.

Such a model is more realistic, since it is able to explain the main aspects of interference starting from the conditions that must be fulfilled to observe this phenomenon. In fact, our everyday experience does not take into account light interference as it is not easily observed using conventional white light sources. Only in 19th century, Thomas Young was able to give an experimental demonstration of light interference after understanding the way of realizing mutually coherent waves, i.e. waves able to realize interference. In other words, we associate the concept of coherence with the ability of waves to produce observable interference. The reason for the difficulty of observing interference using conventional white light sources is the short duration of the wave packets leading to sudden changes of phase, each 10^{-12} s; since the observation is made on a much longer time scale, the phase difference between two light beams made up of many wave packets is not constant in time and gets all the possible values between 0 and 2π, producing an average value of $\langle \cos \phi \rangle = 0$, so that interference cannot be observed.

In order to clarify this concept, it is useful to consider wave packets of the same duration τ_c (*coherence time*), which defines also the *coherence length*:

$$\ell_c = c\tau_c. \tag{1.142}$$

This can be done by considering the *Michelson interferometer* for producing an interference pattern. The system is sketched in Fig. 1.13. A light beam coming from a point source is divided by a beam splitter (BS) in order to produce two sequences of equal wave packets. The two beams [one

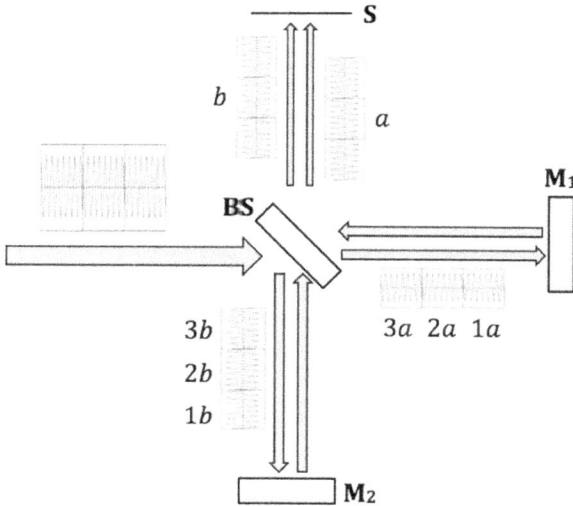

Fig. 1.13. Michelson interferometer. BS — beam splitter; M_1, M_2 — mirrors; S — screen. The two interfering waves are shown as a sequence of corresponding wave packets originated by the amplitude division performed by the beam splitter.

transmitted (a), one reflected (b) by BS] are reflected respectively by mirrors M_1 and M_2 and, after again crossing BS, they can be superimposed, and possibly an interference pattern can be observed on a screen S.

By applying the concept of interference to each wave packet of the sequences (a) and (b), it should be clear from the figure that when the path difference of the two sequences is bigger than ℓ_c each couple of interfering waves presents a different phase difference, owing to the random phase change occurring between different wave packets, and therefore, according to the previous considerations, no interference occurs. In fact, the observation corresponds to superposition of a huge number of wave packets making $\langle \cos \phi \rangle = 0$.

On the contrary, when the path difference is below ℓ_c, each wave packet of the sequence (a) interferes (at least partially) with the corresponding one of the sequence (b), with the phase difference for all of these couples fixed by the path difference, $\phi = 2\pi(L_a - L_b)/\lambda$, and therefore interference will be observed with $\langle \cos \phi \rangle \neq 0$.

In the first case, the two superimposed beam are *mutually incoherent*; in the second, they are *partially or totally coherent*. These two cases are depicted in Fig. 1.14, where the superposition of the two beams is shown after the BS.

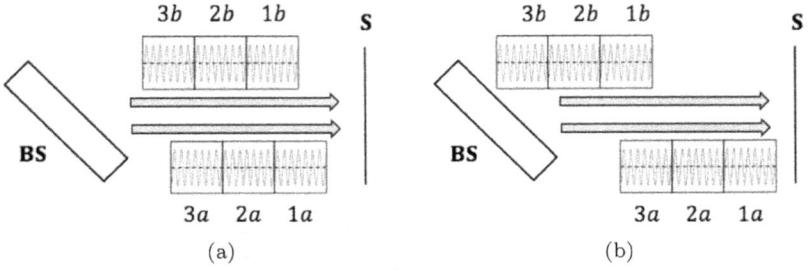

Fig. 1.14. Superposition of light beams in the Michelson interferometer: (a) partially coherent beams; (b) mutually incoherent beams.

Degree of coherence

A more quantitative analysis can be performed by writing the total field in a superposition point P as

$$E_P(t) = E_1(t + \tau) + E_2(t), \qquad (1.143)$$

with τ being the time delay between the two wave signals. Accordingly, the total intensity is

$$I_P(t) = c\langle E_P E_P^* \rangle = I_1 + I_2 + c\langle E_1(t + \tau)E_2^*(t)\rangle + \langle E_1^*(t + \tau)E_2(t)\rangle, \qquad (1.144)$$

where c is a constant dependent on the unit system. We can write

$$I_P(t) = I_1 + I_2 + 2c\Re\{\Gamma_{12}(\tau)\}, \qquad (1.145)$$

where $\Re$ indicates the real part of the *mutual coherence function*, defined as

$$\Gamma_{12}(\tau) = \langle E_1(t + \tau)E_2^*(t)\rangle. \qquad (1.146)$$

In order to obtain a relationship similar to Eq. (1.132), but valid for partially coherent radiation, we define the *complex degree of coherence*:

$$\gamma_{12}(\tau) = \frac{\Gamma_{12}(\tau)}{\sqrt{I_1 I_2}}. \qquad (1.147)$$

Then Eq. (1.145) becomes

$$I_P(\tau) = I_1 + I_2 + 2\sqrt{I_1 I_2}\Re\{\gamma_{12}(\tau)\}. \qquad (1.148)$$

By using the Euler formalism for complex numbers, we can write

$$\gamma_{12}(\tau) = m_{12}(\tau) \exp i\{2\pi\nu_0\tau\}, \qquad (1.149)$$

where $0 \leq m_{12}(\tau) \leq 1$ is a real quantity called the *degree of coherence*. In this way, the total intensity in P becomes

$$I_P(\tau) = I_1 + I_2 + 2\sqrt{I_1 I_2} m_{12}(\tau) \cos\{2\pi\nu_0\tau\}. \qquad (1.150)$$

It is easy to verify that the argument of cosine is the phase delay[6] $\phi = 2\pi\nu_0\tau$, and therefore Eq. (1.150) coincides with Eq. (1.132) when $m_{12}(\tau) = 1$, corresponding to complete coherence between the interacting waves. On the contrary, in the other extreme case of $m_{12}(\tau) = 0$ the interference term disappears.

The interference term depends on the delay τ between the two waves through the product of the slowly varying function $m_{12}(\tau)$ and of the rapidly varying function $\cos\{2\pi\nu_0\tau\}$. The latter is responsible for the appearance of the interference fringes (intensity maxima and minima in the light pattern), while the former affects the amplitude of the fringes, making them weaker and weaker as τ increases. This is easily shown by writing the intensity of the maxima and minima:

$$I_{\max}(t) = I_1 + I_2 + 2\sqrt{I_1 I_2} m_{12}(\tau), \qquad (1.151)$$

and

$$I_{\min}(t) = I_1 + I_2 - 2\sqrt{I_1 I_2} m_{12}(\tau). \qquad (1.152)$$

Using the definition (1.140) for the fringe visibility, we have

$$V = \frac{I_{\max} - I_{\min}}{I_{\max} + I_{\min}} = \frac{2\sqrt{I_1 I_2}}{I_1 + I_2} m_{12}(\tau). \qquad (1.153)$$

When $I_1 = I_2$,

$$V = m_{12}(\tau). \qquad (1.154)$$

In other words, when the light intensities of the interfering waves are the same, a measurement of the visibility is a direct measurement of the degree of coherence.

The dependence of Eq. (1.150) versus the delay time τ is shown in Fig. 1.15 for the common case where the degree of coherence $m_{12}(\tau)$ can be represented by a Gaussian function of τ.

So far we have discussed the *longitudinal coherence* of a light beam, since the Michelson interferometer is a tool for making an amplitude

---

[6]In fact, $\nu_0 = \frac{c}{\lambda_0}$ and then $\phi = 2\pi \frac{c}{\lambda_0} \tau = \frac{2\pi}{\lambda_0} \delta = k\delta$, taking into account that $\delta = c\tau$.

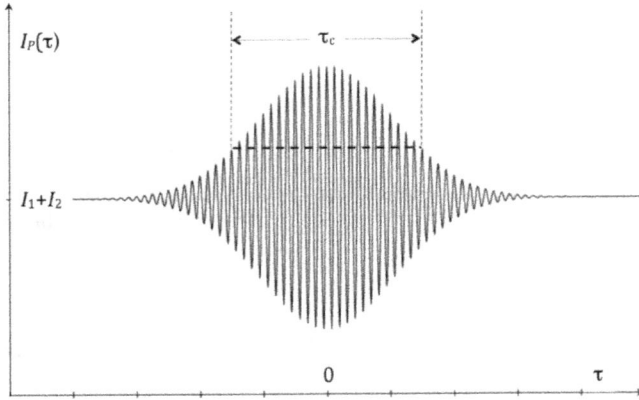

Fig. 1.15.   The function $I_P(\tau)$ versus τ, for a degree of coherence $m_{12}(\tau) = e^{-\frac{4\tau^2}{\tau_c^2}}$.

division creating two beams from a single original beam. In this way, the coherence length is the maximum distance between two different points of the wave that belong to the same sine function, and therefore the phase difference between these two points is determined only by their distance. This is also called *temporal coherence*. The same kind of discussion can be applied to light beams originated by different points in a transversal plane with respect to the wave propagation direction, as can be done by exploiting the Young setup shown in Fig. 1.16.

In this case, the same mathematical approach can be used and the degree of coherence will be expressed as a function of the distance between the considered points P_1 and P_2 instead of the time delay. The final result gives information on the phase correlation between the wave signals at different points of the transversal plane. In this case, we will get full coherence for a point source (when $P_1 = P_2$) and lower and lower coherence by increasing the distance P_1P_2; the minimum distance giving $m_{12}(\tau) = 0$ defines the spatial coherence length.

Actually, temporal and spatial coherence are two aspects of the same phenomenon and can be associated with properties that clearly define the quality of light. In fact, the temporal coherence is strictly related to the wave bandwidth ($\Delta\nu$), and the spatial coherence to the wave spreading ($\Delta\mathbf{k}$).

For what concerns temporal coherence, it can be shown that

$$\Delta\nu = \frac{1}{\tau_c}. \tag{1.155}$$

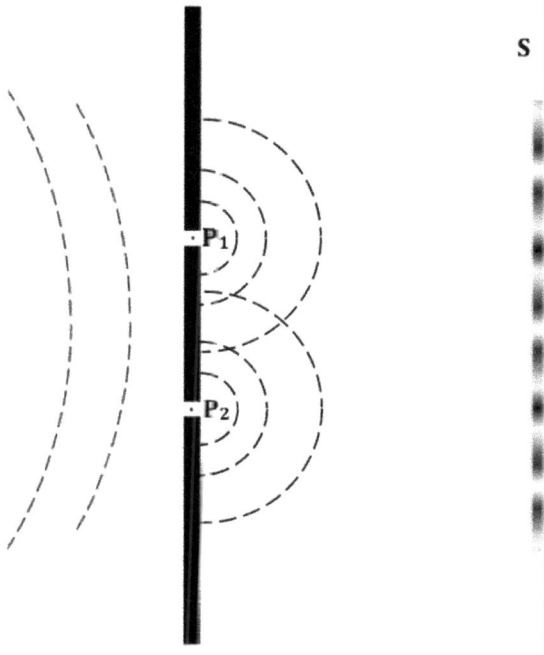

Fig. 1.16. The double-slit interferometer for measurement of the spatial coherence. The interference of the waves originated by the points P_1 and P_2 of a wave front is observed on the screen S.

This relation between coherence time and bandwidth is easily obtained by using the Fourier transform of the wave function. In fact, for a time function $f(t)$, the Fourier transform is

$$g(\nu) = \int_{-\infty}^{+\infty} f(t)e^{-i2\pi\nu t}dt,$$

which for real and even functions leads to

$$g(\nu) = \int_{-\infty}^{+\infty} f(t)\cos\{2\pi\nu t\}dt = 2\int_{0}^{+\infty} f(t)\cos\{2\pi\nu t\}dt$$

and

$$f(t) = \int_{-\infty}^{+\infty} g(\nu)\cos\{2\pi\nu t\}d\nu = 2\int_{0}^{+\infty} g(\nu)\cos\{2\pi\nu t\}d\nu.$$

This means that $g(\nu)$ represents the spectral content of $f(t)$, i.e. the frequencies and their amplitudes included in $f(t)$.

Let us apply this concept to a single wave packet of length τ_c. It can be represented by the following function:

$$f(t) = a \cos\{2\pi\nu_0 t\} \quad \text{when} \quad -\frac{\tau_c}{2} \leq t \leq \frac{\tau_c}{2},$$

with a as constant amplitude, and

$$f(t) = 0 \quad \text{when } t < -\frac{\tau_c}{2} \text{ and } t > \frac{\tau_c}{2}.$$

Then we get

$$g(\nu) = 2 \int_0^{\frac{\tau_c}{2}} a \cos\{2\pi\nu_0 t\} \cos\{2\pi\nu t\} dt$$

$$= a \int_0^{\frac{\tau_c}{2}} [\cos\{2\pi(\nu - \nu_0)t\} + \cos\{2\pi(\nu + \nu_0)t\}] dt$$

$$= \frac{a \sin \pi(\nu - \nu_0)\tau_c}{2\pi(\nu - \nu_0)} + \frac{a \sin \pi(\nu + \nu_0)\tau_c}{2\pi(\nu + \nu_0)} \approx \frac{a \sin \pi(\nu - \nu_0)\tau_c}{2\pi(\nu - \nu_0)},$$

with the approximation $\nu \approx \nu_0$ (then $\nu + \nu_0 \gg \nu - \nu_0$). The functions $f(t)$ and $g(\nu)$ are sketched in Fig. IN1.1.

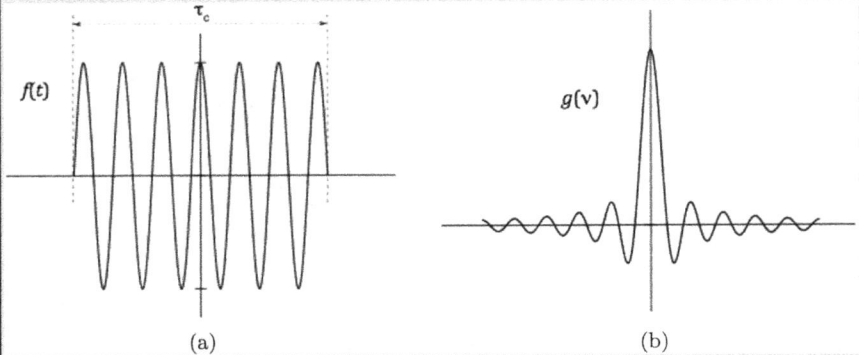

Fig. IN1.1. (a) The wave packet $f(t)$ and (b) the corresponding Fourier transform $g(\nu)$.

The corresponding energy distribution (spectral distribution) is proportional to the square of the field, and therefore to the square of $g(\nu)$:

$$S(\nu) = \left[\frac{a \sin \pi(\nu - \nu_0)\tau_c}{2\pi(\nu - \nu_0)}\right]^2 = \left(\frac{a\tau_c}{2}\right)^2 \left[\frac{\sin \pi(\nu - \nu_0)\tau_c}{\pi(\nu - \nu_0)\tau_c}\right]^2.$$

The term in square brackets called the *sinc function*, determines the frequency dependence. It is shown in Fig. IN1.2.

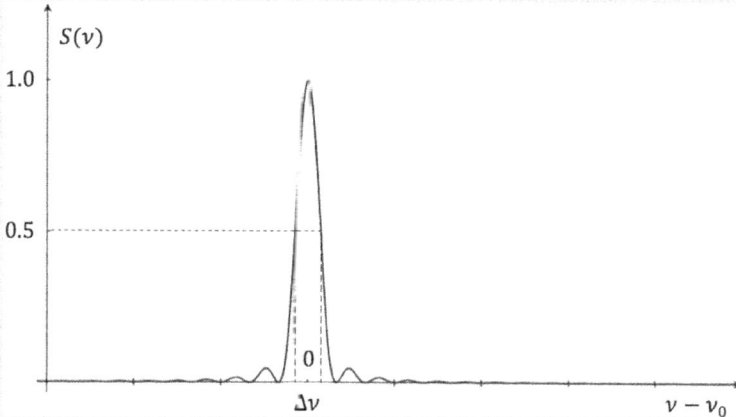

Fig. IN1.2. Normalized spectral distribution of a sinusoidal wave packet.

The first zero of this function occurs at $\pi(\nu_1 - \nu_0)\tau_c = \pi$. Then the width of the first maximum (which includes most of the wave energy) is $2\Delta\nu = 2/\tau_c$, and therefore the bandwidth defined as full width at half maximum is $\Delta\nu = 1/\tau_c$.

This means that a large coherence time corresponds to a narrow bandwidth, i.e. nearly monochromatic light, while a small coherence time means a large bandwidth, i.e. a white spectrum. For a conventional source $\tau_c < 10^{-14}$ s and as a consequence we have a white spectrum. On the contrary, a laser source may have a very large $\tau_c > 10^{-8}$ s and thus a very narrow bandwidth, appearing at a glance as a very definite color of the beam. Since the coherence time affects the shape of the degree of coherence $m_{12}(\tau)$, this will be related to the spectral content of the light.

The mathematical relationship is found by the Wiener–Khinchin theorem linking $\Gamma_{12}(\tau)$ to $S(\nu)$:

$$S(\nu) = \int_{-\infty}^{+\infty} \Gamma_{12}(\tau) e^{-i2\pi\nu\tau} d\tau.$$

By supposing a Gaussian spectral distribution $S(\nu)$, the inverse operation gives $\Gamma_{12}(\tau)$ and

$$\gamma_{12}(\tau) = e^{-\left(\frac{2\tau}{\tau_c}\right)^2} e^{i2\pi\nu_0\tau}.$$

Then

$$m_{12}(\tau) = e^{-\left(\frac{2\tau}{\tau_c}\right)^2}.$$

For what concerns spatial coherence, the beam directionality can be associated with the spatial extension of the coherence along the wave front. A relationship can be easily worked out by exploiting the condition (1.171), as shown in the next section about diffraction.

By considering a limited spatial extension of a wave front, it comes out that it cannot be a planar wave front but it has a minimum half angular aperture given by

$$\alpha = \frac{\lambda}{2a}, \tag{1.156}$$

with a being the spatial extension of the wave front, as sketched in Fig. 1.17.

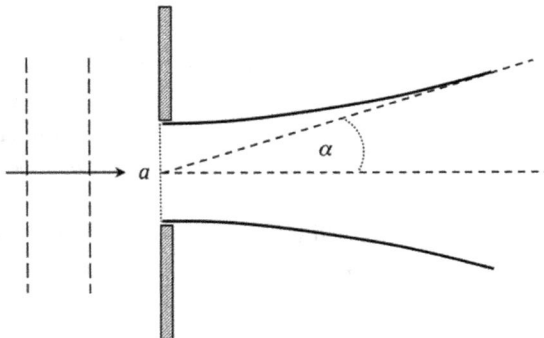

Fig. 1.17.   The minimum divergence of a spatially limited wave front.

Therefore, the minimum value of the transversal component of the wave vector is

$$k_\perp \approx \alpha k_z = \frac{\lambda}{2a} \frac{2\pi}{\lambda} = \frac{\pi}{a},$$ (1.157)

in the approximation $k_\perp \ll k_z$. Then $\tan\alpha \approx \alpha$.

In other words, the transversal key vector distribution has a full width:

$$\Delta k = \frac{2\pi}{a}.$$ (1.158)

This equation is the spatial counterpart of Eq. (1.155), thus connecting the wave vector distribution (directionality) to the spatial extension of the wave front in the same way as Eq. (1.155) connects the frequency bandwidth (monochromaticity) to the average time duration of the wave packets.

1.6. Diffraction

Diffraction is the basic phenomenon characterizing the wave propagation. It occurs any time the wave front is limited by any kind of obstacle, which is often given by an opaque wall with a small aperture. We can easily observe diffraction by looking at elastic waves excited in water. In fact, taking a small tank full of water divided into two parts separated by a wall with an aperture in the central part and exciting a spherical wave in one side, we can observe that when the wave reaches the aperture, it travels as if it were generated in the aperture itself and the wave disturbance will be able to reach all the limiting walls of the other side. This is a diffraction effect typical of wave propagation. On the contrary, by using a similar setup, i.e. a space divided into two parts by a wall with an aperture having a screen somewhere after the aperture and launching a number of beads (particles) from one side in any direction, we will find some beads only in the area of the screen corresponding to the shadow of the aperture, while many other beads will be reflected by the wall (see Fig. 1.18).

The starting point for dealing with wave propagation is *Huygens' principle*. According to it, each point P of the wave front can be considered as the source of a spherical wave whose amplitude is proportional to $\cos\theta$, with θ being the angle between the wave vector **k** of the incoming wave and the half line normal to the wave front originated at P (see Fig. 1.19).

In this way, the wave amplitude in each plane is due to the superposition of each elemental wave originated by all the points of the wave front. This is

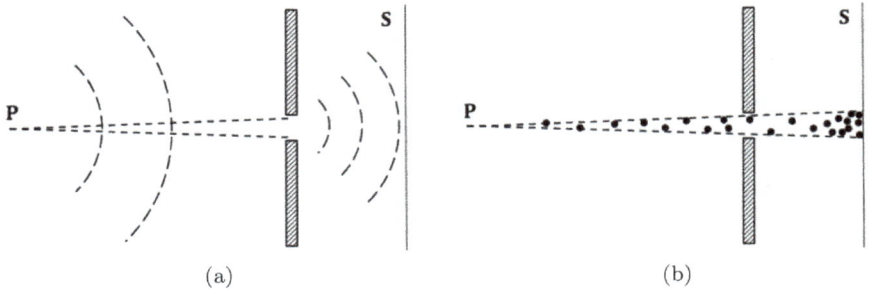

Fig. 1.18.   The difference between wave (a) and particle (b) propagation. Waves undergo diffraction through an aperture.

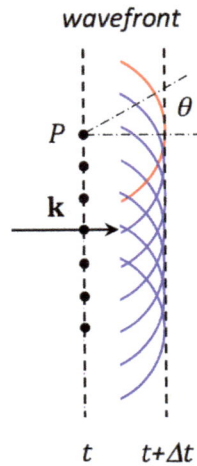

Fig. 1.19.   Application of Huygens' principle to a plane wave — sketch of the formation of the wave front at time $t + \Delta t$ originated by a plane wave at time t.

in agreement with the wave equation, that is a linear equation thus fulfilling the superposition principle.

In order to express mathematically this concept, we write the spherical wave field originated at P as

$$E(\mathrm{P}) = E_0 \frac{e^{ikr}}{r}, \qquad (1.159)$$

and therefore the field at a point P′ beyond the screen is

$$E(\mathrm{P}') = \int_{\mathrm{aperture}} E_0(\varrho) \frac{e^{ikr}}{r} \cos\theta \, d\varrho, \qquad (1.160)$$

where $\varrho(\xi, \eta)$ is a position vector in the plane of the aperture, and the distance $r \equiv r(\mathrm{P}, \varrho)$.

By neglecting the amplitude variation in r and under paraxial approximation ($\cos\theta \sim 1$), we can approximate r in the denominator as the fixed distance z between the aperture and the observation plane:

$$E(\mathrm{P}') = \frac{1}{z} \int_{\text{aperture}} E_0(\varrho) e^{ikr} d\varrho. \tag{1.161}$$

For the phase term, the correct expression of r is required:

$$r = \sqrt{(x - \xi)^2 + (y - \eta)^2 + z^2}$$

$$= R\sqrt{1 - \frac{\xi^2 + \eta^2 - 2(\xi x + \eta y)}{R^2}}, \tag{1.162}$$

where

$$R = \sqrt{x^2 + y^2 + z^2}.$$

All the pertinent coordinates of the problem are depicted in Fig. 1.20.

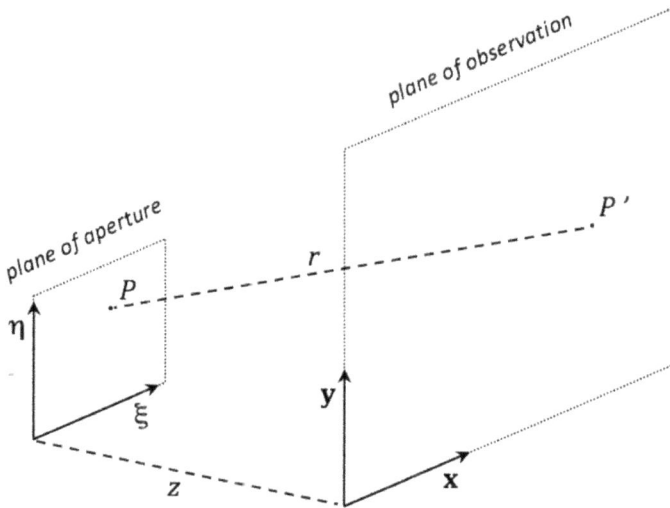

Fig. 1.20. Sketch of the aperture and observation plane with coordinates related to the diffraction problem.

In the case of relatively large distances, $\xi^2 + \eta^2 - 2(\xi x + \eta y) \ll R^2$, a series expansion of Eq. (1.162) gives

$$r \approx R\left(1 + \frac{\xi^2 + \eta^2 - 2(\xi x + \eta y)}{2R^2}\right) = R + \frac{\xi^2 + \eta^2}{2R} - \frac{\xi x + \eta y}{R}.$$

$$(1.163)$$

This expression allows considering the diffraction problems depending on the relative magnitude of the last two terms. We deal with Fraunhofer diffraction when looking at the far field distribution which we get by neglecting $(\xi^2 + \eta^2)/2R$. In the other cases, we deal with Fresnel diffraction. Actually, the Fraunhofer condition corresponds to considering the "far field" distribution, with R being large enough to fulfill the inequality:

$$k\frac{\xi^2 + \eta^2}{2R} \ll 1, \quad \text{i.e. } R \gg k\frac{\xi^2 + \eta^2}{2}. \tag{1.164}$$

This leads to

$$e^{ikr} = e^{ik(R + \frac{\xi^2 + \eta^2}{2R} - \frac{\xi x + \eta y}{R})} \approx e^{ik(R - \frac{\xi x + \eta y}{R})}. \tag{1.165}$$

Then

$$E(x, y) = \frac{e^{ikz}}{z} \int_{\text{aperture}} E_0(\xi, \eta) e^{-i\frac{k}{z}(\xi x + \eta y)} d\xi d\eta, \tag{1.166}$$

where the approximation $R = \sqrt{x^2 + y^2 + z^2} \approx z$ has been used.

In the real cases, in order to observe the far field distribution it is necessary to focus the light beam with a converging lens, and therefore we can write $z = f$, the focal length of the used lens, and

$$E(x, y) = \frac{e^{ikf}}{f} \int_{\text{aperture}} E_0(\xi, \eta) e^{-i\frac{k}{f}(\xi x + \eta y)} d\xi d\eta. \tag{1.167}$$

We know that the observed intensity is proportional to $|E(x, y)|^2$, so we define the term outside the integral as a constant factor C that does not determine the intensity distribution in the observation plane. In this way, the calculation of the diffraction pattern is performed by specifying the aperture geometry:

$$E(x, y) = C \int_{\text{aperture}} E_0(\xi, \eta) e^{-i\frac{k}{f}(\xi x + \eta y)} d\xi d\eta. \tag{1.168}$$

(1) *One-dimensional slit*

A very common case is that of a plane wave of constant amplitude impinging on a one-dimensional slit aperture of width $2a$. In this case Eq. (1.168) reduces to

$$E(x) = C \int_{-a}^{a} E_0 e^{-i\frac{k}{f}(\xi x)} d\xi.$$

Therefore,

$$E(x) = CE_0 \int_{-a}^{a} e^{-i\frac{k}{f}(\xi x)} d\xi$$

$$= -\frac{CE_0 f}{ikx} \left[e^{-i\frac{k}{f}(ax)} - e^{+i\frac{k}{f}(ax)} \right] = 2aCE_0 \frac{\sin\left(\frac{kax}{f}\right)}{\left(\frac{kax}{f}\right)}$$

and the corresponding intensity will be

$$I(x) = K \frac{\sin^2\left(\frac{kax}{f}\right)}{\left(\frac{kax}{f}\right)^2},$$

with K being a dimensional constant, dependent on the unit system.

(2) *Rectangular aperture*

It is easy to show that this result can be extended to a rectangular aperture in the xy plane with side lengths $2a$ and $2b$ to give

$$I(x, y) = K \frac{\sin^2\left(\frac{kax}{f}\right)}{\left(\frac{kax}{f}\right)^2} \frac{\sin^2\left(\frac{kby}{f}\right)}{\left(\frac{kby}{f}\right)^2}.$$

The function $f(x) = \frac{\sin(x)}{x}$ is usually defined as $\mathrm{sinc}(x)$, and therefore the above expressions are often written as

$$I(x) = K\mathrm{sinc}^2\left(\frac{kax}{f}\right) \text{ for a one-dimensional slit,}$$

$$I(x, y) = K\mathrm{sinc}^2\left(\frac{kax}{f}\right) \mathrm{sinc}^2\left(\frac{kby}{f}\right) \text{ for a rectangular aperture slit.}$$

(3) *Circular aperture*

When one is considering a circular aperture of radius a, it is more convenient to employ polar coordinates, using the transformations for the aperture plane $\xi = \zeta \cos \phi$, $\eta = \zeta \sin \phi$, and polar coordinates ρ and θ for the observation plane: $x = \rho \cos \theta$, $y = \rho \sin \theta$.

Then we have

$$E(r,\theta) = C \int_0^a \int_0^{2\pi} E_0 e^{-i\frac{k}{f}(\zeta \cos \phi \rho \cos \theta + \zeta \sin \phi \rho \sin \theta)} \zeta d\zeta d\phi$$

$$= C \int_0^a \int_0^{2\pi} E_0 e^{-i\frac{k}{f}\rho\zeta \cos(\theta-\phi)} \zeta d\zeta d\phi.$$

The integration is performed by using the properties of the Bessel functions defined by

$$J_n(x) = \frac{i^{-n}}{2\pi} \int_0^{2\pi} e^{-ix\cos\gamma} e^{in\gamma} d\gamma,$$

with

$$J_0(x) = \frac{1}{2\pi} \int_0^{2\pi} e^{-ix\cos\gamma} d\gamma \quad \text{and} \quad xJ_1(x) = \int_0^x x' J_0(x') dx'.$$

Then

$$E(\rho,\theta) = 2\pi CE_0 \int_0^a J_0 \left(\frac{k\rho\zeta}{f} \right) \zeta d\zeta,$$

$$E(\rho,\theta) = CE_0 \pi a^2 \frac{2J_1\left(\frac{ka\rho}{f}\right)}{\left(\frac{ka\rho}{f}\right)}.$$

Finally, the intensity distribution is given by

$$I(\rho,\theta) = K \left| \frac{2J_1\left(\frac{ka\rho}{f}\right)}{\left(\frac{ka\rho}{f}\right)} \right|^2 = K \left| \frac{2J_1\left(\frac{2\pi a\rho}{f\lambda}\right)}{\left(\frac{2\pi a\rho}{f\lambda}\right)} \right|^2 = K \left| \frac{2J_1\left(\frac{2\pi a \sin \alpha}{\lambda}\right)}{\left(\frac{2\pi a \sin \alpha}{\lambda}\right)} \right|^2,$$

with K being a dimensional constant, dependent on the unit system, and defining $\rho = f \sin \alpha$, where α is the divergence angle measured with respect to the normal to the aperture. This is the well-known Airy function, whose shape is very similar to the one obtained for a singular slit or for a rectangular aperture.

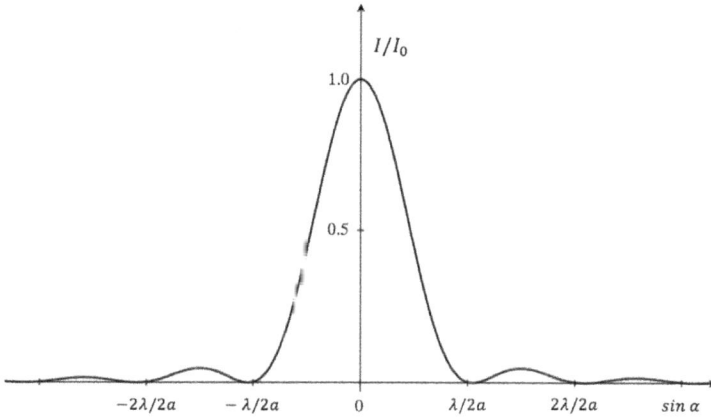

Fig. 1.21. Far field intensity distribution of a plane wave diffracted by a one-dimensional slit of width $2a$.

Let us discuss briefly the intensity distribution shown in Fig. 1.21, originated by a plane wave of wavelength λ crossing a one-dimensional slit of width $2a$:

$$I(x) = I_0 \frac{\sin^2\left(\frac{kax}{f}\right)}{\left(\frac{kax}{f}\right)^2}. \tag{1.169}$$

where $I_0 = I(0)$. By considering the divergence angle α measured with respect to the normal to the aperture, we can write $x = f\sin(\alpha)$, and by substituting $k = 2\pi/\lambda$ we get

$$I(\alpha) = I_0 \left| \frac{\sin\left(\frac{2\pi a \sin\alpha}{\lambda}\right)}{\left(\frac{2\pi a \sin\alpha}{\lambda}\right)} \right|^2. \tag{1.170}$$

Note that the width of distribution depends on the ratio λ/a being wider and wider as this ratio increases. In fact, minima of $I(\alpha)$ correspond to

$$\frac{2\pi a \sin\alpha}{\lambda} = m\pi \rightarrow \sin\alpha = m\frac{\lambda}{2a}, \quad m = 1, 2, 3, \ldots. \tag{1.171}$$

When $\lambda \ll 2a \rightarrow \sin\alpha \approx \alpha$, $\alpha = \frac{\lambda}{2a}$ is the angular width of the brighter central part of the beam. Since we have supposed an impinging plane wave with a **k** vector along the z axis ($\alpha = 0$), this expression gives the minimum divergence induced by the diffraction. It is evident that in the approximation $\lambda \ll 2a$, $\alpha \rightarrow 0$ and we are in the linear propagation regime, typical of geometrical optics, while as soon as a approaches λ the wave

Fig. 1.22.   A diffraction pattern generated by a light beam crossing a one-dimensional slit aperture.

propagation becomes evident through the occurrence of light diffraction leading to increasing divergence of the beam.

An example of a diffraction pattern due to a slit aperture is shown in Fig. 1.22.

These results are very important, since they allow distinguishing wave from particle propagation. On the other hand, they also point out that when $\lambda \ll 2a$ the difference in the behavior of waves and particles is difficult to observe. In this case, we do not need to describe light as waves, but we can account for several optical phenomena just by tracing "rays." The ray is an oriented line whose direction corresponds to the wave vector direction of the associated wave; it undergoes linear propagation and follows the laws of reflection and refraction. This is the regime of *geometrical optics*, which can be used to describe the basic behavior of several optical devices, from simple lenses to optical instruments such as microscopes and telescopes. In this case, any effect not well explained under this approximation is treated as an optical aberration of the system. However, a deeper understanding of the working conditions of an optical system always requires a full electromagnetic treatment.

On the other hand, the occurrence of diffraction when $\lambda \sim 2a$ can be considered the signature of wave propagation. In other words, the observation of a diffraction phenomenon tells us that a wave must be involved in it. This is a very remarkable statement, since it has been a basic demonstration of the wave character of the elemental particles, as we will discuss in the next chapter.

The given example of an aperture limiting the wave front is actually the classical and most evident case for diffraction. However, this phenomenon is a basic one related to wave propagation occurring when any propagation parameter of the wave is locally modified. In order to understand this concept, it is better to write Eq. (1.168) in a more general way:

$$E(x, y) = C \int_{-\infty}^{+\infty} E_0(\xi, \eta) t(\xi, \eta) e^{-i\frac{k}{f}(\xi x + \eta y)} d\xi d\eta, \qquad (1.172)$$

where $E_0(\xi, \eta)$ is the field incident on the plane $\xi\eta$ and $t(\xi, \eta)$ represents the transmission function at this plane. This means that $t(\xi, \eta) = 1$ in the whole plane for a free traveling wave, while in the previous case of an aperture blocking most of the wave front $t(\xi, \eta) = 1$ in the aperture area and $t(\xi, \eta) = 0$ elsewhere, thus reducing Eq. (1.172) to Eq. (1.168). In this way, we can consider Eq. (1.172) the general expression for the Fraunhofer diffraction integral observed at a focal plane of a converging lens whose focal length is f.

We must emphasize that $t(\xi, \eta)$ in general is a complex quantity since the obstacle may induce a change of amplitude and/or a change of phase. In particular, the diffraction integral should be used any time the wave front travels through a region where the optical properties are not uniform. There are many different situations of this kind.

For instance, a light wave that is limited by an obstacle on one side suffers diffraction on that side, diffraction being more easily observed for a sharp obstacle. This effect actually occurs on the sides of any aperture the wave may cross, independently of the size of the aperture.

A different example is given by a wave traveling in a medium of a given refractive index that crosses a small area showing a different index; this will cause a change of phase in the wave front corresponding to this inhomogeneity owing to the different optical path δ traveled ($\delta = \Delta n \, d$, with Δn being the refractive index difference and d the inhomogenety thickness), and the resulting far field can be evaluated by the diffraction integral (1.172).

A typical example is that of a collimated beam partially blocked by a sharp edge obstacle leading to a strong diffraction effect, as shown in Fig. 1.23. The original dimension (before diffraction) of the light beam was limited to the bright central spot.

1.7. Diffraction gratings

Material structures that give rise to a periodic variation of amplitude and/or phase on an incident wave are called *diffraction gratings*. They have a

Fig. 1.23. A diffraction pattern made by a sharp edge obstacle.

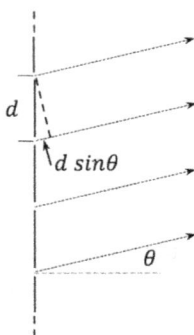

Fig. 1.24.   Wave interference by a diffraction grating made by N slits.

paramount role in many processes involving light–matter interaction, being basic elements of several optical devices and optical instruments. Their properties are a consequence of the effects of diffraction and interference described above.

The simplest diffraction grating is made by a high number (N) of equally spaced slits. Following a classical approach, we consider a plane wave at normal incidence on the grating plane and the waves diffracted by each aperture, which superimpose in the focal plane of a lens. It is worth noting the fundamental role of diffraction in this process: in fact, the origin of waves traveling in different directions is due to diffraction of the incoming plane wave by each aperture. For each direction at an angle θ with respect to the grating normal, we will take into account the sum of N waves, each with a path difference $d\sin\theta$ with respect to the one originated by the next slit, as depicted in Fig. 1.24, with d being the distance between two neighboring slits.

If the incident field amplitude E_0 is constant on each slit, the total field due to the interference of N waves is given by

$$E_{\text{tot}}(\theta) = \sum_1^N E_n = \sum_1^N A(\theta)e^{i(kr_n - \omega t)} = A(\theta)e^{i(kr_1 - \omega t)} \sum_1^N e^{ik(n-1)d\sin\theta},$$

$$(1.173)$$

where $A(\theta)$ is the diffracted amplitude from each slit.

By defining $x = e^{ikd\sin\theta}$, we get

$$E_{\text{tot}}(\theta) = A(\theta)e^{i(kr_1 - \omega t)} \sum_1^N x^{(n-1)} = A(\theta)e^{i(kr_1 - \omega t)} \frac{x^N - 1}{x - 1}. \qquad (1.174)$$

The right-hand side fraction can be rewritten as

$$\frac{x^N - 1}{x - 1} = \frac{e^{ikNd\sin\theta} - 1}{e^{ikd\sin\theta} - 1} = \frac{e^{ik(N/2)d\sin\theta}}{e^{ik(1/2)d\sin\theta}} \frac{e^{ik(N/2)d\sin\theta} - e^{-ik(N/2)d\sin\theta}}{e^{ik(1/2)d\sin\theta} - e^{-ik(1/2)d\sin\theta}}$$

$$= e^{ik((N-1)/2)d\sin\theta} \frac{\sin(k(N/2)d\sin\theta)}{\sin(k(1/2)d\sin\theta)}. \tag{1.175}$$

Then

$$E_{\text{tot}}(\theta) = A(\theta)\frac{\sin(k(N/2)d\sin\theta)}{\sin(k(1/2)d\sin\theta)} e^{i(kr_1 - \omega t)} e^{ik((N-1)/2)d\sin\theta}. \tag{1.176}$$

On the other hand, each slit of width[7] a gives rise to wave diffraction and therefore the far field amplitude distribution will be modulated by the diffraction function and given by

$$E_F(\theta) = E_0 \frac{\sin\left(\frac{ka\sin\theta}{2}\right)}{\frac{ka\sin\theta}{2}} \frac{\sin(k(N/2)d\sin\theta)}{\sin(k(1/2)d\sin\theta)}. \tag{1.177}$$

Then the intensity distribution at the focal plane becomes

$$I_F(\theta) = I_F(0) \left[\frac{\sin\left(\frac{ka\sin\theta}{2}\right)}{\frac{kc\sin\theta}{2}}\right]^2 \left[\frac{\sin\left(\frac{kNd\sin\theta}{2}\right)}{\sin\left(\frac{kd\sin\theta}{2}\right)}\right]^2 \tag{1.178}$$

or

$$I_F(\theta) = I_F(0) \left[\frac{\sin\left(\frac{\pi a\sin\theta}{\lambda}\right)}{\frac{\pi c\sin\theta}{\lambda}}\right]^2 \left[\frac{\sin\left(\frac{\pi Nd\sin\theta}{\lambda}\right)}{\sin\left(\frac{\pi d\sin\theta}{\lambda}\right)}\right]^2. \tag{1.179}$$

The function

$$\left[\frac{\sin\left(\frac{\pi Nd\sin\theta}{\lambda}\right)}{\sin\left(\frac{\pi d\sin\theta}{\lambda}\right)}\right]^2 \quad \text{has a maximum of height } N^2 \text{ each time}$$

the denominator $\sin\left(\frac{\pi d\sin\theta}{\lambda}\right) = 0.$

It occurs for

$$\frac{\pi d\sin\theta}{\lambda} = 0 \pm m\pi \rightarrow \sin\theta = 0 \pm m\frac{\lambda}{d}, \quad m = 1, 2, 3, \dots. \tag{1.180}$$

This function is modulated by the diffraction pattern having minima at

$$\frac{\pi a\sin\theta}{\lambda} = n\pi \rightarrow \sin\theta = n\frac{\lambda}{a}, \quad n = 1, 2, 3, \dots. \tag{1.181}$$

Usually $a < d$, and therefore the intensity pattern has the shape sketched in Fig. 1.25.

---

[7]It must be taken into account that in the former section the slit width has been defined as $2a$. This explains the formal difference between Eq. (1.181) and Eq. (1.171).

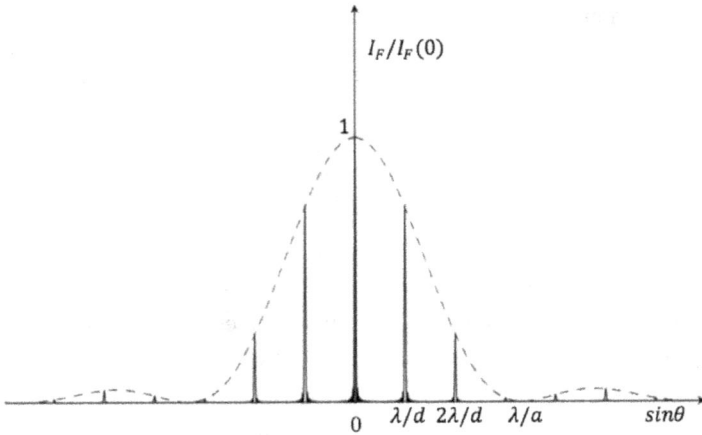

Fig. 1.25. A normalized intensity pattern created by a diffraction grating made by N slits. The dashed line represents the sinc function given by the first term on the right of Eq. (1.179), which modulates the second term made by equally spaced peaks (λ/d). The total intensity is normalized dividing the maximum by N^2.

In the intensity pattern due to a diffraction grating, the position of maxima is remarkably affected by the light wavelength. Since for small angles we can approximate $\sin\theta \approx \theta$, Eq. (1.180) shows a linear dependence of the angular dispersion of maxima with wavelength. This property makes diffraction gratings very useful as dispersive devices for many applications, from spectrometers to tunable lasers.

By neglecting the central maximum at $\theta = 0$, it is useful to reverse Eq. (1.180) as

$$d\sin\theta = m\lambda, \quad m = 1, 2, 3, \ldots . \tag{1.182}$$

which is the typical grating equation found under many different physical conditions. It defines the location of the maxima, where the integer m defines the *order of diffraction*.

Figure 1.26 shows how the number of slits affects the width of the diffracted peaks. In it is plotted the normalized intensity of the first order ($m = 1$) diffraction peak occurring at $\sin\theta_1 = \lambda/d$ for an increasing number of slits: $N = 10, 100, 1000$. The strong spectral narrowing suggests the use of grating with a high value of N to enhance the wavelength selectivity, as discussed in the inset on the next page.

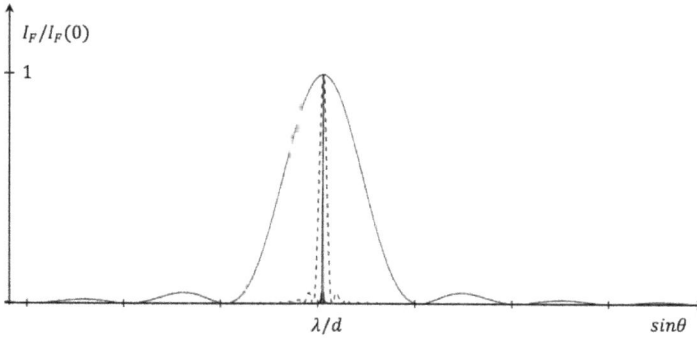

Fig. 1.26. A normalized intensity pattern of the first order diffracted beam for gratings with an increasing number of slits. The broad continuous curve corresponds to $N = 10$; the dashed line corresponds to $N = 100$; the narrow peak corresponds to $N = 1000$.

An important issue related to application of a diffraction grating as a dispersive device is the *resolving power*, i.e. the ability to separate maxima originated by different wavelengths. It is defined as $\lambda/\Delta\lambda$, where $\Delta\lambda$ is the wavelength separation around a specific value λ.

A common criterion states that maxima are resolved if the one corresponding to the next wavelength falls in the position of a minimum of the other wavelength.

Minima are found from Eq. (1.179) to be at

$$\sin\left(\frac{\pi N d \sin\theta}{\lambda}\right) = 0 \rightarrow \frac{\pi N d \sin\theta}{\lambda} = m\pi \rightarrow \sin\theta = m\frac{\lambda}{Nd},$$

$$m = 1, 2, 3, \ldots.$$

Since for $m = 0$ we have the first maximum, the quantity λ/Nd is the separation between maxima and minima of the same order of diffraction.

On the other hand, maxima are determined by the condition (1.182), that for the wavelength $\lambda + \Delta\lambda$ becomes

$$d(\sin\theta)_{\lambda+\Delta\lambda} = m(\lambda + \Delta\lambda).$$

Thus, for a variation of wavelength from λ to $\lambda + \Delta\lambda$ the corresponding shift of maxima is

$$\Delta(\sin\theta) = \frac{m\Delta\lambda}{d}.$$

The maxima are resolved if the one at $\lambda + \Delta\lambda$ falls in the position of the minimum for the wavelength λ for the same order of diffraction m.

Then the maximum shift must be equal to the separation between maxima and minima for the wavelength λ:

$$\frac{m\Delta\lambda}{d} = \frac{\lambda}{Nd}.$$

Finally, the resolving power is obtained as

$$\frac{\lambda}{\Delta\lambda} = mN.$$

Therefore, the resolving power at a given order of diffraction depends on the total number of slits (which are grooves in practical devices).

As an example, we can evaluate the value of N required to separate lines which are 1 Å apart around the center of the visible spectrum at 5500 Å. For the first order $(m = 1)$ it simply gives $N = 5500$, and for the second order $N = 11,000$.

The problem of a diffraction grating could have been addressed in a straightforward way starting from Eq. (1.172), by taking into account the proper transmission function $t(\xi, \eta)$, which in the case of a sequence of N slits would be a periodic step function with alternate values 0 and 1.

Therefore, starting from that equation, we can discuss different physical situations described by a specific periodic function $t(\xi, \eta)$ affecting the amplitude and phase of the incoming wave. However, sometimes alternative analytic methods allow achieving the most important results, overcoming some problems related to integration of Eq. (1.172) following approximations based on the experimental observations.

Actually, different kinds of gratings besides the one made by N slits are considered and widely used in technology. For application in spectroscopy, rather than a *transmission grating* like that previously discussed, the most popular one is the *reflection grating* made by a reflecting surface whose reflectivity is periodically modulated by a large number of grooves. The same treatment as the one developed above can be applied to this case, considering the superposition of the wave diffracted by each successive grove having the same direction and a relative phase shift dependent on the incident angle and grove inclination, as depicted in Fig. 1.27.

Images of dispersion by diffraction gratings are given in Fig. 1.28. A broad band "white beam" is considered to be impinging on a diffraction grating to highlight the dispersion properties of this optical device.

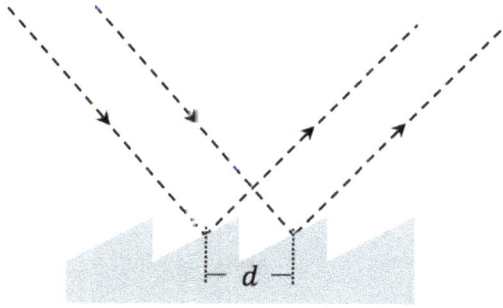

Fig. 1.27. Wave interference by an N-groove reflection grating.

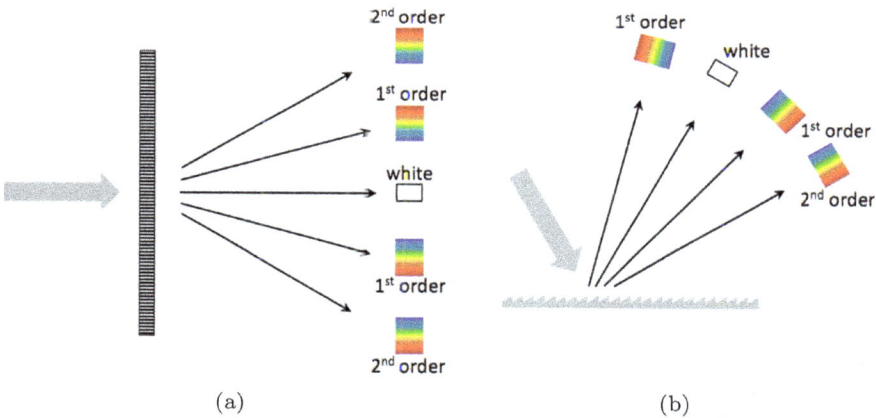

(a)   (b)

Fig. 1.28. Wavelength dispersion in (a) transmission grating; (b) reflection grating.

The examples considered above concern diffraction gratings realized by the modulation of one optical property (transmittivity or reflectivity) of a single surface. In many other cases, the periodic modulation of the optical properties is the consequence of a wave traveling through the volume of a material, whose physical properties change periodically in space; in this case, we have a three-dimensional diffraction grating. As outlined in Section 1.6, the role of diffraction is important when the wavelength is close to the period of modulation of a given physical property. For instance, this is the case for solid crystals, where the atomic distribution efficiently diffracts X-rays, as we will discuss in Chapter 3. However, this volume periodic modulation can be present or be induced in materials also at visible wavelengths.

Spectrometer

The reflection grating described above is the basic element of a spectrometer. This is an instrument for measuring the wavelength and intensity of light over a specific range of the electromagnetic spectrum. It is widely used for spectroscopic investigation and material characterization. One basic scheme is shown in Fig. IN1.3.

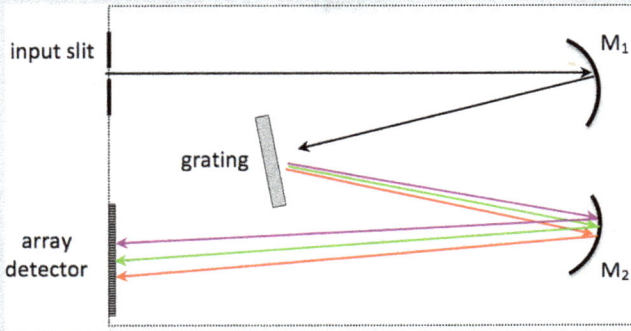

Fig. IN1.3. Scheme of a grating spectrometer.

The light enters through the input slit and is collected by a spherical mirror M_1 producing a quasi-plane wave on the grating. The diffracted beams are collected by a second mirror M_2, which reflects them on the array detector plane. The light detector measures the relative intensity corresponding to each wavelength component, recording the whole spectrum at the same time. In an older design (classical scheme), the second mirror focuses the light on an exit slit and each wavelength is analyzed separately by rotating the grating.

In the most general case, the periodic modulation in the material structure concerns both the real and the imaginary part of the refractive index, and thus it affects both the amplitude and the phase of the traveling wave. However, in several cases, absorption may be neglected and the modulation concerns only the real part of n; in these cases, we speak of a *phase grating* opposite to an *amplitude grating* occurring when the most important modulation concerns the imaginary part of n.

We recall here some basic properties of phase gratings, owing to the very important role they play in light–matter interaction and in many applications, while a more detailed discussion on optical gratings can be found in several specialized books.

Thin phase grating

Let us consider first a thin slab of material that presents a spatial modulation of the refractive index. In this case, the situation is very similar to the one already discussed for surface gratings. The result is the observation of multiple beams diffracted by the structure when a single plane wave impinges on it. The solution of the analytical problem can be addressed by writing the output field as the product of the incident field times the slab transmission function:

$$E_{out} = E_o t(x). \tag{1.183}$$

We now consider an impinging plane wave traveling along $\mathbf{z}$ and a sinusoidal modulation of the refractive index along the $\mathbf{x}$ direction with a modulation δn:

$$n(x) = n_o + \frac{\delta n}{2} \cos\left(\frac{2\pi}{\Lambda}x\right), \tag{1.184}$$

where Λ is the period of modulation, i.e. the grating pitch. This means that the refractive index oscillates between a maximum n_M and a minimum n_m,

$$n_M = n_o + \frac{\delta n}{2}, \quad n_m = n_o - \frac{\delta n}{2}, \tag{1.185}$$

as depicted in Fig. 1.29.

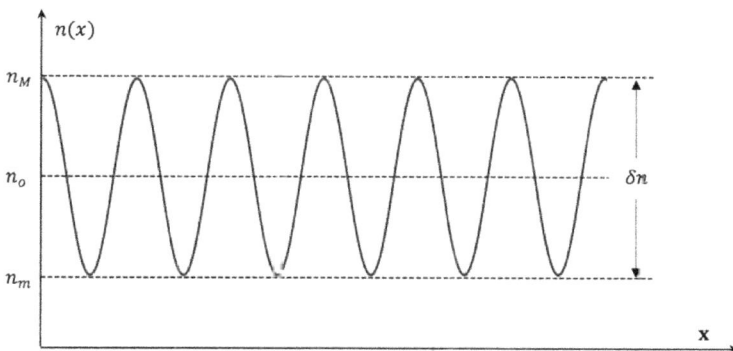

Fig. 1.29. Plot of $n(x)$ given by Eq. (1.184), showing the sinusoidal modulation of the refractive index around the value n_o.

Since after traveling through a length z the original amplitude of a wave is multiplied by the exponential factor e^{ikz}, the transmission factor of a slab of thickness d with the index modulation given by Eq. (1.184) is

$$t(x) = e^{ikd} = e^{ik_0 d(n_o + \frac{\delta n}{2} \cos(\frac{2\pi}{\Lambda} x))} = e^{ik_0 d n_o} e^{ik_0 d \frac{\delta n}{2} \cos(\frac{2\pi}{\Lambda} x)} \qquad (1.186)$$

or

$$t(x) = e^{ik_0 d n_o} e^{i\xi \cos \beta x}, \qquad (1.187)$$

by defining

$$\xi = k_0 d \frac{\delta n}{2}, \quad \beta = \frac{2\pi}{\Lambda}. \qquad (1.188)$$

Now we can exploit the identity given by the Jacobi–Anger expansion:

$$e^{i\xi \cos \beta x} = \sum_{-\infty}^{+\infty} i^m J_m(\xi) e^{im\beta x}, \qquad (1.189)$$

where $J_m(\xi)$ are the Bessel functions. Then the transmission function becomes

$$t(x) = e^{ik_0 d n_o} \sum_{-\infty}^{+\infty} i^m J_m(\xi) e^{im\frac{2\pi}{\Lambda} x} \qquad (1.190)$$

and the output field is

$$E(x, d) = E_0 e^{ik_0 d n_o} \sum_{-\infty}^{+\infty} i^m J_m(\xi) e^{im\frac{2\pi}{\Lambda} x}, \quad m = 0, 1, 2, \ldots . \qquad (1.191)$$

This is a linear superposition of plane waves, each with a transversal component of the **k** vector increasing with the order of diffraction m:

$$k_x = m \frac{2\pi}{\Lambda}. \qquad (1.192)$$

This means that we have m waves diffracted at angles:

$$\theta_m \approx m \frac{k_x}{n_o k_0} = m \frac{2\pi}{\Lambda} \frac{\lambda_0}{n_o 2\pi} = m \frac{\lambda}{\Lambda}. \qquad (1.193)$$

The intensity distribution of the diffraction pattern among the different orders is easily found by squaring Eq. (1.191). By defining the diffraction

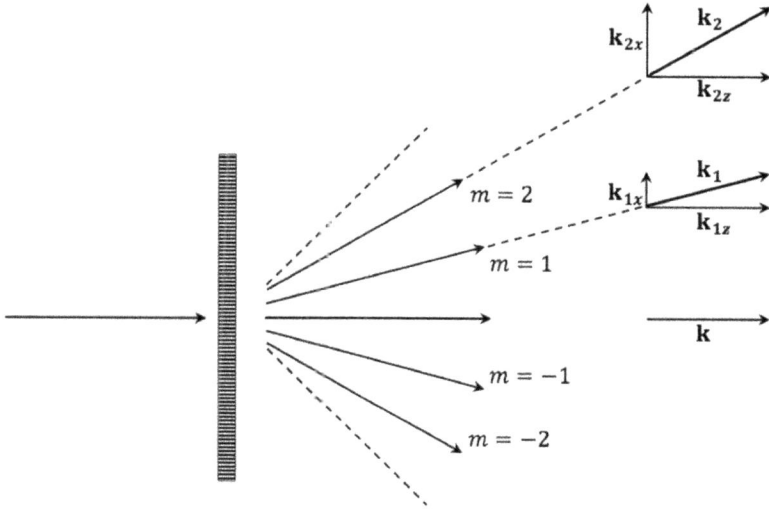

Fig. 1.30.   Diffraction of a single slab phase grating and wave **k** vectors.

efficiency η_m as the ratio of the intensity of the mth order diffracted beam
to the incident intensity, we get

$$\eta_m = \frac{|E_0|^2|J_m(\xi)|^2}{|E_0|^2} = |J_m(\xi)|^2. \tag{1.194}$$

A sketch of diffraction by a phase grating made by a single slab is shown
in Fig. 1.30.

Bessel functions

The Bessel functions $J_m(x)$ are the solution to the differential equation

$$\frac{d^2 J_m(x)}{dx^2} + \frac{1}{x}\frac{dJ_m(x)}{dx} + \left(1 - \frac{m^2}{x^2}\right)J_m(x) = 0.$$

There are several ways of representing the Bessel functions. The
integral representation is

$$J_m(x) = \frac{1}{2\pi}\int_{-\pi}^{\pi} e^{-mi\xi + ix\sin\theta}d\theta = \frac{1}{\pi}\int_0^{\pi}\cos(m\theta - x\sin\theta)d\theta.$$

A plot of the Bessel functions for $m = 0, 1, 2, 3$ is shown in Fig. IN1.4.

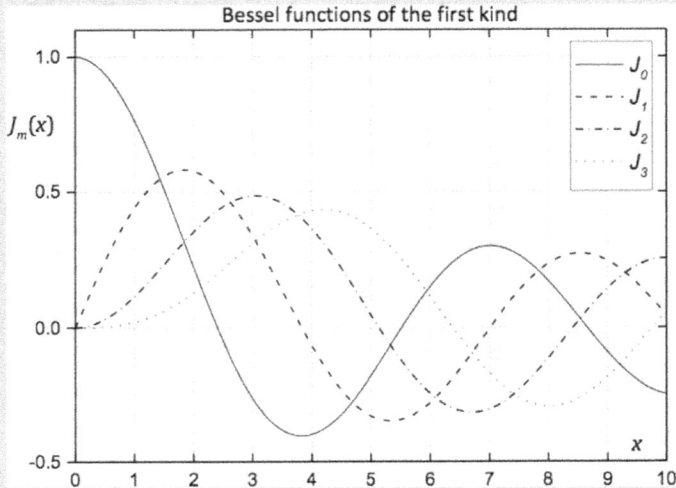

Fig. IN1.4. Plot of the Bessel functions $J_m(x)$ for $m = 0, 1, 2, 3$.

According to Eq. (1.194), the maximum diffraction efficiency is found on the first order with a maximum value of 0.33, and therefore most of the energy remains in the undiffracted beam. Differences occur in the case of modulation with a different functional behavior, such as a steplike function.

By increasing the thickness, propagation through the grating structure affects also the diffracted beams and it is not possible to approximate the overall effect on the diffraction of a single slab. However, even if the grating should be regarded as being made by a number of successive layers with a modulated index, it is possible to neglect multiple diffraction if the path difference between the beam diffracted by the first layer and the one diffracted by the last layer is much lower than the light wavelength λ. This guarantees that they can still interfere constructively in the far field without modification of the diffraction pattern.

Looking at Fig. 1.31, we see that the path difference between the ray (b) diffracted by the last layer and the ray (a) diffracted by the first layer is

$$\Delta s = CA - BA = \frac{d}{\cos \varphi} - \frac{d}{\cos \varphi} \cos \theta = \frac{d(1 - \cos \theta)}{\cos \varphi}. \qquad (1.195)$$

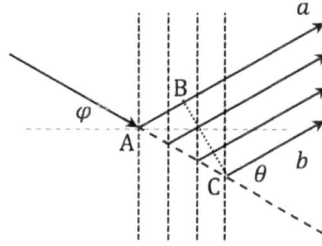

Fig. 1.31. Diffraction by multiple layers. The path difference between the beams (b) and (a) is given by CA–BA.

In the approximation $\cos\theta \approx 1 - \theta^2/2$ and $\cos\varphi \approx 1 - \varphi^2/2$, we can write the corresponding phase difference as

$$\Delta\phi = \frac{2\pi}{\lambda}\Delta s = \frac{2\pi}{\lambda}\frac{d(1 - \cos\theta)}{\cos\varphi} = \frac{2\pi}{\lambda}\frac{d\theta^2/2}{1 - \varphi^2/2}. \tag{1.196}$$

Considering the first order diffraction maximum for a single layer grating at $\theta = \lambda/\Lambda$, according to Eq. (1.193), the phase difference becomes

$$\Delta\phi = \frac{\pi}{\Lambda^2}\frac{d\lambda}{1 - \varphi^2/2}. \tag{1.197}$$

At normal incidence $\varphi = 0$, and therefore, in order to keep the constructive interference condition for the rays (b) and (a), their phase difference must be small:

$$\Delta\phi = \frac{\pi}{\Lambda^2}\frac{d\lambda}{1} \ll 1. \tag{1.198}$$

That is to say,

$$d \ll \frac{\Lambda^2}{\lambda\pi} \quad \text{(thin grating condition)}. \tag{1.199}$$

In other words, the grating follows the same behavior as the one made by a single diffracting plane grating.

On the other hand, we speak of a thick grating when the opposite occurs:

$$d \gg \frac{\Lambda^2}{\lambda\pi} \quad \text{(thick grating condition)}. \tag{1.200}$$

Thick phase grating

A thick phase grating can be represented by a sequence of a large number of planes where the incident wave feels the refractive index change. The

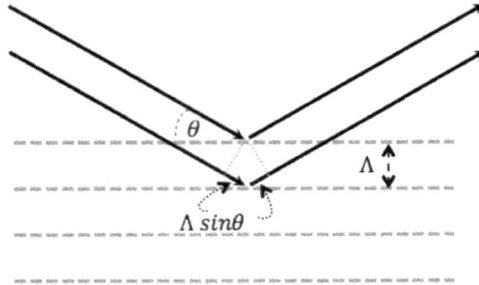

Fig. 1.32.   A simple demonstration of Bragg's law of diffraction.

waves scattered by each plane having the same direction can give rise to constructive interference if their path difference is a multiple integer of the light wavelength.

Looking at Fig. 1.32, we consider the wave reflected by a couple of successive layers. Calling θ the angle between the wave direction and the diffracting plane, it is easy to see that the path difference between the considered waves is $2\Lambda \sin\theta$, being the same for each couple of waves diffracted by the sequence of couples of planes. All these waves give rise to a diffraction maximum if

$$2\Lambda \sin\theta = m\lambda. \tag{1.201}$$

That is the well-known *Bragg's law* of diffraction.

The meaning of Eq. (1.201) is evident: if the wave impinges on the grating plane at an angle fulfilling Bragg's law, it will be diffracted with the same inclination angle, otherwise it will be transmitted with negligible diffraction. As a consequence, in the case of thick gratings we have only one diffracted beam existing only if Eq. (1.201) is fulfilled. For this reason, the theoretical problem is in this case simplified by taking into account the propagation of only two waves: one corresponding to the incident wave, the other being the diffracted wave. This is developed in the coupled two-wave theory, which allows computing the diffraction efficiency of the grating usually called the *Bragg grating*.

We will not go into the details of the theory, but will recall some optical properties of this type of gratings owing to their important role in photonic devices and applications.

First of all, one has to recognize two kinds of gratings: *transmission grating* and *reflection grating*. These definitions are linked to the effect of the structure on the impinging wave: if the diffracted beam travels in the

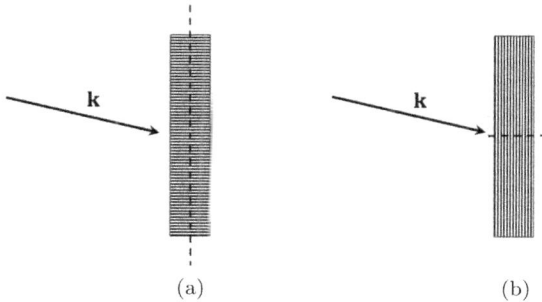

Fig. 1.33. (a) Transmission grating; (b) reflection grating. The dashed line is perpendicular to the grating planes.

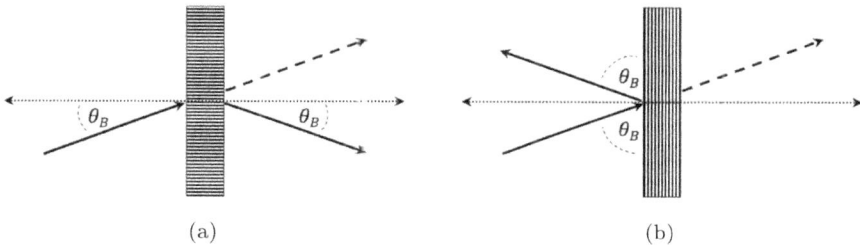

Fig. 1.34. (a) Diffraction by a transmission grating; (b) diffraction by a reflection grating. In both cases it occurs only for $\theta \approx \theta_B$; for $\theta > \theta_B$ or $\theta < \theta_B$ the wave is transmitted unaffected by the grating.

half-space of the transmitted beam, we speak of a transmission grating, while if it travels in the half-space of the incident beam, we speak of a reflection grating. We can roughly state that the first case occurs for incidence directions at angles $\theta < \pi/4$ with respect to the grating planes and the second case when $\theta > \pi/4$, as depicted in Fig. 1.33.

The optical effect of these gratings is sketched in Fig. 1.34. If a wave with wavelength λ is incident in the transmission grating geometry, by varying the incidence angle θ, the beam will be always transmitted except when $\theta \approx \theta_B$, where θ_B is the angle fulfilling Eq. (1.201). In this case the beam will be diffracted [see Fig. 1.34(a)], and we should note here that in this case the diffraction efficiency can approach 100%. In the case of a reflection grating, a similar behavior is observed [see Fig. 1.34(b)].

Looking at the first order of diffraction ($m = 1$), we note that Eq. (1.201) states that at each wavelength there corresponds a different Bragg angle θ_B, and therefore if we fix the incidence angle and allow varying the wavelength of the incident wave, we get diffraction only for the wavelength fulfilling Eq. (1.201) while all the other waves will be transmitted. In this way, the thick grating acts as a narrow band filter transmitting all the wavelengths except for $\lambda_B = 2\Lambda \sin \theta_B$.

An example of experimental detailed analysis of the angular dependence of the diffraction efficiency (the ratio between the intensity of the diffracted beam and of the incident beam) for a transmission grating is shown in Fig. 1.35(a). As expected, high efficiency is measured in a narrow range around θ_B. Similarly, the wavelength dependence is shown in Fig. 1.35(b),

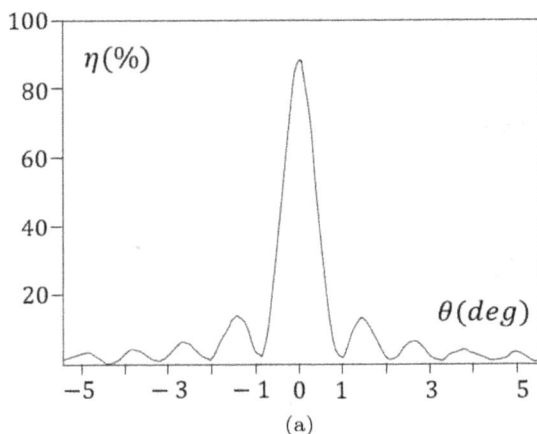

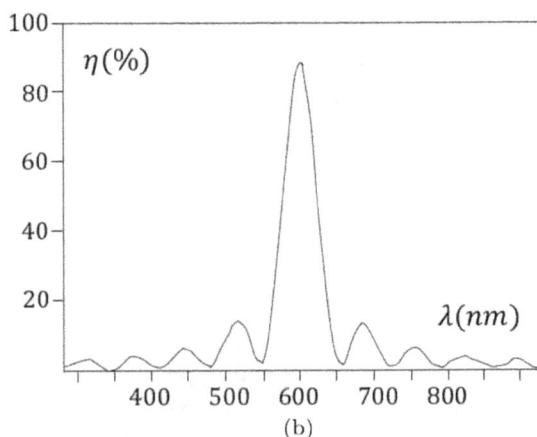

Fig. 1.35.   Diffraction efficiency of a transmission grating (a) by varying the incidence angle at a fixed wavelength; (b) by varying the wavelength at a fixed incidence angle.

where, at a fixed incidence angle, strong diffraction is found only for a narrow band around λ_B.

In particular, the reflection Bragg grating becomes a very selective reflector for a narrow wavelength range. This property is widely exploited in many optical devices.

Chapter 2

The Basics of Quantum Mechanics

2.1. The old quantum theory

We know that the physical sciences have been developing continuously since the setup of novel experimental methods paved the way for new research areas, and at the same time the theoretical explanation of physical phenomena has been pushing scientists to perform new experiments. For this reason, a good scientist should be ready to change his or her point of view concerning the interpretation of phenomena. Thanks to this open-mindedness, Galilei and Newton were able to change the approach to natural phenomena that had lasted many centuries since the time of the Greek philosophers. More recently, Planck, Einstein, Bohr, de Broglie, Heisenberg, Born, Dirac and many more clever scientists have been able to overcome the crisis of classical physics by developing the theory of quantum mechanics.

The crisis occurred at the end of 1800 and the beginning of 1900, owing to a number of experimental facts which could not be explained following classical mechanics and classical electromagnetism. These experimental facts could fall neither into the frame of Newton's mechanics nor into the frame of Maxwell's equations. They made questionable the sharp border between the wave description of electromagnetic radiation and the corpuscular description of elementary particles.

The main observations leading to the new point of view for the microscopic world concerned the light–matter interaction and the atomic structure.

The explanation given by Max Planck for the spectrum of *blackbody radiation* is historically the beginning of this story, since he introduced for the first time the idea of the energy packet, opening the way to the

74

concept of a *quantum of energy*, a basic one for the new theory. It is well known that all materials emit electromagnetic radiation with a spectral distribution dependent on the temperature of the emitting body. While at room temperature emission is usually in the infrared, by increasing the temperature to some hundreds of degrees Celsius emission is visible, giving rise to incandescence which occurs at a temperature dependent on the particular material. At the same time, the emitted power increases with temperature. This process can be studied by considering a hollow box with walls maintained at a constant temperature. The radiation inside the box is in thermal equilibrium with the walls and can be analyzed through a small hole that allows the exit of radiation. Since external radiation entering through the hole will be completely absorbed by the walls after a large number of reflections, the box is called a *blackbody*, which acts like a black surface absorbing all the impinging radiation. Therefore, the radiation emitted from the hole is known as blackbody radiation.

The mentioned equilibrium of the radiation in the blackbody can be described by a universal function of frequency and temperature $u(\nu, T)$ giving the energy density in the frequency range between ν and $\nu + d\nu$, which can be defined by

$$U(T) = \int_0^\infty u(\nu, T) d\nu, \tag{2.1}$$

with $U(T)$ being the total energy density at a fixed temperature T.

A very important fact is that G.R. Kirchoff in 1859 was able to demonstrate that the ratio between the energy flux emitted and absorbed is determined only by the temperature and therefore black body radiation at temperature T corresponds to the radiation emitted by any body at the same temperature. As a consequence, the energy density $u(\nu, T)$ is

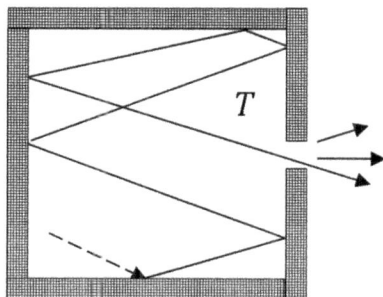

Fig. 2.1. Sketch of radiation emitted by a blackbody.

proportional to the emitted energy flux. For this reason, by detecting the energy irradiated by the hole of the above-described box, it is possible to determine $u(\nu, T)$.

The first careful determination of the blackbody spectrum was made by O. Lummer and E. Pringsheim in 1899, in terms of $u(\lambda, T)$, which can be transformed into $u(\nu, T)$ taking into account that $\lambda = c/\nu$.[1] The actual shape of $u(\lambda, T)$ is shown in Fig. 2.2(a), while $u(\nu, T)$ is shown in Fig. 2.2(b), for increasing values of T. Arbitrary units are used in these plots.

These curves allow finding two important phenomenological rules. The first is the Stefan–Boltzmann law:

$$U(T) = \int_0^\infty u(\lambda, T)d\lambda = \sigma T^4. \tag{2.2}$$

Namely, the total energy density grows as the fourth power of the body temperature. The second is Wien's law of displacement of the peak at wavelength λ_{max} with temperature:

$$\lambda_{\text{max}} T = \text{const} = 2.9 \cdot 10^{-3}\,\text{mK}. \tag{2.3}$$

However, the explanation of the functional behavior of $u(\lambda, T)$ is the basic problem. Following classical arguments, Rayleigh and Jeans were able to describe the second part of the curves shown in the Fig. 2.2 by the formula

$$u(\lambda, T) = \frac{8\pi}{\lambda^4} k_B T, \tag{2.4}$$

where $k_B = 1.38064852 \cdot 10^{-23}$ J/K, the Boltzmann constant. The expression (2.4) is not satisfactory, since it does not explain the maximum present in the spectrum and contains the "unphysical" behavior of a diverging energy density at low wavelengths — the so-called *ultraviolet catastrophe*.

Planck in 1900 was able to fit theoretically the curves of Fig. 2.2 by introducing into the classical model of Rayleigh and Jeans a new revolutionary concept: the exchange of energy between electromagnetic radiation and matter in the black body walls can occur through a number of a minimum integer quantity $h\nu$.

---

[1]Transformation of $u(\lambda, T)$ into $u(\nu, T)$ must take into account that $u(\lambda, T)d\lambda = -u(\nu, T)d\nu$. Since $d\lambda = d(\frac{c}{\nu}) = -\frac{c}{\nu^2}d\nu$, we have $u(\lambda, T) = \frac{\nu^2}{c}u(\nu, T) = \frac{c}{\lambda^2}u(\nu, T)$.

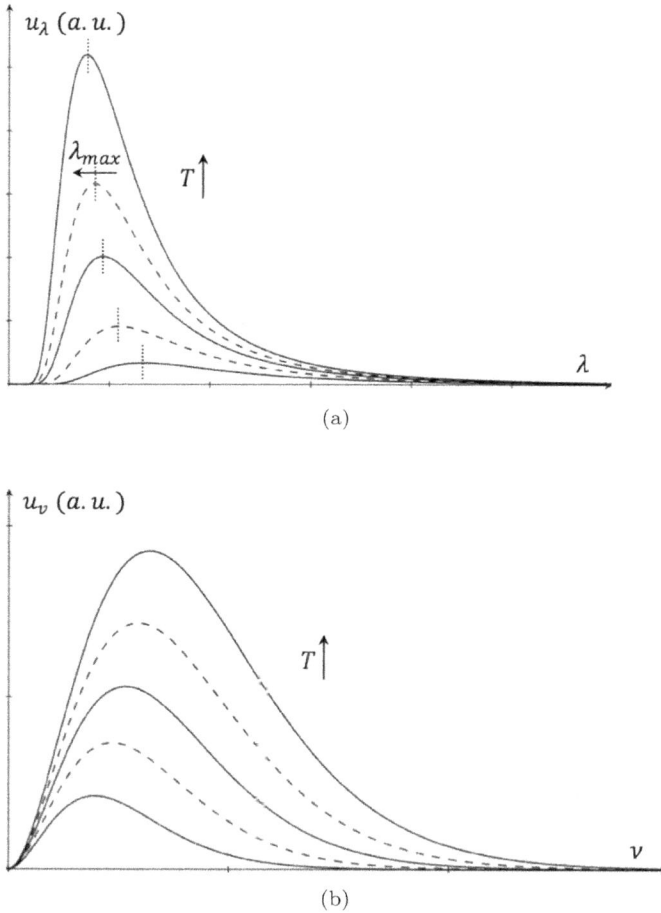

u_λ (a. u.)

λ_{max}

T

λ

(a)

u_ν (a. u.)

T

ν

(b)

Fig. 2.2. Blackbody spectra for increasing values of temperature: (a) $u(\lambda, T)$; (b) $u(\nu, T)$.

In this way, Planck was able to obtain the expression

$$u(\nu, T) = \frac{8\pi n^3}{c^3} \frac{h\nu^3}{e^{\frac{h\nu}{k_B T}} - 1}; \tag{2.5}$$

n is the refractive index of the medium where the electromagnetic wave is traveling and h is a constant with the dimensions of an "action" (energy multiplied by time, i.e. dimensions of angular momentum):

$$h = 6.626 \cdot 10^{-34} \text{ J} \cdot \text{s}. \tag{2.6}$$

This value is extremely small in the SI system, accounting for the fact that this "quantization" concept can be effective only when interaction on the microscopic scale plays a major role.[2]

Of course, Planck's hypothesis was against the classical law and it was just an intuition used to work out a theoretical expression able to fit the experimental data. However, the explanation of other observations strengthened this idea, leading to the need for formulating a new theory necessary for describing the microscopic world.

Five years later, in 1905, Albert Einstein used this quantization concept to explain the *photoelectric effect*. He actually took a step further, stating that $h\nu$ is the minimum amount of energy of the electromagnetic wave that must be considered as an integer number of quanta $h\nu$.

The photoelectric effect is the emission of electrons by a metal surface when visible or ultraviolet radiation impinges on it. These electrons can be collected by an anode and the consequent current can be measured under a different applied voltage. The behavior that resulted as being the most amazing to the first observers is shown in Fig. 2.3, where the induced current is reported versus the frequency of the impinging radiation.

Here we can see that a threshold frequency ν_{th} exists below which no effect is induced, irrespective of the intensity or power of the radiation.

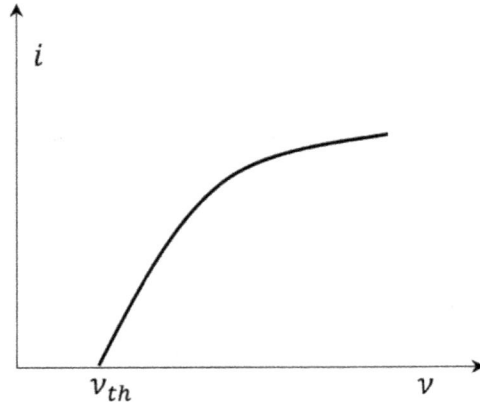

Fig. 2.3. The electric current induced by the photoelectric effect versus the frequency of the impinging light.

---

[2]According to footnote 1, we get $u(\lambda, T) = \dfrac{8\pi c}{\lambda^5} \dfrac{h}{e^{\frac{hc}{k_B \lambda T}} - 1}$.

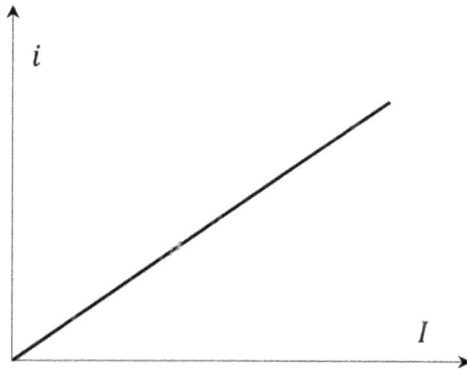

Fig. 2.4. The electric current induced by the photoelectric effect versus the intensity of the impinging light at $\nu > \nu_{th}$.

Only if $\nu > \nu_{th}$ does the current linearly increase with light intensity, as shown in Fig. 2.4.

In summary, the main features of the photoelectric effect are:

- Emission occurs only above a threshold ν_{th} for the radiation frequency;
- The kinetic energy of the emitted electrons is proportional to the radiation frequency ν;
- The number of electrons emitted per second and unit surface at fixed frequency is proportional to the light intensity.

These facts could not be explained by the classical theory of the electromagnetic field. In particular, the threshold in frequency rather than in intensity was difficult to understand. Nevertheless, all these features were explained by Einstein in a simple way just by considering the energy of the light field made by packets of an integer number of the minimum quantity $h\nu$.

This is actually a step forward with respect to Planck's hypothesis, where this energy quantum was concerned with exchange of energy between radiation and matter, without considering it as a property of light itself. On the contrary, in the Einstein model the existence of quanta $h\nu$ is not connected with the possible interaction processes but is a property of radiation. This quantum is what we call a *photon* and the photoelectric effect can be considered a collision process between photons and electrons.

Let us suppose that one electron will be able to exit from the metal surface if the energy absorbed from the photon is at least equal to the extraction potential W, i.e. the work necessary for extracting the electron from the metal. In this case, when the impinging photon has energy $\epsilon = h\nu \geq W$ it will be able to extract one electron, otherwise if $h\nu < W$ this process does not take place. In this way, the threshold behavior is easily understood to be:

$$\nu_{th} = \frac{W}{h}. \tag{2.7}$$

When extracted the electron has kinetic energy

$$K_e = h\nu - W = h\nu - h\nu_{\text{th}} = h(\nu - \nu_{\text{th}}), \tag{2.8}$$

thus explaining the proportionality between kinetic energy and frequency over the threshold.

Finally, the proportionality between the number of emitted electrons and the number of impinging photons explains the current dependence on light intensity.

Another major step toward the formulation of the new theory of quantum mechanics was Bohr's model for the hydrogen atom. It succeeded in explaining the spectroscopic observations made by different researchers on the emission and absorption of elements in the atomic state. In 1885, Balmer identified the basic series of six lines in the emission of the hydrogen and Rydberg found the following empirical expression for the spectral position of these lines:

$$\frac{1}{\lambda_m} = R\left(\frac{1}{n_1^2} - \frac{1}{n_2^2}\right), \tag{2.9}$$

where $n_1 < n_2$ are integers and R is the universal Rydberg constant.

The main lines of Balmer's spectrum of hydrogen are shown in Fig. 2.5.

The discrete lines spectrum in absorption or emission was found for all the atomic vapors investigated, following rules similar to the one given by Eq. (2.9).

Bohr was able to account for this behavior by a semiclassical model for the hydrogen atom, introducing the quantization of the energy of the electron. He started from Rutherford's model consequent to the observations

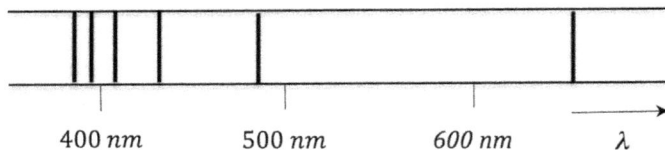

Fig. 2.5. Balmer's emission spectrum of hydrogen. The vertical lines represent the spectral location of the light emitted by excited atoms of hydrogen.

of scattering of α particles (helium positive nuclei). According to it, the atom was modeled as a positive nucleus surrounded by a cloud of electrons. By assuming the number of electrons in a neutral atom to be equal to the atomic number Z (the number of protons in the nucleus), Bohr represented the hydrogen atom as one electron with charge $-e$ moving around a fixed proton with charge $+e$, since the proton mass is about 2000 times that of the electron. By considering the interaction due to the Coulomb force between the two opposite charges, this planetary model could not be formulated in the frame of a pure classical electromagnetic theory. In fact, an electron rotating around the atomic nucleus, being an accelerated charge, should emit radiation according to Larmor's law, thus losing energy very quickly and falling on the nucleus in a spiral motion. The only possibility for such an atom to exist is to allow stable orbits where the electron does not emit any radiation, keeping constant its energy. Thus, Bohr's hypothesis was about the existence of stable orbits for the electron corresponding to the ones where the angular moment of the electron in the revolutionary movement around the proton was an integer number of $\hbar = h/2\pi$:

$$L_n = n\hbar, \quad n = 1, 2, 3, \ldots \tag{2.10}$$

Following this statement, it is straightforward to calculate the electron energy in the Coulomb field of the proton as

$$W_n = -\frac{1}{n^2} \frac{2\pi^2 m e^4 Z^2}{h^2} \quad \text{(Gauss units).} \tag{2.11}$$

In order to get Eq. (2.11), one starts from the Coulomb force between the electron and the nucleus of a generic atom with atomic number Z:

$$F = \frac{e^2 Z}{a^2} \quad \text{(Gauss units),}$$

with a being the electron–nucleus distance. By applying Newton's law of dynamics, we have

$$F = \frac{e^2 Z}{a^2} = mA = \frac{mv^2}{a} \quad (A \equiv \text{acceleration}).$$

Therefore,

$$v^2 = \frac{e^2 Z}{ma}$$

and the kinetic energy is

$$K = \frac{1}{2}mv^2 = \frac{e^2 Z}{2a}.$$

The potential energy originated by the electrostatic interaction is written as

$$V = -\frac{e^2 Z}{a} \quad (V = 0 \text{ when } a \to \infty).$$

Therefore, $V = -2K$ and the total energy is

$$W = K + V = -K.$$

On the other hand, we can also write

$$K = \frac{1}{2}mv^2 = \frac{1}{2}ma^2\omega^2 = \frac{1}{2}ma^2(2\pi f)^2,$$

with ω being the angular velocity of the circular motion and f its frequency.

Then we can write

$$K = \frac{1}{2}(2\pi f)^2 m \left(\frac{e^2 Z}{2K}\right)^2.$$

Now we use the hypothesis (2.10), taking into account that $L = ma^2 2\pi f = n\hbar$. Hence

$$K = \frac{1}{2}ma^2(2\pi f)^2 = \frac{1}{2}ma^2 2\pi f(2\pi f) = \frac{1}{2}n\hbar(2\pi f) = \frac{1}{2}nhf \to f = \frac{2K}{nh},$$

$$K = 2\pi^2 m \left(\frac{2K}{nh}\right)^2 \left(\frac{e^2 Z}{2K}\right)^2 = \frac{2\pi^2 me^4 Z^2}{n^2 h^2} = -W,$$

which is Eq. (2.11).

Thus, the only allowed energies for the electron are

$$W_1 = -\frac{2\pi^2 me^4 Z^2}{h^2}, \quad W_2 = -\frac{1}{4}\frac{2\pi^2 me^4 Z^2}{h^2}, \quad W_2 = -\frac{1}{9}\frac{2\pi^2 me^4 Z^2}{h^2}, \cdots \tag{2.12}$$

The additional, most radical assumption is that radiation is emitted when the electron jumps from one energy level to another; the emission produces a photon whose energy is the difference between the energies of the two involved levels:

$$h\nu_{ij} = W_i - W_f,$$ (2.13)

where i and f indicate the initial and the final state of the transition. By using Eq. (2.11) we get

$$\nu_{if} = \frac{W_i - W_f}{h} = \frac{2\pi^2 m e^4 Z^2}{h^3} \left(\frac{1}{n_i^2} - \frac{1}{n_f^2} \right),$$ (2.14)

which is formally equal to Eq. (2.9) if

$$R = \frac{2\pi^2 m e^4}{ch^3} = 1.09677 \cdot 10^5 \, \text{cm}^{-1},$$ (2.15)

since $1/\lambda = \nu/c$.

Again, we have an "ad hoc" statement that is outside the frame of classical theory, which allows explaining in a very satisfactory way several spectroscopic observations related to light absorption and emission of atomic vapors, additionally verifying the value of Rydberg's constant.

Moreover, Bohr's theory allows the calculation of the radius a_n of the stable electronic orbits:

$$a_n = \frac{h^2}{4\pi^2 m e^2} \frac{n^2}{Z}.$$ (2.16)

Using the known values of the physical constants in the Gauss units system,

$$m = 9.11 \cdot 10^{-28} \text{g}; \quad e = 4.806 \cdot 10^{-10} \text{esu}; \quad h = 6.626 \cdot 10^{-27} \text{erg} \cdot \text{s},$$

we get

$$a_n = 0.529 \cdot 10^{-8} \frac{n^2}{Z} \, (\text{cm}).$$ (2.17)

With $Z = 1$ and $n = 1$ we have the radius of the first orbit of the hydrogen atom, the so-called Bohr's radius:

$$a_{\text{Bohr}} = 0.529 \cdot 10^{-8} \text{cm} = 0.529 \, \text{Å}.$$ (2.18)

Bohr's quantization of electronic states was very successful in explaining several spectroscopic observations, introducing an atomic model accounting for them and really novel concepts that will find their complete justification in the complete quantum-mechanical theory. Let us briefly summarize them:

1. The only possible trajectories for the circular orbits of the electron are the ones where the angular momentum of the electron is

$$L_n = n\hbar,$$

and thus the radius fulfills the relation (2.17),

$$a_n = a_{\text{Bohr}} \frac{n^2}{Z},$$

and corresponds to different energy states for the electron:

$$W_n = -\frac{1}{n^2} \frac{2\pi^2 m e^4 Z^2}{h^2}.$$

2. In these orbits, the electron does not follow the Larmor principle, which states that an accelerated charge emits electromagnetic waves, losing energy owing to an emitted power $P = 2e^2 A^2/3c^3$ proportional to the acceleration A. In this way, the electron can keep the energy stable, otherwise the very existence of the atom would not be possible owing to the fast dropping of the electron on the nucleus due to light irradiation. Only by contradicting the Larmor principle can the electron have a steady state.

3. The emission (or absorption) of energy by the electron occurs through jumping between two different stable states of energy W_2 and W_1, through electromagnetic waves with the frequency given by

$$\nu = \frac{W_2 - W_1}{h}.$$

The photon

The quantum of electromagnetic radiation was introduced by Einstein in 1905, but only in 1926 was it designated as the *photon*. Moreover, the quasiparticle representation of the photon was completed by the discovery of the Compton effect in 1927. Photons have no mass, but they bring energy and linear momentum, and may bring an angular momentum linked to their polarization state:

$$\epsilon = h\nu = \hbar\omega = h\frac{c}{\lambda} \quad \text{(energy)},$$

$$\vec{p} = \hbar\vec{k}, \quad p = \frac{h}{\lambda} = \frac{h\nu}{c} \quad \text{(linear momentum)},$$

$$\vec{l} = \pm\hbar\hat{k} \quad \text{(angular momentum for circular polarization)}.$$

It should be noted that the photon mass must be zero, since according to relativistic mechanics the energy of a particle is written as

$$\epsilon = \frac{mc^2}{\sqrt{1 - \frac{v^2}{c^2}}}.$$

Since the photon in vacuum travels at a speed $v = c$ in order for the energy to be finite, the numerator must be zero, and then $m = 0$.

It is quite interesting to underline that even if the mass is zero, photons are able to transfer both linear and angular momentum to matter. This concept is currently widely exploited in optical trapping and manipulation of microsize and nanosize objects using the "optical tweezers" realized by a strongly focused laser beam.

2.2. de Broglie's hypothesis and the wave description of matter

In the previous section, we recalled that the explanation of several phenomena related to microscopic properties of matter required the introduction of "ad hoc" statements about the existence of quantization of different physical quantities. Among them, the quantum of electromagnetic energy, the photon, shows particle-like properties through the Compton and the photoelectric effect. The duality of the photon appearing in the *old quantum theory* can be considered a key factor in the development of the theory of quantum mechanics, since simple arguments arising from this property lead to the basic assumptions that have to be made when one is dealing with the microscopic world.

It is remarkable that the quantum nature of photons requires a different approach in the observation of physical phenomena on the microscopic scale: namely, it is necessary to discard the deterministic view of classical mechanics while adopting a statistical one that introduces the concept of probability into the result of a measurement. The discussion on a simple ideal experiment about the polarization of photons reported by Paul A.M. Dirac in his fundamental essay "The Principles of Quantum Mechanics" is very clear in highlighting this concept. It explains at the same time the *principle of superposition of states*. On the other hand, the extension of this duality to particles with mass introduced by Louis de Broglie led to the necessity of formulating the *uncertainty principle* and the *complementarity principle*. Thus, it was very soon clear to the founders of the new quantum physics that a new approach was necessary to account for phenomena occurring on the microscale. In order to clarify this concept, we will follow a reasoning line similar to the mentioned Dirac' argument.

We know that an electromagnetic wave can exist in a definite linear polarization state if the field vector keeps stable its direction during oscillation. Moreover, by considering a linearly polarized wave (in any

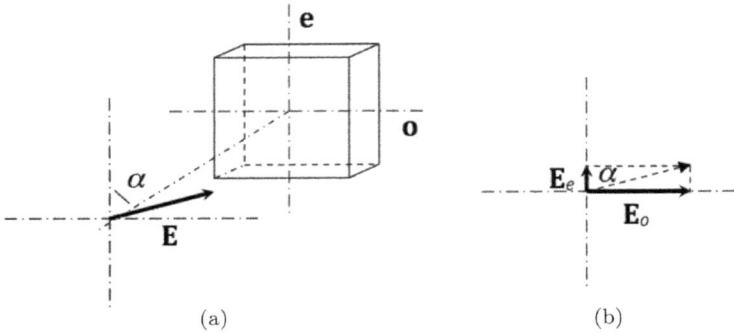

Fig. 2.6. (a) **E** is the electric field of the electromagnetic wave oscillating along a polarization direction, making the angle α with respect to the optic axis of the crystal slab; (b) $\mathbf{E}_o$ and $\mathbf{E}_e$ are the components of the electric field of the transmitted waves.

direction) we can always consider it as being due to superposition of two vector components in two directions orthogonal to each other, arbitrarily chosen in a plane normal to the propagation wave vector (see Section 1.3). On the other hand, the photoelectric effect shows strong dependence on the light polarization, thus demonstrating that the property of polarization must also be given to photons.

Let us suppose this linearly polarized wave, represented by the electric field **E**, impinging at normal incidence on a polarizer crystal slab, with optic axis **e** parallel to its surface, thus allowing propagation of light only with polarization vectors parallel and perpendicular to **e** (as discussed in Section 3.13). If the field **E** is oriented at an angle α with respect to **e**, it will be broken into two components $\mathbf{E}_e$ and $\mathbf{E}_o$ respectively parallel and perpendicular to **e** (see Fig. 2.6). Calling E_i the amplitude of the incoming field, the amplitude of the two orthogonal components transmitted by the crystal polarizer will be $E_e = E_i \cos\alpha$ and $E_o = E_i \sin\alpha$. As a consequence, the corresponding measured intensities will be $I_e = I_i \cos^2\alpha$ and $I_o = I_i \sin^2\alpha$, with I_i being the intensity of the incident wave. This is the well-known classical Malus law. Since the property of polarization belongs also to photons and since intensity is proportional to energy, this means that I_i corresponds to N incident photons, while after passing through the crystal a number $N\cos^2\alpha$ is polarized along **e** (a direction at an angle α with respect to the initial polarization) and a number $N\sin^2\alpha$ is polarized along **o** (a direction at an angle $\pi/2 - \alpha$ with respect to the initial polarization).

Now, the problem arises if we consider a single photon impinging on the polarizer. In fact, the energy of the photon cannot be divided into smaller parts, and therefore the transmitted photon can be polarized either along **e** or along **o**.

As a consequence, the expected result of the experiment is not known "*a priori*": in other words, we lose the deterministic knowledge of the final state of the photon. On the other hand, when we have a large number N of photons we know that we will get a fraction $N \cos^2 \alpha$ polarized along **e** and a fraction $N \sin^2 \alpha$ polarized along **o** ($N \cos^2 \alpha + N \sin^2 \alpha = N$). Since the frequency of occurrence of a single event approaches its probability as the number of events increases, it is quite obvious to consider the factor ($\cos^2 \alpha$) as the probability that a single photon impinging on the crystal slab with polarization at an angle α with respect to the crystal axis **e** is transmitted with polarization along **e**, while $\sin^2 \alpha$ will be the probability to be transmitted with polarization along **o**. The total probability is obviously 1 ($\cos^2 \alpha + \sin^2 \alpha = 1$).

Therefore, when we are dealing with a single photon, once the directions of the crystal optic axis and of light polarization are fixed, we cannot foresee the final state of polarization of the photon, but are only able to say what the probability is of finding the photon in each of the possible polarization states when transmitted by the crystal.

At the same time, this allows considering the initial state of the photon as a superposition of the two possible final states, each with its probability, and the measurement (i.e. the transmission through the crystal polarizer) pushes the photon to choose one of them. This *principle of superposition of states* leads also to the concept that a measurement affects the state of a microscopic system, thus preventing us from achieving a complete knowledge of its dynamical quantities in the same way as classical mechanics does.

In 1924, Louis de Broglie extended the concept of the wave–particle duality to any physical object, stating the following relationship between the modulus p of momentum of a particle and the associated wavelength λ:

$$\lambda = \frac{h}{p}, \tag{2.19}$$

with h being Planck's constant. This is actually the relation that holds for the photon. In fact, we know that $E = h\nu, \lambda = c/\nu$ and the photon momentum can be written as $p = E/c$. Then, combining these equalities, we get Eq. (2.19), which can also be written as

$$p = \hbar k. \tag{2.20}$$

As a matter of fact, the hypothesis of de Broglie was to associate with any particle with mass m and velocity v a plane wave with the wavelength given by Eq. (2.19) and the frequency $\nu = E/h$ to be described by

$$\psi(r,t) = A[\exp i(\mathbf{k} \cdot \mathbf{r} - \omega t)] = A\left[\exp \frac{i}{\hbar}(\mathbf{p} \cdot \mathbf{r} - Et)\right], \qquad (2.21)$$

thus traveling in the $\mathbf{p}$ direction.

On the other hand, the energy of a free particle is just $E = \frac{1}{2}mv^2 = p^2/2m$.

In the next section, we will exploit this wave–particle duality in the energy expression in order to get intuitively the basic equation of quantum mechanics. It is interesting to recall the idea of de Broglie that the stable energy levels of Bohr's atom should correspond to stationary waves in the electronic orbits. In fact, in order to get stationary waves in a circular orbit with ray r, the length of the orbit $2\pi r$ should correspond to an integer number n of wavelengths:

$$2\pi r = n\lambda. \qquad (2.22)$$

Taking into account Eq. (2.19),

$$2\pi r = n\frac{h}{p}, \qquad (2.23)$$

which can be written as

$$pr = n\hbar. \qquad (2.24)$$

Since $pr = L$ is the angular momentum, we again get

$$L = n\hbar. \qquad (2.25)$$

This is Bohr's "ad hoc" statement introduced to find the atom energy levels.

This intuition receives a stronger theoretical background from the hypothesis of de Broglie.

The question now is how to keep the concept of spatial localization of a particle needed to describe the particle's motion while describing it as a wave. For this aim (in one dimension for the sake of simplicity), we try to describe a particle using a wave function $A(x,t)$ made by the superposition of several plane waves of the same amplitude A, having different wave vectors with value $-\Delta k < k < \Delta k$:

$$A(x,t) = \int_{-\Delta k}^{\Delta k} A \exp[i(kx - \omega t)]dk, \qquad (2.26)$$

where k and ω are not independent, but linked through the dispersion relation which for a free particle can be obtained by the mentioned duality of the energy expression:

$$E = \frac{p^2}{2m} = \hbar\omega \quad \rightarrow \quad \omega = \frac{\hbar k^2}{2m}. \tag{2.27}$$

By defining $k = k_0 + (k - k_0)$, we get the Taylor expansion for ω:

$$\omega(k) = \omega_0 + \left(\frac{d\omega}{dk}\right)_{k=k_0} (k - k_0) + \cdots, \tag{2.28}$$

where $\omega_0 = \omega(k_0)$.

By substitution in Eq. (2.26), we get at first order in k

$$A(x,t) = A\exp[i(k_0 x - \omega_0 t)]$$

$$\times \int_{-\Delta k}^{\Delta k} \exp i\left[(k - k_0)x - (k - k_0)\left(\frac{d\omega}{dk}\right)_{k=k_0} t\right] d(k - k_0) \tag{2.29}$$

Calculation of the integral is easy, since it corresponds to a basic integral:

$$\int_{\xi_1}^{\xi_2} \exp[i\alpha\xi]d\xi = \left[\left(\frac{1}{i\alpha}\right)\exp[i\alpha\xi]\right]_{\xi_1}^{\xi_2},$$

where $\alpha = x - \left(\frac{d\omega}{dk}\right)_{k=k_0} t$ and $\xi = k - k_0$.

Then

$$\int_{-\Delta k}^{\Delta k} \exp i\left[(k - k_0)x - (k - k_0)\left(\frac{d\omega}{dk}\right)_{k=k_0} t\right] d(k - k_0)$$

$$= \left[\frac{i}{\left(\frac{d\omega}{dk}\right)_{k=k_0} t - x}\exp\left\{i(k - k_0)\left[x - \left(\frac{d\omega}{dk}\right)_{k=k_0} t\right]\right\}\right]_{-\Delta k}^{\Delta k}$$

$$= \left[\frac{i}{\left(\frac{d\omega}{dk}\right)_{k=k_0} t - x}\right]\left\{\exp i\Delta k\left[x - \left(\frac{d\omega}{dk}\right)_{k=k_0} t\right]\right.$$

$$\left. - \exp i(-\Delta k)\left[x - \left(\frac{d\omega}{dk}\right)_{k=k_0} t\right]\right\}$$

$$= \left[\frac{i}{\left(\frac{d\omega}{dk}\right)_{k=k_0} t - x}\right] \left\{2i \sin \Delta k \left[x - \left(\frac{d\omega}{dk}\right)_{k=k_0} t\right]\right\}$$

$$= \left[\frac{2 \sin \Delta k \left[\left(\frac{d\omega}{dk}\right)_{k=k_0} t - x\right]}{\left(\frac{d\omega}{dk}\right)_{k=k_0} t - x}\right].$$

Therefore, we finally have

$$A(x,t) = A \exp[i(k_0 x - \omega_0 t)] \left[\frac{2 \sin \Delta k \left[\left(\frac{d\omega}{dk}\right)_{k=k_0} t - x\right]}{\left(\frac{d\omega}{dk}\right)_{k=k_0} t - x}\right]. \qquad (2.30)$$

The expression (2.30) corresponds to a plane wave with wave vector k_0 and frequency ω_0, modulated by a function in the form of $2(\sin \Delta k \zeta / \zeta)$, where $\zeta = x_M - x$, as shown in Fig. 2.7. The maximum of the amplitude is at

$$x_M = \left(\frac{d\omega}{dk}\right)_{k=k_0} t. \qquad (2.31)$$

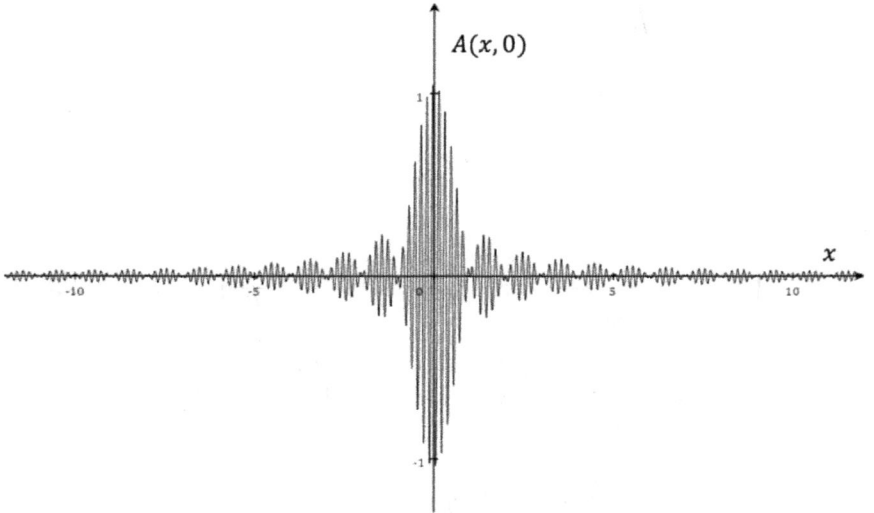

Fig. 2.7. Plot of Eq. (2.30) for $x_M = 0$, showing the modulated high frequency oscillations.

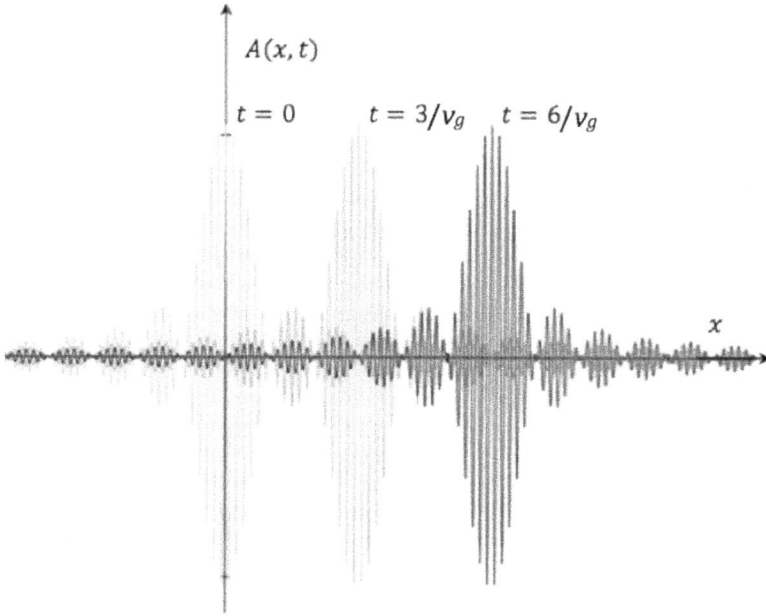

Fig. 2.8.   Plot of Eq. (2.30) shcwing that x_M is moving with time at speed v_g.

The expression (2.31) shows that x_M moves with a speed v_g given by

$$v_g = \left(\frac{d\omega}{dk}\right)_{k=k_0},\qquad(2.32)$$

which is the "group velocity" of the wave packet; in fact, by increasing t we get just a translation of the curve $A(x,t)$ toward higher x. This is clearly shown in Fig. 2.8, where the same function (the real part which is equal to the wave packet amplitude) is plotted for different times corresponding to $x_M = 0, x_M = 3, x_M = 6$. On the other hand, under de Broglie's assumption for a particle, we have from Eq. (2.27) $\omega = \hbar k^2/2m$. Then

$$\frac{d\omega}{dk} = \frac{\hbar k}{m} = \frac{p}{m} = v \ !$$

Therefore, the group velocity is exactly equal to the particle velocity.

At the same time, this property makes it natural to consider x_M as the "most probable" position of the particle described by the wave packet.

The width of the peak can be measured by the distance between the two values of x corresponding to zero amplitude on both sides of the peak.

It can be easily calculated at $t = 0$ as

$$\sin(\Delta k x_{0-}) = 0 \quad \text{and} \quad \sin(\Delta k x_{0+}) = 0;$$
$$\text{then} \quad x_{0-} = -\frac{\pi}{\Delta k} \quad x_{0+} = \frac{\pi}{\Delta k}.$$

Therefore,

$$\Delta x = \frac{2\pi}{\Delta k}. \tag{2.33}$$

The meaning of Eq. (2.33) becomes clear when one looks at the function $F(x, \Delta k)$ modulating the plane wave in Eq. (2.30):

$$F(x, \Delta k) = \frac{2 \sin \Delta k \left[\left(\frac{d\omega}{dk} \right)_{k=k_0} t - x \right]}{\left(\frac{d\omega}{dk} \right)_{k=k_0} t - x}, \tag{2.34}$$

which is plotted in Fig. 2.9 for increasing values of Δk expressed in arbitrary units, for the same range of x values at $t = 0$. One can see that increasing Δk means to narrow the peak around the maximum, and therefore the uncertainty around it rapidly decreases. In fact, the quantity Δx represents the indetermination in the particle location, which can be given as $x_M \pm \Delta x$, and Eq. (2.33) shows that it is dependent on the indetermination of the particle momentum $\Delta p = \hbar \Delta k$ in such a way that the wider the indetermination of the momentum is, the smaller the error on the particle location will be, and vice versa.

This leads to a very important relationship that introduces the concept of the *principle of uncertainty* as being intrinsically linked to the wave representation of a particle:

$$\Delta x = \frac{h}{\Delta p} \quad \text{or} \quad \Delta x \Delta p = h. \tag{2.35}$$

Thus, the indetermination product has the Planck constant h as the minimum value, and therefore the more general expression will be

$$\Delta x \Delta p \geq h. \tag{2.36}$$

This shows that position and momentum are conjugated quantities which are not possible to know at the same time with infinitely low indetermination: if the experimental error in the particle position is extremely low, the value of the particle momentum will be almost undetermined, and vice versa.

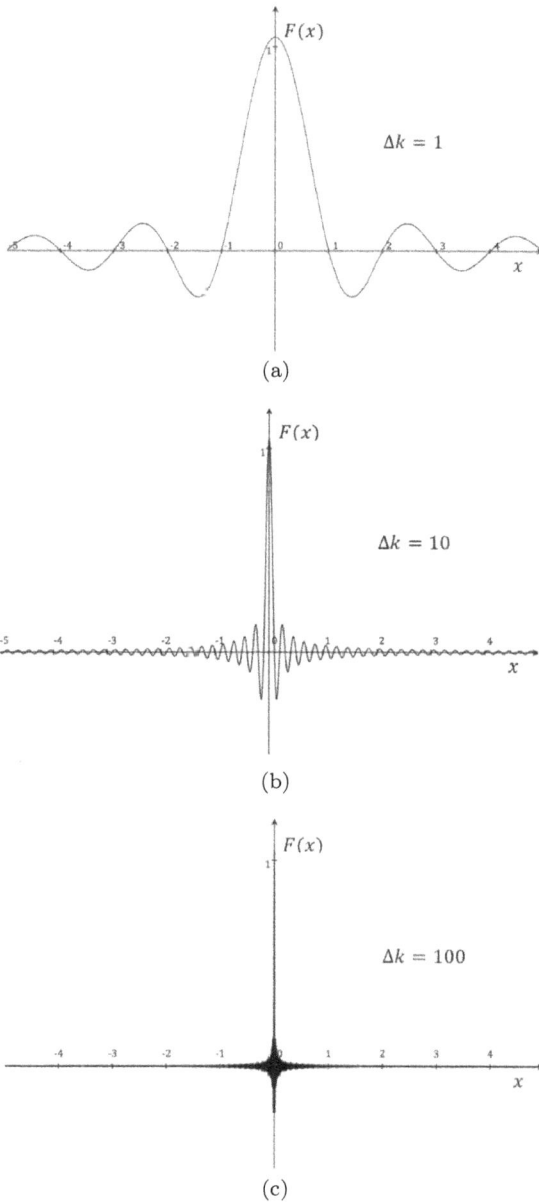

Fig. 2.9. Plot of Eq. (2.34) showing the peak narrowing by increasing the k vector width of the wave packet: (a) $\Delta k = 1$, (b) $\Delta k = 10$, (c) $\Delta k = 100$, in arbitrary units.

This result is linked to the *principle of complementarity*, formulated by Bohr. Complementarity is a property of conjugated physical quantities such as energy and time, the angular momentum components, etc., which obey the principle of indetermination. This property is caused by the limited knowledge of the microscopic world we can have because of the perturbations we introduce at the moment of performing measurements. The small value of the Planck constant guarantees that this effect is negligible in the macroscopic world. In the frame of the complete theory of quantum mechanics, we will meet the most important conjugated physical quantities.

It is amazing that the few arguments coming from the introduction of the concept of the photon and from wave–particle representation after de Broglie's hypothesis led to formulating some basic principles: *uncertainty, complementarity, superposition of states.* It is also clear that the deterministic view of classical mechanics must be abandoned to adopt a statistical approach to the result of any measurement. Of course, this was the most difficult step to be taken, and in this view it was also difficult to find the basic quantity linked to the *wave function* which must represent the probability of finding the particle in a specific location. The answer came from experiments showing diffraction of electrons. The famous experiment reported in 1927 by C. Davisson and L.H. Germer[3] was one of the first experimental demonstrations of de Broglie's hypothesis. They sent an electron beam on a crystal and measured the scattered electrons at different angles, obtaining a distribution showing maxima and minima similar to the one obtained by X-ray diffraction. A similar experiment was performed by G.P. Thomson and A. Reid.[4] They sent the electron beam against a very thin target able to transmit the impinging electrons. These were finally detected on a screen producing the distribution that is typical of diffraction of electromagnetic waves. Figure 2.10 shows the scheme of the experiment and the corresponding measurement.

Since the measured intensity I is proportional to the square of the amplitude, the measured quantity will be

$$I = \text{const} \times |A(x,t)|^2.$$

---

[3]C. Davisson and L.H. Germer, *Phys. Rev.* **30**, 705 (1927).
[4]G.P. Thomson and A. Reid, *Nature* **119**, 890 (1927).

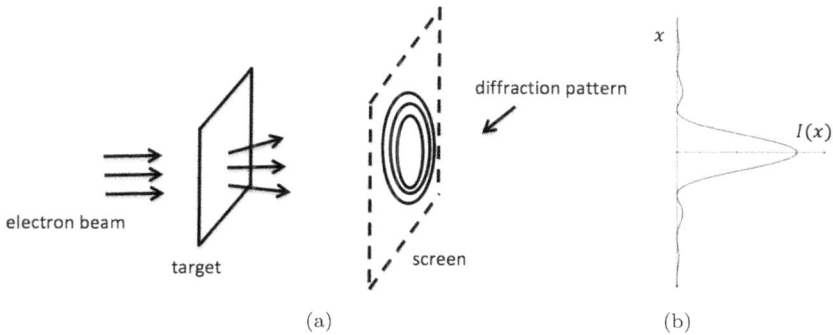

(b)

Fig. 2.10. (a) Scheme of the experiment by Thomson and Reid; (b) intensity distribution of electrons on the screen plane.

On the other hand, also in this case we could argue about the scattering of a single electron in a way similar to the one we followed about the photon passing through the crystal polarizer. A single electron would be diffracted in any of the possible positions on the screen; in making the experiment with many electrons, each of them will be detected in a particular position of the screen, giving rise to the measured diffraction pattern.

Therefore, also in this case, this distribution represents the probability for the position of a single electron that cannot be known in advance. For this reason, we find the physical meaning of $|A(x,t)|^2$ as being proportional to the probability of finding the electron in the position x at time t.

This statistical approach is the most radical change with respect to the classical deterministic point of view. In Newton's mechanics, the second law of dynamics describes the dynamic conditions of a particle and, together with the initial conditions of motion, it allows determining all the characteristic physical quantities.

On the contrary, the quantum point of view means identifying the wave function that describes the motion of a particle under specific dynamical conditions, which include the boundary conditions, and getting from it the expected physical quantities as most probable among a number of possible values. In order to follow this logic, one needs a starting equation that all the wave functions must obey, playing a role that in classical mechanics is played by Newton's equation $\mathbf{F} = m\mathbf{a}$. This role is played by the Schrödinger equation which must be integrated to find the wave function describing the particle in any specific physical condition. From it one should be able to find the probability distribution of the particle position and of the other relevant dynamical quantities.

2.3. The Schrödinger equation

To get wave functions describing many different physical situations requires a basic equation valid in any case, in order to solve the general problem as integration of such an equation. In fact, a simple wave packet created by superposition of plane waves makes understandable how a wave description can be adopted for a particle, but it is not suitable for treating most of the cases, mainly because of dispersion, i.e. the different phase velocity ω/k of the superimposed waves — an effect that leads to a rapid modification of the wave envelope.

The arguments including both mathematics and physics which are used to get the Schrödinger equation can be found in specialized textbooks on quantum mechanics. We will show here how it is possible to get it for a free particle obeying de Broglie's hypothesis. Of course, it is not a demonstration of the equation that works as a principle in quantum mechanics since its demonstration lies in the success of its expectations when compared to experimental results.

We have already pointed out that the energy of a free wave–particle can be written in the following way:

$$E = \hbar\omega = \frac{\hbar^2 k^2}{2m}.$$

Describing the particle as a plane wave, we get

$$\psi(x,t) = A\exp[i(kx - \omega t)]. \tag{2.37}$$

Then

$$\frac{\partial\psi}{\partial t} = -i\omega\psi, \tag{2.38}$$

$$\frac{\partial^2\psi}{\partial x^2} = -k^2\psi. \tag{2.39}$$

Multiplying Eq. (2.38) by $i\hbar$ and Eq. (2.39) by $-\hbar^2/2m$, we get

$$i\hbar\frac{\partial\psi}{\partial t} = \hbar\omega\psi, \tag{2.40}$$

$$-\frac{\hbar^2}{2m}\frac{\partial^2\psi}{\partial x^2} = \frac{\hbar^2 k^2}{2m}\psi. \tag{2.41}$$

Since $\hbar\omega = \hbar^2 k^2/2m$, the right-hand side of Eqs. (2.40) and (2.41) are equal and we can write

$$-\frac{\hbar^2}{2m}\frac{\partial^2\psi}{\partial x^2} = i\hbar\frac{\partial\psi}{\partial t}. \tag{2.42}$$

That is the differential equation for ψ we were looking for. In the three-dimensional case, in Eq. (2.37) we must replace kx with $k_x x + k_y y + k_z z$ and, as a consequence, in Eq. (2.39) we must replace $\partial^2 \psi/\partial x^2$ with

$$\left(\frac{\partial^2}{\partial x^2} + \frac{\partial^2}{\partial y^2} + \frac{\partial^2}{\partial z^2} \right) \psi \equiv \nabla^2 \psi,$$

with ∇^2 being the Laplace operator.

Finally, we get the *Schrödinger equation* for a free particle:

$$-\frac{\hbar^2}{2m} \nabla^2 \psi = i \frac{\partial \psi}{\partial t}. \tag{2.43}$$

The first term of the equation is given by the operator $-(\hbar^2/2m)\nabla^2$ applied to the wave function ψ. Looking at Eq. (2.41), we recognize that when applied to ψ this operator gives the kinetic energy of the particle times ψ. For this reason, it is named the *kinetic energy operator*.

In the case where the particle is not free but moves under the action of forces, giving rise to a potential energy $\mathcal{V}(\mathbf{r}, t)$, the total energy of the particle will be

$$E = \frac{\hbar^2 k^2}{2m} + \mathcal{V}(\mathbf{r}, t). \tag{2.44}$$

Therefore, a *total energy operator* or *Hamiltonian operator* will be defined as

$$\mathcal{H} \equiv -\frac{\hbar^2}{2m} \nabla^2 + \mathcal{V}(\mathbf{r}, t). \tag{2.45}$$

Accordingly, we can replace the first term operator of Eq. (2.43) with the Hamiltonian operator to get the *general form of the Schrödinger equation*,

$$\left[-\frac{\hbar^2}{2m} \nabla^2 + \mathcal{V}(\mathbf{r}, t) \right] \psi = i\hbar \frac{\partial \psi}{\partial t}, \tag{2.46}$$

or

$$\mathcal{H}\psi = i\hbar \frac{\partial \psi}{\partial t}. \tag{2.47}$$

This is the fundamental equation that any wave function must fulfill. The general problem is the integration of this equation given the potential energy $\mathcal{V}(\mathbf{r}, t)$ and the boundary conditions that define the physical problem.

Equation (2.47) can be simplified if the potential energy is not dependent on time, i.e. $\mathcal{V}(\mathbf{r}, t) = \mathcal{V}(\mathbf{r})$. In this case, it is possible to make a preliminary statement writing the wave function as the product of a function of time and a function of space coordinates:

$$\psi(\mathbf{r}, t) = f(t)\phi(\mathbf{r}). \tag{2.48}$$

Then the Schrödinger equation becomes

$$f(t)\left[-\frac{\hbar^2}{2m}\nabla^2 + \mathcal{V}(\mathbf{r}, t)\right]\phi(\mathbf{r}) = i\hbar\phi(\mathbf{r})\frac{\partial f(t)}{\partial t}, \tag{2.49}$$

i.e.

$$\frac{1}{\phi(\mathbf{r})}\left[-\frac{\hbar^2}{2m}\nabla^2 + \mathcal{V}(\mathbf{r}, t)\right]\phi(\mathbf{r}) = \frac{i\hbar}{f(t)}\frac{\partial f(t)}{\partial t}. \tag{2.50}$$

As the first term is dependent only on $\mathbf{r}$ and the second only on t, each of them must be equal to the same constant E. Then

$$\frac{i\hbar}{f(t)}\frac{\partial f(t)}{\partial t} = E \quad \rightarrow \quad \frac{\partial f(t)}{f(t)} = -\frac{iE}{\hbar}\partial t. \tag{2.51}$$

This can be easily integrated to give

$$f(t) = C\exp\left(-\frac{iE}{\hbar}t\right). \tag{2.52}$$

Then, from Eq. (2.48), we write

$$\psi(\mathbf{r}, t) = \phi(\mathbf{r})\exp\left(-\frac{iE}{\hbar}t\right), \tag{2.53}$$

where the integration constant C has been included in $\phi(\mathbf{r})$. The constant E can be easily identified as the particle's total energy corresponding to the wave function $\psi(\mathbf{r}, t)$. In fact, we can recognize the coefficient of the exponential function as the leading frequency of wave the packet described by $\psi(\mathbf{r}, t)$:

$$\psi(\mathbf{r}, t) = \phi(\mathbf{r})\exp(-\omega_0 t). \tag{2.54}$$

Since the energy of the particle can be expressed as $E = \hbar\omega_0$, if we combine Eqs. (2.53) and (2.54) it coincides with the constant appearing in Eq. (2.51).

Using Eq. (2.53) in Eq. (2.46) and dividing by $\exp\left(-\frac{iE}{\hbar}t\right)$, we get

$$\left[-\frac{\hbar^2}{2m}\nabla^2 + \mathcal{V}(\mathbf{r}, t)\right]\phi(\mathbf{r})\exp\left(-\frac{iE}{\hbar}t\right) = i\hbar\frac{\partial}{\partial t}\phi(\mathbf{r})\exp\left(-\frac{iE}{\hbar}t\right)$$

$$\rightarrow \left[-\frac{\hbar^2}{2m}\nabla^2 + \mathcal{V}(\mathbf{r}, t)\right]\phi(\mathbf{r}) = i\hbar\left(-\frac{iE}{\hbar}\right)\phi(\mathbf{r}), \tag{2.55}$$

leading to the *time-independent Shrödinger equation,*

$$\left[-\frac{\hbar^2}{2m}\nabla^2 + \mathcal{V}(\mathbf{r}, t) \right] \phi(\mathbf{r}) = E\phi(\mathbf{r}), \qquad (2.56)$$

or

$$\mathcal{H}\phi(\mathbf{r}) = E\phi(\mathbf{r}). \qquad (2.57)$$

We have already pointed out that the meaning of the wave function is given by a statistical approach. Then we can define the probability $dP(\mathbf{r}, t)$ of finding the particle described by $\psi(\mathbf{r}, t)$ on time t in a small volume $dV = dx\, dy\, dz$ around the point $\mathbf{r}$,

$$dP(\mathbf{r}, t) = |\psi(\mathbf{r}, t)|^2 dV, \qquad (2.58)$$

or, in one dimension,

$$dP(x, t) = |\psi(x, t)|^2 dx, \qquad (2.59)$$

giving the probability of finding the particle in the range between x and $x + dx$. An example is given in Fig. 2.11, where a function $|\psi(x, t)|^2$ is reported versus x. The dashed area represents the quantity

$$P_{x_1}^{x_2}(t) = \int_{x_1}^{x_2} |\psi(x, t)|^2 dx,$$

which is the probability of finding the particle in the range $x_1 \leq x \leq x_2$. Then the general expression (in the one-dimensional case) for the probability that the position of the particle will be in the range (a, b) is given by

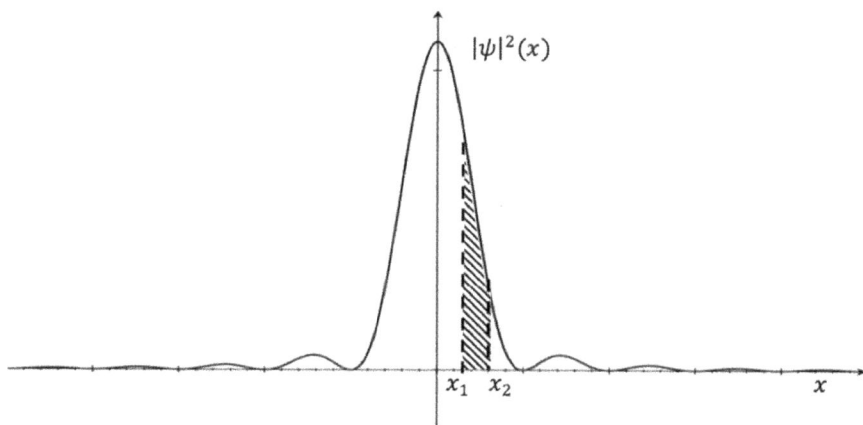

Fig. 2.11. An example of the plot of $|\psi(x, t)|^2$ versus x showing the probability of finding the particle in the range $x_1 \leq x \leq x_2$ (dashed area).

the area limited by the function $|\psi(x,t)|^2$ and the x axis:

$$P_a^b(t) = \int_a^b |\psi(x,t)|^2 dx. \tag{2.60}$$

Since the probability of finding the particle somewhere is 1, i.e. in any place along the x axis, we obviously get

$$\int_{-\infty}^{+\infty} |\psi(x,t)|^2 dx = 1. \tag{2.61}$$

or, for the three-dimensional case,

$$\int_{-\infty}^{+\infty} |\psi(\mathbf{r},t)|^2 dV = 1. \tag{2.62}$$

This is the *normalization condition* for the wave function $\psi(\mathbf{r},t)$.

It is worth underlining that if $\psi(\mathbf{r},t)$ is a solution to the Schrödinger equation any other function $A\psi(\mathbf{r},t)$ is also a solution, with A being a constant quantity. Therefore, Eq. (2.62) allows determining the constant A to fix the specific wave function describing the particle.

It is necessary that Eq. (2.62) is fulfilled by a solution to the Schrödinger equation, in order to represent the particle dynamics. For this reason, one says that the possible states are given by the *square-integrable* solutions.

In this context, it is very relevant that Schrödinger equation allows holding the normalization condition with time: if a wave function is normalized at $t = 0$, it holds the normalization at any time. In one dimension it means that

$$\frac{d}{dt} \int_{-\infty}^{+\infty} |\psi(x,t)|^2 dx = 0. \tag{2.63}$$

Demonstration is straightforward:

$$\frac{d}{dt} \int_{-\infty}^{+\infty} |\psi(x,t)|^2 dx = \int_{-\infty}^{+\infty} \frac{\partial}{\partial t} |\psi(x,t)|^2 dx.$$

On the other hand,

$$\frac{\partial}{\partial t} |\psi|^2 = \frac{\partial}{\partial t}(\psi\psi^*) = \frac{\partial \psi^*}{\partial t}\psi + \psi^*\frac{\partial \psi}{\partial t}.$$

Then we write the Schrödinger equation for ψ and for ψ^*:

$$\frac{\partial \psi}{\partial t} = \frac{i\hbar}{2m}\frac{\partial^2 \psi}{\partial x^2} - \frac{i}{\hbar}\mathcal{V}\psi, \qquad \frac{\partial \psi^*}{\partial t} = -\frac{i\hbar}{2m}\frac{\partial^2 \psi^*}{\partial x^2} + \frac{i}{\hbar}\mathcal{V}\psi^*.$$

Therefore,

$$\frac{\partial}{\partial t}|\psi|^2 = \frac{i\hbar}{2m}\left(\psi^*\frac{\partial^2 \psi}{\partial x^2} - \frac{\partial^2 \psi^*}{\partial x^2}\psi\right) = \frac{\partial}{\partial x}\left[\frac{i\hbar}{2m}\left(\psi^*\frac{\partial \psi}{\partial x} - \frac{\partial \psi^*}{\partial x}\psi\right)\right],$$

$$\int_{-\infty}^{+\infty}\frac{\partial}{\partial t}|\psi(x,t)|^2 dx = \frac{i\hbar}{2m}\left[\psi^*\frac{\partial \psi}{\partial x} - \frac{\partial \psi^*}{\partial x}\psi\right]_{-\infty}^{+\infty} \to 0.$$

Since ψ and ψ^* must vanish when $x \to \pm\infty$ (otherwise they could not be normalized),

$$\frac{d}{dt}\int_{-\infty}^{+\infty}|\psi(x,t)|^2 dx = 0.$$

This result can be easily extended to the more general three-dimensional case.

2.4. Physical quantities, operators and expectation values

In the previous section, we have introduced the Hamiltonian operator $\mathcal{H} \equiv -\frac{\hbar^2}{2m}\nabla^2 + \mathcal{V}(\mathbf{r}, t)$, associating it with the total energy. It is actually a general procedure in quantum mechanics to have a correspondence between physical quantities and operators. Starting from the kinetic energy $-\frac{\hbar^2}{2m}\nabla^2$ operator, we can easily find a way to get other operators. In fact, since we can write the kinetic energy as $\frac{p^2}{2m}$, we can associate the operator $\frac{\hbar}{i}\nabla$ with the momentum p, with ∇ being the gradient operator:

$$\nabla \equiv \left(\frac{\partial}{\partial x}, \frac{\partial}{\partial y}, \frac{\partial}{\partial z}\right).$$

In this way, we correctly get

$$\frac{p^2}{2m} \to -\frac{\hbar^2}{2m}\nabla^2.$$

Then we have the association

$$\mathbf{p} \quad \rightarrow \quad \frac{\hbar}{i}\nabla \tag{2.64}$$

as the *momentum operator* and

$$\mathbf{L} = \mathbf{r} \times \mathbf{p} \quad \rightarrow \quad \mathbf{r} \times \frac{\hbar}{i}\nabla \tag{2.65}$$

as the *angular momentum operator*.

One basic assumption of quantum mechanics is that each dynamical quantity related to the motion of the particle can be represented by a linear operator.[5] The meaning and the usefulness of associating a mathematical operator with each physical quantity rely on the way the theory is able to evaluate the value of the physical quantity which can be measured, allowing a comparison between theory and experiments. In this way, a real foundation for the theory is given. This meaning is actually linked to the statistical approach that is required when one is dealing with the wave function.

This concept can be highlighted starting from the basics of statistical treatment of experimental data. By measuring N times (where N should be large) a physical quantity by a very sensitive instrument, one gets a statistical distribution of data, depending on how many times each datum has been obtained. In these conditions, one considers the average of all the data as the value of the physical quantity under examination:

$$\langle \xi \rangle = \frac{\sum_1^N \xi_i}{N} = \sum_1^K \frac{n_j \xi_j}{N}, \tag{2.66}$$

with ξ_i being the result of a single measurement, $K \leq N$ the set of different values obtained in the total set of N measurements, and n_j the number of times each value ξ_j has been found.

It is well known that if N is large the probability of the event coincides with the repetition frequency of the occurrence of the event, namely

$$P(\xi_j) \equiv P_j = \frac{n_j}{N}.$$

Then, from Eq. (2.66), the mean value can be written as

$$\langle \xi \rangle = \sum_1^K P_j \xi_j. \tag{2.67}$$

---

[5]A linear operator Ω fulfills the equation $\Omega(a_1\psi_1 + a_2\psi_2) = a_1\Omega\psi_1 + a_2\Omega\psi_2$ for any complex number a_i and wave function ψ_i.

In other words, the mean value is the summation of each measured value times the probability of its occurrence. In the case of a continuous range of values, we must replace P_j with $\rho(\zeta)d\zeta$, where $\rho(\zeta)$ is the probability density for a measurement of the physical quantity ζ, to give a value falling in the range $\zeta' \leq \zeta \leq \zeta' + d\zeta'$:

$$P_j \quad \rightarrow \quad \rho(\zeta')d\zeta.$$

As a consequence, the average result of a measurement on ζ will be given by[6]

$$\langle \zeta \rangle = \int_{\zeta_1}^{\zeta_2} \zeta\rho(\zeta)d\zeta \tag{2.68}$$

if the possible values are limited to the range $(\zeta_1\zeta_2)$.

In quantum mechanics, this is called the *expectation value*. It is the mathematical expectation of the measurement of the corresponding physical quantity.

Let us apply these concepts to the most-used physical quantities, starting from the particle position in one dimension. We know the probability of finding the particle in the range between x and $x + dx$ as $|\psi(x,t)|^2dx$. Therefore, applying the definition (2.68) for the expectation value of x, we must write

$$\langle x \rangle = \int_{-\infty}^{+\infty} x|\psi(x,t)|^2dx \tag{2.69}$$

if the wave function is already normalized.

In the most general three-dimensional case, we must replace the interval dx with the volume $dV = dx\,dy\,dz$ around the point $\mathbf{r}(x,y,z)$. Therefore, the expectation values of the three-point coordinates are

$$\langle x \rangle = \int_{-\infty}^{+\infty} x|\psi(\mathbf{r},t)|^2dV, \quad \langle y \rangle = \int_{-\infty}^{+\infty} y|\psi(\mathbf{r},t)|^2dV,$$

$$\langle z \rangle = \int_{-\infty}^{+\infty} z|\psi(\mathbf{r},t)|^2dV, \tag{2.70}$$

---

[6]For a continuous quantity, the integral must replace summation over the range of possible values.

which can be written as

$$\langle x \rangle = \int_{-\infty}^{+\infty} \psi^*(\mathbf{r}, t) x \psi(\mathbf{r}, t)\, dV, \quad \langle y \rangle = \int_{-\infty}^{+\infty} \psi^*(\mathbf{r}, t) y\, \psi(\mathbf{r}, t) dV,$$

$$\langle z \rangle = \int_{-\infty}^{+\infty} \psi^*(\mathbf{r}, t) z\, \psi(\mathbf{r}, t) dV. \tag{2.71}$$

There is no difference between the expressions (2.70) and (2.71) since the position operators x, y, z correspond just to multiplication by the wave function.

In the general case, the correct form of the expectation value is the one given in Eq. (2.71). For instance, this can be pointed out by considering the time derivative of $\langle x \rangle$ starting from Eq. (2.69) to work out the expectation value of momentum. Taking into account the Schrödinger equation,[7] we get

$$\frac{d\langle x \rangle}{dt} = \int_{-\infty}^{+\infty} x \frac{\partial}{\partial t} |\psi(x,t)|^2 dx = \frac{i\hbar}{2m} \int_{-\infty}^{+\infty} x \frac{\partial}{\partial x} \left(\psi^* \frac{\partial \psi}{\partial x} - \frac{\partial \psi^*}{\partial x} \psi \right) dx. \tag{2.72}$$

Integration by parts gives

$$\frac{d\langle x \rangle}{dt} = \frac{i\hbar}{2m} \int_{-\infty}^{+\infty} \left(\psi^* \frac{\partial \psi}{\partial x} - \frac{\partial \psi^*}{\partial x} \psi \right) dx, \tag{2.73}$$

where we have used the condition $\psi \to 0$ when $x \to \pm\infty$. A second integration allows writing

$$\frac{d\langle x \rangle}{dt} = -\frac{i\hbar}{m} \int_{-\infty}^{+\infty} \psi^* \frac{\partial \psi}{\partial x} dx. \tag{2.74}$$

Assuming the expectation value of the velocity to be[8]

$$\langle \nu_x \rangle = \frac{d\langle x \rangle}{dt}, \tag{2.75}$$

the expression (2.74) gives the expectation value $\langle \nu_x \rangle$. By introducing the linear momentum $p_x = m\nu_x$, we get

$$\langle p_x \rangle = m \frac{d\langle x \rangle}{dt} = -i\hbar \int_{-\infty}^{+\infty} \psi^* \frac{\partial \psi}{\partial x} dx = \int_{-\infty}^{+\infty} \psi^* \left(\frac{\hbar}{i} \frac{\partial}{\partial x} \right) \psi dx, \tag{2.76}$$

---

[7]See calculation on page 101.
[8]This statement comes from the Ehrenfest theorem, which will be demonstrated in the next section.

which corresponds to Eq. (2.71) when we are considering the momentum operator in place of the position operator. In the three-dimensional case, we have

$$\langle \mathbf{p} \rangle = \int_{-\infty}^{+\infty} \psi^*(\mathbf{r}, t) \left(\frac{\hbar}{i} \nabla \right) \psi(\mathbf{r}, t) dV, \qquad (2.77)$$

which means that

$$\langle p_x \rangle = \int_{-\infty}^{+\infty} \psi^*(\mathbf{r}, t) \left(\frac{\hbar}{i} \frac{\partial}{\partial x} \right) \psi(\mathbf{r}, t) dV,$$

$$\langle p_y \rangle = \int_{-\infty}^{+\infty} \psi^*(\mathbf{r}, t) \left(\frac{\hbar}{i} \frac{\partial}{\partial y} \right) \psi(\mathbf{r}, t) dV,$$

$$\langle p_z \rangle = \int_{-\infty}^{+\infty} \psi^*(\mathbf{r}, t) \left(\frac{\hbar}{i} \frac{\partial}{\partial z} \right) \psi(\mathbf{r}, t) dV. \qquad (2.78)$$

Therefore, as a general rule, the expectation value of any physical quantity expressed by the operator Ω is given by

$$\langle \Omega \rangle = \int_{-\infty}^{+\infty} \psi^*(\mathbf{r}, t) \, \Omega \, \psi(\mathbf{r}, t) dV. \qquad (2.79)$$

For instance, we have

$$\langle \mathcal{V} \rangle = \int_{-\infty}^{+\infty} \psi^*(\mathbf{r}, t) \mathcal{V} \psi(\mathbf{r}, t) dV \qquad (2.80)$$

for the potential energy, and

$$\langle \mathbf{L} \rangle = \int_{-\infty}^{+\infty} \psi^*(\mathbf{r}, t) \left[\mathbf{r} \times \frac{\hbar}{i} \nabla \right] \psi(\mathbf{r}, t) dV \qquad (2.81)$$

for the angular momentum, which means that[9]

$$\langle L_x \rangle = \frac{\hbar}{i} \int_{-\infty}^{+\infty} \psi^*(\mathbf{r}, t) \, [\mathbf{r} \times \nabla]_{\mathbf{x}} \, \psi(\mathbf{r}, t) dV,$$

$$\langle L_y \rangle = \frac{\hbar}{i} \int_{-\infty}^{-\infty} \psi^*(\mathbf{r}, t) [\mathbf{r} \times \nabla]_{\mathbf{y}} \psi(\mathbf{r}, t) dV,$$

$$\langle L_z \rangle = \frac{\hbar}{i} \int_{-\infty}^{+\infty} \psi^*(\mathbf{r}, t) [\mathbf{r} \times \nabla]_{\mathbf{z}} \psi(\mathbf{r}, t) dV. \qquad (2.82)$$

In summary, in quantum mechanics each physical quantity has its operator counterpart that allows evaluating the expectation value that

---

[9]We recall that $[\mathbf{r} \times \nabla]_{\mathbf{x}} = y\frac{\partial}{\partial z} - z\frac{\partial}{\partial y}$, $[\mathbf{r} \times \nabla]_{\mathbf{y}} = z\frac{\partial}{\partial x} - x\frac{\partial}{\partial z}$, $[\mathbf{r} \times \nabla]_{\mathbf{z}} = x\frac{\partial}{\partial y} - y\frac{\partial}{\partial x}$.

corresponds to the result expected from the measurement. This is one of the basic links, between theory and experiment.

2.5. Hermitian operators, commutator brackets and the correspondence principle

It is now important to underline a property of several operators representing physical quantities. First of all, we define the *adjoint* Ω^+ of an operator Ω through the expression

$$\int_{-\infty}^{+\infty} f^*(\mathbf{r})\Omega g(\mathbf{r})dV = \int_{-\infty}^{+\infty} [\Omega^+ f(\mathbf{r})]^* g(\mathbf{r})dV, \qquad (2.83)$$

with $f(\mathbf{r})$ and $g(\mathbf{r})$ being two complex functions.

The operator is defined as *Hermitian* or *self-adjoint* if $\Omega^+ = \Omega$, thus fulfilling the property

$$\int_{-\infty}^{+\infty} f^*(\mathbf{r})\Omega g(\mathbf{r})dV = \int_{-\infty}^{+\infty} [\Omega f(\mathbf{r})]^* g(\mathbf{r})dV. \qquad (2.84)$$

It is important to show that an operator representing a physical quantity is *Hermitian*. In fact, for this type of operator the expectation value must be a real quantity since it corresponds to the result of a measurement, and therefore $\langle \Omega \rangle = \langle \Omega \rangle^*$,

$$\langle \Omega \rangle = \int_{-\infty}^{+\infty} \psi^* \Omega \psi dV = \int_{-\infty}^{+\infty} [\Omega \psi]^* \psi dV = \langle \Omega \rangle^*, \qquad (2.85)$$

corresponding to Eq. (2.84).[10]

Exploiting this property, it is possible to get a very important equation giving the temporal evolution of the expectation value of an operator describing a physical quantity. Starting from the definition (2.85) of the expectation value, we can write

$$\frac{d\langle \Omega \rangle}{dt} = \frac{d}{dt} \int_{-\infty}^{+\infty} \psi^* \, \Omega \, \psi dV, \qquad (2.86)$$

and performing the calculation detailed below we get

$$\frac{d\langle \Omega \rangle}{dt} = \left\langle \frac{\partial \Omega}{\partial t} \right\rangle + \frac{i}{\hbar} \int_{-\infty}^{+\infty} \{\psi^* \mathcal{H}\Omega\psi - \psi^* \Omega\mathcal{H}\psi\}dV. \qquad (2.87)$$

---

[10]Complex conjugation requires the reversal of order of the factors in the integral.

In fact,

$$\frac{d\langle\Omega\rangle}{dt} = \frac{d}{dt}\int_{-\infty}^{+\infty}\psi^*\Omega\psi dV$$

$$= \int_{-\infty}^{+\infty}\left\{\psi^*\frac{\partial}{\partial t}[\Omega\psi] + \frac{\partial\psi^*}{\partial t}[\Omega\psi]\right\}dV$$

$$= \int_{-\infty}^{+\infty}\left\{\psi^*\left[\left(\frac{\partial\Omega}{\partial t}\right)\psi + \Omega\frac{\partial\psi}{\partial t}\right] + \frac{\partial\psi^*}{\partial t}[\Omega\psi]\right\}dV.$$

One can exploit the Schrödinger equation to write $\frac{\partial\psi}{\partial t} = -\frac{i}{\hbar}\mathcal{H}\psi$ and obtain

$$\frac{d\langle\Omega\rangle}{dt} = \int_{-\infty}^{+\infty}\left\{\psi^*\left(\frac{\partial\Omega}{\partial t}\right)\psi - \frac{i}{\hbar}\psi^*\Omega\mathcal{H}\psi + \frac{i}{\hbar}[\mathcal{H}\psi]^*\Omega\psi\right\}dV.$$

The first term under integral operation is the expectation value of $\frac{\partial\Omega}{\partial t}$; moreover, we can apply to $\mathcal{H}$ the property (2.85) of being *self-adjoint* to get

$$\frac{d\langle\Omega\rangle}{dt} = \left\langle\frac{\partial\Omega}{\partial t}\right\rangle + \frac{i}{\hbar}\int_{-\infty}^{+\infty}\{-\psi^*\Omega\mathcal{H}\psi + \psi^*\mathcal{H}\Omega\psi\}dV,$$

which is (2.87).

Looking at Eq. (2.87), we recognize that the first term under the integral sign is the expectation value of the operator $\Omega\mathcal{H}$, while the second term is the expectation value of $\mathcal{H}\Omega$, namely the same operators applied to the wave function in reverse order. In general, the order is important and the result of the operations will be different, depending on it. Only if the operations given by Ω and $\mathcal{H}$ fulfill the commutation property is the result of $\Omega\mathcal{H}\psi$ equal to the result of $\mathcal{H}\Omega\psi$. Therefore, it is useful to define the *commutator bracket* operator as

$$[\mathcal{H}, \Omega] = \mathcal{H}\Omega - \Omega\mathcal{H}, \tag{2.88}$$

which in the general case of two operators Ω_1 and Ω_2 is

$$[\Omega_1, \Omega_2] = \Omega_1\Omega_2 - \Omega_2\Omega_1. \tag{2.89}$$

In the case where the two operators fulfill the commutation property, the commutator bracket is obviously zero. With this definition Eq. (2.87) can be written as

$$\frac{d\langle\Omega\rangle}{dt} = \left\langle\frac{\partial\Omega}{\partial t}\right\rangle + \frac{i}{\hbar}\int_{-\infty}^{+\infty}\{\psi^*(\mathcal{H}\Omega - \Omega\mathcal{H})\psi\}dV$$

$$= \left\langle\frac{\partial\Omega}{\partial t}\right\rangle + \frac{i}{\hbar}\int_{-\infty}^{+\infty}\{\psi^*[\mathcal{H},\Omega]\psi\}dV, \qquad (2.90)$$

and therefore the second term on the right is just the expectation value of the commutator brackets:

$$\frac{d\langle\Omega\rangle}{dt} = \left\langle\frac{\partial\Omega}{\partial t}\right\rangle + \frac{i}{\hbar}\langle[\mathcal{H},\Omega]\rangle. \qquad (2.91)$$

In this way, we get a very compact expression for the time evolution of the expectation value of an operator representing a physical quantity. If the operator is not explicitly dependent on time, its time derivative is directly proportional to the commutator brackets $[\mathcal{H},\Omega]$, this means that if Ω commutes with $\mathcal{H}$, the corresponding physical quantity is a *constant of motion*:

$$[\mathcal{H},\Omega] = 0 \quad\rightarrow\quad \frac{d\langle\Omega\rangle}{dt} = 0 \quad\rightarrow\quad \langle\Omega\rangle = \text{constant of motion}. \qquad (2.92)$$

Of course, this is true if the commutation property holds with time, i.e. if $\mathcal{H}$ is also a constant of motion.

This result is also important for showing that quantum mechanics is somehow compatible with classical mechanics. In fact, it is possible to demonstrate that the equations of classical mechanics holding for the physical quantities describing the motion of a particle are still valid in quantum mechanics for the expectation values of those quantities. This is the *principle of correspondence*. It also guarantees that the same conservation laws can be applied to quantum-mechanical systems. It means that the results of analysis coming from the statistical approach of quantum mechanics correspond to those of the deterministic approach of classical mechanics when measurement is based on a high number of single events.

The mathematical demonstration of this principle is given by *Ehrenfest's theorem*. We will first give the demonstration of Eq. (2.75), showing that for each component of the velocity the expectation value is the time derivative of the corresponding position coordinate, and get

$$\langle v_x\rangle = \frac{d\langle x\rangle}{dt}, \quad \langle v_y\rangle = \frac{d\langle y\rangle}{dt}, \quad \langle v_z\rangle = \frac{d\langle z\rangle}{dt}. \qquad (2.93)$$

Let us do the calculation for the x component,

$$\frac{d\langle x\rangle}{dt} = \frac{d}{dt}\int_{-\infty}^{+\infty}\psi^* x\psi dV = \int_{-\infty}^{+\infty}\left(\psi^* x\frac{\partial\psi}{\partial t} + \frac{\partial\psi^*}{\partial t}x\psi\right)dV,$$

taking into account that inside the integral x appears as a time-independent operator. From the Schrödinger equation,

$$\frac{\partial\psi}{\partial t} = -\frac{i}{\hbar}\left[-\frac{\hbar^2}{2m}\nabla^2 + \mathcal{V}\right]\psi.$$

Then

$$\frac{d\langle x\rangle}{dt} = \int_{-\infty}^{+\infty}\left\{\psi^* x\left[\frac{i\hbar}{2m}\nabla^2\psi - \frac{i}{\hbar}\mathcal{V}\psi\right]\right.$$

$$\left. + \left[-\frac{i\hbar}{2m}\nabla^2\psi^* + \frac{i}{\hbar}\mathcal{V}\psi^*\right]x\psi\right\}dV.$$

The ∇^2 operator is self-adjoint:

$$\int_{-\infty}^{+\infty}\nabla^2\psi^* x\psi dV = \int_{-\infty}^{+\infty}\psi^*\nabla^2 x\psi dV.$$

Then

$$\frac{d\langle x\rangle}{dt} = \frac{i\hbar}{2m}\int_{-\infty}^{+\infty}\{\psi^* x\nabla^2\psi - \psi^*\nabla^2 x\psi\}dV.$$

On the other hand,

$$\nabla^2(x\psi) = x\nabla^2\psi + 2\frac{\partial\psi}{\partial x}.$$

Therefore,

$$\frac{d\langle x\rangle}{dt} = -\frac{i\hbar}{m}\int_{-\infty}^{+\infty}\psi^*\frac{\partial\psi}{\partial x}dV$$

$$= \frac{1}{m}\int_{-\infty}^{+\infty}\psi^*\left(-i\hbar\frac{\partial}{\partial x}\right)\psi dV$$

$$= \frac{1}{m}\int_{-\infty}^{+\infty}\psi^*\hat{p}\psi dV = \frac{\langle p\rangle_x}{m},$$

which coincides with the first of Eqs. (2.93)

Therefore, we can apply to expectation values the usual definition of velocity used in classical mechanics:

$$\langle \mathbf{v} \rangle = \frac{\langle \mathbf{p} \rangle}{m} = \frac{d\langle \mathbf{r} \rangle}{dt}. \tag{2.94}$$

Another fundamental equation is the second law of classical dynamics, $\mathbf{F} = m\mathbf{a}$, or

$$-\nabla \mathcal{V} = \frac{d\mathbf{p}}{dt},$$

which we can demonstrate to be fulfilled by the expectation values of the corresponding physical quantities. This can be done by using Eq. (2.91) to calculate the time derivative of the linear momentum $\mathbf{p}$.

According to Eq. (2.91), we write

$$\frac{d\langle p_x \rangle}{dt} = \left\langle \frac{\partial p_x}{\partial t} \right\rangle + \frac{i}{\hbar} \langle [\mathcal{H}, p_x] \rangle.$$

Since p_x is not explicitly dependent on t,

$$\left\langle \frac{\partial p_x}{\partial t} \right\rangle = 0.$$

Then

$$\frac{d\langle p_x \rangle}{dt} = \frac{i}{\hbar} \langle [\mathcal{H}, p_x] \rangle = \frac{i}{\hbar} \int_{-\infty}^{+\infty} \left\{ \psi^* \left[-\frac{\hbar^2}{2m} \nabla^2 + \mathcal{V}, -i\hbar \frac{\partial}{\partial x} \right] \psi \right\} dV.$$

On the other hand,

$$\left[-\frac{\hbar^2}{2m} \nabla^2 + \mathcal{V}, -i\hbar \frac{\partial}{\partial x} \right] = \left[-\frac{\hbar^2}{2m} \nabla^2, -i\hbar \frac{\partial}{\partial x} \right] + \left[\mathcal{V}, -i\hbar \frac{\partial}{\partial x} \right],$$

$$\left[-\frac{\hbar^2}{2m} \nabla^2, -i\hbar \frac{\partial}{\partial x} \right] \psi = i\frac{\hbar^3}{2m} \left(\nabla^2 \frac{\partial \psi}{\partial x} - \frac{\partial}{\partial x} \nabla^2 \psi \right)$$

$$= i\frac{\hbar^3}{2m} \left(\frac{\partial^3 \psi}{\partial x^3} + \frac{\partial^2}{\partial y^2} \frac{\partial \psi}{\partial x} + \frac{\partial^2}{\partial z^2} \frac{\partial \psi}{\partial x} \right.$$

$$\left. - \frac{\partial^3 \psi}{\partial x^3} - \frac{\partial}{\partial x} \frac{\partial^2 \psi}{\partial y^2} - \frac{\partial}{\partial x} \frac{\partial^2 \psi}{\partial z^2} \right) = 0,$$

since the derivation order is not relevant. As a consequence,

$$\frac{d\langle p_x \rangle}{dt} = \frac{i}{\hbar} \left\langle \left[\mathcal{V}, -i\hbar \frac{\partial}{\partial x} \right] \right\rangle$$

$$= \frac{i}{\hbar} \left\langle -i\hbar \mathcal{V} \frac{\partial}{\partial x} + i\hbar \frac{\partial}{\partial x} \mathcal{V} \right\rangle = \left\langle \mathcal{V} \frac{\partial}{\partial x} - \frac{\partial}{\partial x} \mathcal{V} \right\rangle,$$

i.e.

$$\frac{d\langle p_x \rangle}{dt} = \int_{-\infty}^{+\infty} \left\{ \psi^* \left(\mathcal{V} \frac{\partial}{\partial x} - \frac{\partial}{\partial x} \mathcal{V} \right) \psi \right\} dV$$

$$= \int_{-\infty}^{+\infty} \left\{ \psi^* \mathcal{V} \frac{\partial \psi}{\partial x} - \psi^* \frac{\partial}{\partial x} (\mathcal{V}\psi) \right\} dV$$

$$= \int_{-\infty}^{+\infty} \left\{ \psi^* \mathcal{V} \frac{\partial \psi}{\partial x} - \psi^* \frac{\partial \mathcal{V}}{\partial x} \psi - \psi^* \mathcal{V} \frac{\partial \psi}{\partial x} \right\} dV$$

$$= \int_{-\infty}^{+\infty} -\psi^* \frac{\partial \mathcal{V}}{\partial x} \psi \, dV \equiv -\left\langle \frac{\partial \mathcal{V}}{\partial x} \right\rangle.$$

This calculation can be repeated for the other components to give

$$\frac{d\langle p_x \rangle}{dt} = -\left\langle \frac{\partial \mathcal{V}}{\partial x} \right\rangle, \quad \frac{d\langle p_y \rangle}{dt} = -\left\langle \frac{\partial \mathcal{V}}{\partial y} \right\rangle, \quad \frac{d\langle p_z \rangle}{dt} = -\left\langle \frac{\partial \mathcal{V}}{\partial z} \right\rangle.$$

Therefore, we get

$$\frac{d\langle \mathbf{p} \rangle}{dt} = -\langle \nabla \mathcal{V} \rangle. \tag{2.95}$$

This is a very important result of the theory, since it gives the link between the microscopic statistical approach of quantum mechanics and the macroscopic deterministic approach of Newtonian mechanics.

2.6. Eigenfunctions and eigenvalues

Looking at the Schrödinger equation independent of time (2.57),

$$\mathcal{H}\phi(\mathbf{r}) = E\phi(\mathbf{r}),$$

on the left we have the Hamiltonian operator applied to the wave function $\phi(\mathbf{r})$ and on the right a number E which corresponds to the total energy of the system described by $\phi(\mathbf{r})$ multiplied by the same wave function. In other words, we can say that the result of applying $\mathcal{H}$ to $\phi(\mathbf{r})$ is the value of energy times $\phi(\mathbf{r})$. In such a case, we call all the functions $\phi(\mathbf{r})$ fulfilling this equation *eigenfunctions* of $\mathcal{H}$ and we call *eigenvalues* the corresponding values of E.

This type of equation may be written for any other operator Ω describing physical quantities (of course, with different eigenfunctions and eigenvalues), and it is called an *eigenvalue equation* for the operator Ω:

$$\Omega\phi(\mathbf{r}) = \alpha\phi(\mathbf{r}), \tag{2.96}$$

with $\phi(\mathbf{r})$ being the eigenfunctions and α the corresponding eigenvalues.

The eigenvalue equation allows determining the set of eigenfunctions $\phi(\mathbf{r})$ and the corresponding eigenvalues for a given operator Ω when the boundary conditions are known.

The very important property of the eigenvalues is that they correspond to the results of the measurement of the physical quantity described by Ω when the system is in the *eigenstate* described by the eigenfunction of Ω. This means that if the wave function of the system is an eigenfunction for a specific operator, then the measurement of the corresponding physical quantity gives the eigenvalue of that eigenfunction. Thus, with $\phi_j(\mathbf{r})$ being a specific eigenfunction of Ω, its eigenvalue α_j is equal to the expectation value $\langle\Omega\rangle$ in the eigenstate $\phi_j(\mathbf{r})$. This can be easily shown, since

$$\Omega\phi_j(\mathbf{r}) = \alpha_j\phi_j(\mathbf{r}), \tag{2.97}$$

$$\langle\Omega\rangle = \int_{-\infty}^{+\infty} \phi_j^*\Omega\phi_j dV. \tag{2.98}$$

Then

$$\langle\Omega\rangle = \int_{-\infty}^{+\infty} \phi_j^*\Omega\phi_j dV = \alpha_j \int_{-\infty}^{+\infty} \phi_j^*\phi_j dV = \alpha_j \tag{2.99}$$

if the wave function is normalized to give $\int_{-\infty}^{+\infty} \phi_j^*\phi_j dV = 1$.

We can say that $\phi_j(\mathbf{r})$ is a state where the physical quantity represented by Ω has a definite value α_j, and we call it the *eigenstate*. In other words, if the system is in the state $\phi_j(\mathbf{r})$ a measurement of the *observable*[11] Ω gives as the result α_j.

This is a key point of the whole quantum-mechanical description of a microscopic physical system. In fact, a measurement of a dynamical quantity represented by Ω has always as a result one of the eigenvalues α_j of Ω, i.e. the measurement process necessarily introduces a perturbation

---

[11]We call a physical quantity that is possible to measure an "observable."

in the physical system, forcing it in the eigenstate $\phi_j(\mathbf{r})$. That is to say, one of the eigenvalues α_j of any observable is the only possible result of a measurement of the physical quantity represented by Ω. Since we do not know in advance what the expected result of the measurement is, but only its probability, this means that the initial state of the system can be represented by a superposition of the set of eigenstates.

Therefore, it is also assumed that all the eigenfunctions of any observable constitute a complete set of functions in such a way that any wave function can be expressed as a linear combination of the eigenfunctions:

$$\psi = \sum_j c_j \phi_j. \tag{2.100}$$

Let us focus now on the *energy eigenfunctions*, namely the solutions to the time-independent Schrödinger equation, to highlight some fundamental properties.

They satisfy the property of *orthonormality* expressed by the condition

$$\int_{-\infty}^{+\infty} \phi_m^* \phi_k dV = 0, \quad \text{with} \ k \neq m, \tag{2.101}$$

coupled to the normalization condition:

$$\int_{-\infty}^{+\infty} \phi_m^* \phi_k dV = 1, \quad \text{with} \ k = m. \tag{2.102}$$

With ϕ_m and ϕ_k as energy eigenfunctions, we have

$$\mathcal{H}\phi_k = E_k \phi_k,$$
$$\mathcal{H}\phi_m^* = E_m \phi_m^*.$$

Multiplying the first equation by ϕ_m^* and the second equation by ϕ_k and subtracting, we get

$$\phi_m^* \mathcal{H}\phi_k - \phi_k \mathcal{H}\phi_m^* = E_k \phi_m^* \phi_k - E_m \phi_k \phi_m^*,$$

i.e.

$$-\frac{\hbar^2}{2m}(\phi_m^* \nabla^2 \phi_k - \phi_k \nabla^2 \phi_m^*) + \mathcal{V}(\phi_m^* \phi_k - \phi_k \phi_m^*) = (E_k - E_m)\phi_k \phi_m^*,$$

where the second term vanishes.

Integration is performed exploiting Green's theorem:

$$\int_V (f\nabla^2 g - g\nabla^2 f)dV = \int_S (f\nabla g - g\nabla f) \cdot \mathbf{n}\, dS,$$

with S being the surface closing the volume V, and $\mathbf{n}$ the unit vector normal to S. In this way,

$$\int_V (\phi_m^* \nabla^2 \phi_k - \phi_k \nabla^2 \phi_m^*) dV = \int_S (\phi_m^* \nabla \phi_k - \phi_k \nabla \phi_m^*) \cdot \mathbf{n} \, dS.$$

Then

$$\int_S (\phi_m^* \nabla \phi_k - \phi_k \nabla \phi_m^*) \cdot \mathbf{n} \, dS = (E_k - E_m) \int_V \phi_k \phi_m^* dV.$$

The surface integral on the left can be shown to vanish by using periodic boundary conditions.[12] Therefore,

$$(E_k - E_m) \int_V \phi_k \phi_m^* dV = 0,$$

i.e.

$$\int_V \phi_k \phi_m^* dV = 0 \quad \text{when} \quad E_k - E_m \neq 0,$$

corresponding to Eq. (2.101).

We can use the Kronecker δ symbol to specify the orthonormal set of eigenfunctions,

$$\delta_{km} = 0 \quad \text{if} \quad k \neq m, \tag{2.103}$$
$$\delta_{km} = 1 \quad \text{if} \quad k = m,$$

so that

$$\int_{-\infty}^{+\infty} \phi_m^* \phi_k dV = \delta_{km}. \tag{2.104}$$

This result is important for understanding the meaning of the coefficient c_j of the expansion (2.100). In fact, we can consider the normalization condition

$$\int_{-\infty}^{+\infty} \psi^* \psi dV = 1,$$

---

[12]See: L.I. Schiff, *Quantum Mechanics*, Chapter 3 (McGraw-Hill, 1955).

i.e.

$$\int_{-\infty}^{+\infty} \left(\sum_m c_m^* \phi_m^* \right) \left(\sum_k c_k \phi_k \right) dV = 1. \qquad (2.105)$$

Then

$$\sum_m \sum_k c_m^* c_k \int_{-\infty}^{+\infty} \phi_m^* \phi_k \, dV = 1. \qquad (2.106)$$

Owing to orthonormality, only the terms with $k = m$ do not vanish. Then

$$\sum_k |c_k|^2 = 1. \qquad (2.107)$$

Therefore, each coefficient c_k gives the probability $|c_k|^2$ that the system described by the wave function $\psi = \sum_j c_j \phi_j$ is found in the eigenstate represented by ϕ_k, with the expectation value of the energy given by E_k. In the case where the system is known to be in one specific eigenstate ϕ_m, it means that all the coefficients $c_j = 0$ for $j \neq m$ and $c_m = 1$.

2.7. Commutator brackets and the uncertainty principle

Let us consider two different operators Ω_1 and Ω_2 that fulfill the eigenvalue equations with the same eigenfunctions:

$$\Omega_1 \phi_j = \alpha_{1j} \phi_j, \qquad (2.108)$$

$$\Omega_2 \phi_j = \alpha_{2j} \phi_j.$$

We can calculate the commutator brackets as

$$[\Omega_1, \Omega_2]\phi_j = (\Omega_1 \Omega_2 - \Omega_2 \Omega_1)\phi_j = (\alpha_{1j}\alpha_{2j} - \alpha_{2j}\alpha_{1j})\phi_j = 0, \qquad (2.109)$$

i.e.

$$[\Omega_1, \Omega_2] = 0. \qquad (2.110)$$

The meaning of Eqs. (2.108) is that the physical quantities represented by Ω_1 and Ω_2 can be measured at the same time without any mutual limitation on error determination and as a consequence we have found the condition (2.110) for the commutator brackets. On the other hand, if two operators satisfy this condition it is possible to find a set of eigenfunctions of Ω_1 which are eigenfunctions of Ω_2.

These concepts become clear when we are considering explicitly the uncertainty that we have about the result of a measurement. For a statistical set of data related to the measurement of a specific physical quantity, the uncertainty is given by the *standard deviation*, which is a measure of the dispersion of the set of data around the average value: the more the data are spread apart, the higher the standard deviation. The standard deviation is defined as the square root of the *variance*, which is calculated as the average squared deviation from the average. For a discrete set of data ξ_i, the variance is given by

$$\sigma^2 = \langle (\Delta\xi)^2 \rangle = \langle (\xi_i - \langle\xi\rangle)^2 \rangle, \tag{2.111}$$

where the square procedure is necessary since obviously $\langle\Delta\xi\rangle = 0$.

Accordingly, the standard deviation is

$$\sigma = \sqrt{\sigma^2} = \sqrt{\langle (\xi_i - \langle\xi\rangle)^2 \rangle}. \tag{2.112}$$

By transferring these statistical concepts to quantum mechanics, it is possible to demonstrate that any pair of observables represented by the operators A and B satisfy the following equation involving the product of the variances on the measurement of the physical quantities represented by A and B:

$$\sigma_A^2 \sigma_B^2 \geq \left\{ \frac{1}{2i} \langle [A, B] \rangle \right\}^2. \tag{2.113}$$

This is the general form of the Heisenberg *uncertainty principle*. According to it, one cannot specify simultaneously the values of particular pairs of physical variables with unlimited sensitivity. Namely, if the error of the measurement on *one of the pair* is infinitely low ($\sigma_A \to 0$), the value of the other physical quantity will be completely undetermined ($\sigma_B \to \infty$). In other words, the achievable information on the system is limited for the pairs of *conjugated* quantities.

Applying Eq. (2.111) to a quantum-mechanical continuous system, we should write

$$\sigma_A^2 = \int \psi^* |A - \langle A \rangle|^2 \psi dV = \int \psi^* (A - \langle A \rangle)^* (A - \langle A \rangle) \psi \, dV,$$

$$\sigma_B^2 = \int \psi^* |B - \langle B \rangle|^2 \psi dV = \int \psi^* (B - \langle B \rangle)^* (B - \langle B \rangle) \psi \, dV.$$

Taking into account the property (2.85), we have

$$\sigma_A^2 = \int [\psi(A - \langle A \rangle)]^*(A - \langle A \rangle)\psi \, dV$$

$$= \int [(A - \langle A \rangle)^*\psi]^*(A - \langle A \rangle)\psi \, dV$$

$$= \int (A - \langle A \rangle)\psi^*(A - \langle A \rangle)\psi \, dV,$$

$$\sigma_B^2 = \int [\psi(B - \langle B \rangle)]^*(B - \langle B \rangle)\psi \, dV$$

$$= \int [(B - \langle B \rangle)^*\psi]^*(B - \langle B \rangle)\psi \, dV$$

$$= \int (B - \langle B \rangle)\psi^*(B - \langle B \rangle)\psi \, dV.$$

We now exploit Schwarz's inequality,

$$\left| \int_a^b f^*g \, dV \right|^2 \leq \int_a^b |f|^2 dV \int_a^b |g|^2 dV,$$

and define

$$f \equiv (A - \langle A \rangle)\psi \quad \text{and} \quad g \equiv (B - \langle B \rangle)\psi.$$

The obtained expressions for the standard deviations can be written as

$$\sigma_A^2 = \int |f|^2 dV,$$

$$\sigma_B^2 = \int |g|^2 dV.$$

Therefore, Schwarz's inequality reads

$$\sigma_A^2 \sigma_B^2 \geq \left| \int (A - \langle A \rangle)\psi^*(B - \langle B \rangle)\psi \, dV \right|^2.$$

On the other hand, for any complex number z we can write

$$|z|^2 \geq |\Im m(z)|^2 = \left| \frac{1}{2i}(z - z^*) \right|^2.$$

Then we can replace the integral with its imaginary part and the inequality still holds

$$\sigma_A^2 \sigma_B^2 \geq \left| \frac{1}{2i} \left(\int (A - \langle A \rangle) \psi^* (B - \langle B \rangle) \psi \, dV \right. \right.$$

$$\left. \left. - \int (B - \langle B \rangle) \psi^* (A - \langle A \rangle) \psi \, dV \right) \right|^2 .$$

We have

$$\int (A - \langle A \rangle) \psi^* (B - \langle B \rangle) \psi \, dV = \int A \psi^* B \psi \, dV - \int \langle A \rangle \psi^* B \psi \, dV$$

$$- \int A \psi^* \langle B \rangle \psi \, dV + \langle A \rangle \langle B \rangle \int \psi^* \psi \, dV,$$

$$\int (B - \langle B \rangle) \psi^* (A - \langle A \rangle) \psi \, dV = \int B \psi^* A \psi \, dV - \int \langle B \rangle \psi^* A \psi \, dV$$

$$- \int B \psi^* \langle A \rangle \psi \, dV + \langle A \rangle \langle B \rangle \int \psi^* \psi \, dV.$$

Finally,

$$\sigma_A^2 \sigma_B^2 \geq \left| \frac{1}{2i} \left(\int A \psi^* B \psi \, dV - \int B \psi^* A \psi \, dV \right) \right|^2 .$$

Since A and B are self-adjoint operators,

$$\int A \psi^* B \psi \, dV - \int B \psi^* A \psi \, dV$$

$$= \int \psi^* AB \psi \, dV - \int \psi^* BA \psi \, dV = \langle AB \rangle - \langle BA \rangle.$$

Therefore,

$$\sigma_A^2 \sigma_B^2 \geq \left| \frac{1}{2i} \langle [A, B] \rangle \right|^2 ,$$

which is (2.113).

The meaning of the uncertainty principle can be understood by looking at the most important cases of conjugated quantities.

Let us consider the position x and the component p_x of the linear momentum. The commutator brackets are

$$[A, B] = \left[x, -i\hbar \frac{\partial}{\partial x} \right],$$ (2.114)

which can be calculated as applied to any wave function ψ:

$$\left[x, -i\hbar \frac{\partial}{\partial x} \right] \psi = -i\hbar \left(x \frac{\partial}{\partial x} - \frac{\partial}{\partial x} x \right) \psi$$

$$= -i\hbar \left(x \frac{\partial \psi}{\partial x} - \frac{\partial}{\partial x} [x\psi] \right)$$

$$= -i\hbar \left(x \frac{\partial \psi}{\partial x} - x \frac{\partial \psi}{\partial x} - \psi \right) = i\hbar\psi.$$ (2.115)

Then

$$\left[x, -i\hbar \frac{\partial}{\partial x} \right] = [x, p_x] = i\hbar.$$ (2.116)

Using (2.113) we get

$$\sigma_x^2 \sigma_{p_x}^2 \geq \left\{ \frac{1}{2i} i\hbar \right\}^2 = \frac{\hbar^2}{4},$$ (2.117)

$$\sigma_x \sigma_{p_x} \geq \frac{\hbar}{2}.$$ (2.118)

In the case of a Gaussian distribution of data, the uncertainty associated with the average value is often considered as the standard deviation times $\sqrt{2}$, i.e. the width of the distribution[13] taken where the data value is $1/e$ times the maximum (which coincides with the average). That is to say, $\Delta x = \sqrt{2}\sigma_x$ and $\Delta p_x = \sqrt{2}\sigma_{p_x}$, and the uncertainty product

---

[13]The Gauss distribution is characteristic of several sets of experimental data when casual errors are dominant. It is written as

$$f(x) = Ae^{-\frac{(x-x_0)^2}{2\sigma^2}},$$

with A being a normalization factor. It is straightforward to show that $f(x_0) = A$ and $f(\tilde{x}) = A/e$ if $\tilde{x} = x_0 \pm \Delta x$, with $\Delta x = \sqrt{2}\sigma$.

becomes

$$\Delta x \Delta p_x \geq \hbar. \qquad (2.119)$$

The same calculation can be done for the y and z components, leading to

$$\Delta y \Delta p_y \geq \hbar, \qquad (2.120)$$

$$\Delta z \Delta p_z \geq \hbar. \qquad (2.121)$$

This means the same components of the position and momentum cannot be determined simultaneously with unlimited precision. In fact, from (2.119)–(2.121) we have

$$\Delta x \geq \frac{\hbar}{\Delta p_x}, \quad \Delta y \geq \frac{\hbar}{\Delta p_y}, \quad \Delta z \geq \frac{\hbar}{\Delta p_z}, \qquad (2.122)$$

therefore the higher the uncertainty Δp_i is on the component of the momentum, the lower it will be on the corresponding component of the position, and vice versa. We see that the uncertainty product is $\hbar$, which is a very small number on a macroscopic scale, and thus this principle is actually effective only on a microscopic scale.

It is useful to underline that the same uncertainty does not apply if we consider one component of the position and a different one of the momentum. As an example, let us calculate the commutator brackets for x and p_y:

$$\left[x, -i\hbar \frac{\partial}{\partial y} \right] \psi = -i\hbar \left(x \frac{\partial}{\partial y} - \frac{\partial}{\partial y} x \right) \psi$$

$$= -i\hbar \left(x \frac{\partial \psi}{\partial y} - \frac{\partial}{\partial y} [x \psi] \right) = -i\hbar \left(x \frac{\partial \psi}{\partial y} - x \frac{\partial \psi}{\partial y} \right)$$

$$= 0 \quad \text{since} \quad \frac{\partial x}{\partial y} = 0. \qquad (2.123)$$

Then

$$\left[x, -i\hbar \frac{\partial}{\partial y} \right] = [x, p_y] = 0, \qquad (2.124)$$

and by applying (2.113) for a Gaussian distribution we get

$$\Delta x \Delta p_y \geq 0. \qquad (2.125)$$

This means that there is no limit to the simultaneous measurement with high sensitivity of both quantities. In summary, we can write

$$[x, p_x] = [y, p_y] = [z, p_z] = i\hbar, \tag{2.126}$$

$$[x, p_y] = [x, p_z] = [y, p_x] = [y, p_z] = [z, p_x] = [z, p_y] = 0. \tag{2.127}$$

Another important conjugate pair is energy and time. The commutator calculation is made easy by taking into account the Schrödinger equation,

$$\mathcal{H}\psi = i\hbar \frac{\partial \psi}{\partial t},$$

showing the Hamiltonian operator $\mathcal{H}$ corresponding to the operator $i\hbar \frac{\partial}{\partial t}$. Then

$$[\mathcal{H}, t] \equiv \left[i\hbar \frac{\partial}{\partial t}, t \right]. \tag{2.128}$$

Applying to a wave function ψ, we have

$$\left[i\hbar \frac{\partial}{\partial t}, t \right] \psi = i\hbar \left(\frac{\partial}{\partial t}[t\psi] - t \frac{\partial \psi}{\partial t} \right) = i\hbar \left(t \frac{\partial \psi}{\partial t} + \psi - t \frac{\partial \psi}{\partial t} \right). \tag{2.129}$$

Therefore,

$$[\mathcal{H}, t] = i\hbar, \tag{2.130}$$

which, following the arguments mentioned above, gives the uncertainty product:

$$\Delta E \Delta t \geq \hbar. \tag{2.131}$$

This means that a state of the system whose energy is determined with uncertainty equal to ΔE must last for a time duration $\Delta t \geq \hbar/\Delta E$. On the other hand, if the lifetime of an energy level is Δt, the same energy has an uncertainty $\Delta E \geq \hbar/\Delta t$. This result has consequences for the spectroscopic determination of energy levels and their lifetimes for electrons in atoms, giving rise to the minimum *intrinsic* linewidth of an atomic transition.

As a last example, let us consider the angular momentum operator $\mathbf{r} \times (\hbar/i)\nabla$, namely

$$L_x = \frac{\hbar}{i} \left(y \frac{\partial}{\partial z} - z \frac{\partial}{\partial y} \right), \quad L_y = \frac{\hbar}{i} \left(z \frac{\partial}{\partial x} - x \frac{\partial}{\partial z} \right), \quad L_z = \frac{\hbar}{i} \left(x \frac{\partial}{\partial y} - y \frac{\partial}{\partial x} \right). \tag{2.132}$$

Each couple of angular momentum components constitutes a pair of conjugated physical quantities with an uncertainty product:

$$[L_x, L_y] = i\hbar L_z, \quad [L_y, L_z] = i\hbar L_x, \quad [L_z, L_x] = i\hbar L_y. \tag{2.133}$$

The calculation of the first commutator is easily performed:

$$[L_x, L_y]\psi = -\hbar^2 \left\{ \left(y\frac{\partial}{\partial z} - z\frac{\partial}{\partial y} \right) \left(z\frac{\partial}{\partial x} - x\frac{\partial}{\partial z} \right) \right.$$

$$\left. - \left(z\frac{\partial}{\partial x} - x\frac{\partial}{\partial z} \right) \left(y\frac{\partial}{\partial z} - z\frac{\partial}{\partial y} \right) \right\} \psi$$

$$= -\hbar^2 \left\{ y\frac{\partial}{\partial z}\left(z\frac{\partial}{\partial x} \right) - yx\frac{\partial^2}{\partial z^2} - z^2\frac{\partial^2}{\partial y\partial x} + xz\frac{\partial^2}{\partial y\partial z} \right\} \psi$$

$$+ \hbar^2 \left\{ yz\frac{\partial^2}{\partial z\partial x} - z^2\frac{\partial^2}{\partial y\partial x} - yx\frac{\partial^2}{\partial z^2} + x\frac{\partial}{\partial z}\left(z\frac{\partial}{\partial y} \right) \right\} \psi$$

$$= -\hbar^2 \left\{ y\frac{\partial}{\partial z}\left(z\frac{\partial}{\partial x} \right) + xz\frac{\partial^2}{\partial y\partial z} - yz\frac{\partial^2}{\partial z\partial x} - x\frac{\partial}{\partial z}\left(z\frac{\partial}{\partial y} \right) \right\} \psi$$

$$= -\hbar^2 \left\{ yz\frac{\partial^2}{\partial z\partial x} + y\frac{\partial}{\partial x} + xz\frac{\partial^2}{\partial y\partial z} \right.$$

$$\left. - yz\frac{\partial^2}{\partial z\partial x} - xz\frac{\partial^2}{\partial z\partial y} - x\frac{\partial}{\partial y} \right\} \psi$$

$$= -\hbar^2 \left\{ y\frac{\partial}{\partial x} - x\frac{\partial}{\partial y} \right\} \psi = \hbar^2 \frac{L_z}{\frac{\hbar}{i}}\psi = i\hbar L_z\psi.$$

A similar calculation can be done for the other pairs of components.

On the contrary, each component commutes with the square of the total angular momentum:

$$[L^2, L_j] = 0, \quad j = x, y, z. \tag{2.134}$$

In fact,

$$[L^2, L_z] = [L_x^2 + L_y^2 + L_z^2, L_z] = [L_x^2, L_z] + [L_y^2, L_z] + [L_z^2, L_z]$$

$$= L_x[L_x, L_z] + [L_x, L_z]L_x + L_y[L_y, L_z] + [L_y, L_z]L_y$$

$$= L_x(-i\hbar L_y) + (-i\hbar L_y)L_x + L_y(i\hbar L_x) + (i\hbar L_x)L_y = 0,$$

where the following properties of commutator brackets have been exploited:

$$[A + B, C] = [A, C] + [B, C],$$

$$[AB, C] = A[B, C] + [A, C]B.$$

2.8. The free particle

Let us consider the motion of a free particle under the quantum mechanics approach. According to the Schrödinger equation (2.57) in one dimension, we write

$$-\frac{\hbar^2}{2m}\frac{d^2}{dx^2}\phi(x) = E\phi(x), \qquad (2.135)$$

i.e.

$$\frac{d^2}{dx^2}\phi(x) + \frac{2m}{\hbar^2}E\phi(x) = 0. \qquad (2.136)$$

By choosing as the solution to Eq. (2.136) a plane wave traveling in the positive **x** direction, we write

$$\phi(x) = Ae^{ikx} \qquad (2.137)$$

and get

$$-k^2\phi(x) + \frac{2m}{\hbar^2}E\phi(x) = 0. \qquad (2.138)$$

Then

$$k = \sqrt{\frac{2mE}{\hbar^2}}, \qquad (2.139)$$

which actually corresponds to the expected expression of the kinetic energy of the particle:

$$E = \frac{\hbar^2 k^2}{2m}.$$

The absence of boundary conditions allows the particle to have a continuous spectrum of energy values, and thus there is no actual quantization. However, this description has something completely different from classical mechanics, because by considering the wave functions (2.137) we can easily show that they are eigenfunctions also of momentum,

$$\hat{p}\phi = \frac{\hbar}{i}\frac{d}{dx}Ae^{ikx} = \hbar k Ae^{ikx} = p\phi, \qquad (2.140)$$

since de Broglie's condition gives $p = \hbar k$.

The result given by Eq. (2.140) means that energy eigenfuntions are also momentum eigenfunctions, and therefore the exact value of momentum is determined. According to Heisenberg's principle expressed by (2.119), it means that we have no information about the particle position. On the other hand, in Section 2.2 we have shown that to make meaningful the particle position in the wave description we must use a wave packet, namely use a nonzero distribution of wave vectors Δk in order to allow determination of the particle position within a range given by Δx, in such a way that, in agreement with (2.119),

$$\Delta x \geq \frac{\hbar}{\Delta p} = \frac{1}{\Delta k}.$$

2.9. Localized wave functions: the potential well

In the next chapters our main interest will be in localized wave functions describing a particle whose position is restricted to a limited volume by a specific potential energy $\mathcal{V}(\mathbf{r})$. The most simple boundary conditions define the wave functions different from zero only on a cubic box with volume $V = a^3$, with a being the cube side.

The very simple example of this kind is the one-dimensional potential well, showing that these boundary conditions lead to quantization of energy states and eigenvalues. Therefore, let us consider the one-dimensional problem where a particle is confined in the range $-\frac{a}{2} < x < \frac{a}{2}$ by a constant potential energy that becomes infinite at $x = -\frac{a}{2}, \frac{a}{2}$ (see Fig. 2.12).

Inside the well we can arbitrarily choose $\mathcal{V}(x) = 0$, and thus the particle wave function in the range $-\frac{a}{2} < x < \frac{a}{2}$ must fulfill the Schrödinger equation of the free particle:

$$-\frac{\hbar^2}{2m} \frac{d^2\phi(x)}{dx^2} = E\phi(x), \tag{2.141}$$

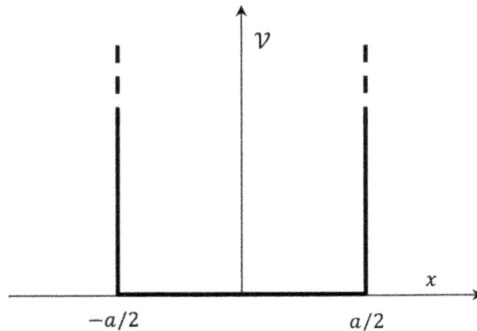

Fig. 2.12.   Infinite potential well.

while the boundary conditions are given by

$$\phi\left(-\frac{a}{2}\right) = \phi\left(\frac{a}{2}\right) = 0. \tag{2.142}$$

The general solution to Eq. (2.141) can be given as

$$\phi(x) = A\sin(\alpha x) + B\cos(\alpha x). \tag{2.143}$$

In this case, the expression (2.143) is more convenient than the general solution expressed through exponential functions,

$$\phi(x) = Ce^{ikx} + De^{-ikx},$$

which will be used in the next section.
    The two solutions are coincident provided that

$$A = i(C - D),$$
$$B = C + D.$$

By substitution in Eq. (2.141) we get

$$\frac{\partial}{\partial x}\phi(x) = A\alpha\cos(\alpha x) - B\alpha\sin(\alpha x),$$

$$\frac{\partial^2}{\partial x^2}\phi(x) = -A\alpha^2\sin(\alpha x) - B\alpha^2\cos(\alpha x).$$

Then

$$\frac{\partial^2}{\partial x^2}\phi(x) = -\alpha^2\phi(x), \quad \text{i.e.} \quad \frac{\hbar^2}{2m}\alpha^2\phi(x) = E\phi(x),$$

which gives

$$\alpha = \sqrt{\frac{2mE}{\hbar^2}}, \tag{2.144}$$

the same as Eq. (2.139), where $\alpha \equiv k$. The boundary conditions (2.142) become

$$A\sin\left(\alpha\frac{a}{2}\right) + B\cos\left(\alpha\frac{a}{2}\right) = 0,$$

$$-A\sin\left(c\frac{a}{2}\right) + B\cos\left(\alpha\frac{a}{2}\right) = 0. \tag{2.145}$$

By neglecting the trivial solution $A = B = 0$ leading to $\phi(x) \equiv 0$, we get either

$$A = 0, \quad \cos\left(\alpha\frac{a}{2}\right) = 0 \tag{2.146}$$

or

$$B = 0, \quad \sin\left(\alpha\frac{a}{2}\right) = 0. \tag{2.147}$$

In both cases,

$$\alpha\frac{a}{2} = n\frac{\pi}{2},$$

with the integer n being an odd number in the first case and an even number in the second case.

Then

$$\alpha = n\frac{\pi}{2}\frac{2}{a} = n\frac{\pi}{a}. \tag{2.148}$$

In this way, we are able to write the wave functions as

$$\phi_n(x) = B\cos\left(n\frac{\pi x}{a}\right), \quad \text{with } n \text{ integer odd number;} \tag{2.149}$$

$$\phi_n(x) = A\sin\left(n\frac{\pi x}{a}\right), \quad \text{with } n \text{ integer even number.} \tag{2.150}$$

The constants A and B can be determined by the normalization condition

$$\int_{-\frac{a}{2}}^{\frac{a}{2}} |\phi_n(x)|^2 dx = 1. \tag{2.151}$$

This means that

$$\int_{-\frac{a}{2}}^{\frac{a}{2}} \left|B\cos\left(n\frac{\pi x}{a}\right)\right|^2 dx = 1 \quad \text{or} \quad \int_{-\frac{a}{2}}^{\frac{a}{2}} \left|A\sin\left(n\frac{\pi x}{a}\right)\right|^2 dx = 1.$$

If $n = 1$,

$$\phi_1(x) = B\cos\left(\frac{\pi x}{a}\right).$$

Then Eq. (2.151) gives

$$\int_{-\frac{a}{2}}^{\frac{a}{2}} B^2\cos^2\left(\frac{\pi x}{a}\right) dx = 1,$$

It can be easily calculated:

$$\int_{-\frac{a}{2}}^{\frac{a}{2}} B^2 \cos^2\left(\frac{\pi x}{a}\right) dx = \frac{B^2 a}{\pi} \int_{-\frac{a}{2}}^{\frac{a}{2}} \cos^2\left(\frac{\pi x}{a}\right) d\left(\frac{\pi x}{a}\right)$$

$$= \frac{B^2 a}{\pi} \int_{-\frac{\pi}{2}}^{\frac{\pi}{2}} \cos^2\xi d\xi,$$

defining $\xi = \frac{\pi x}{a}$.

The integral has a well-known solution:

$$\int_{-\frac{\pi}{2}}^{\frac{\pi}{2}} \cos^2\xi d\xi = \left[\frac{\xi}{2} + \frac{\sin 2\xi}{4}\right]_{-\frac{\pi}{2}}^{\frac{\pi}{2}} = \frac{\pi}{2}.$$

Then

$$\int_{-\frac{a}{2}}^{\frac{a}{2}} B^2 \cos^2\left(\frac{\pi x}{a}\right) dx = \frac{B^2 a}{\pi}\frac{\pi}{2} = 1.$$

leading to

$$B = \sqrt{\frac{2}{a}}.$$

If $n = 2$, we can perform a similar calculation:

$$\phi_2(x) = A \sin\left(\frac{2\pi x}{a}\right),$$

which gives

$$\int_{-\frac{a}{2}}^{\frac{a}{2}} A^2 \sin^2\left(\frac{2\pi x}{a}\right) dx = 1,$$

It can be easily calculated:

$$\int_{-\frac{a}{2}}^{\frac{a}{2}} A^2 \sin^2\left(\frac{2\pi x}{a}\right) dx = \frac{A^2 a}{2\pi} \int_{-\frac{a}{2}}^{\frac{a}{2}} \sin^2\left(\frac{2\pi x}{a}\right) d\left(\frac{2\pi x}{a}\right)$$

$$= \frac{A^2 a}{2\pi} \int_{-\frac{\pi}{2}}^{\frac{\pi}{2}} \sin^2\xi d\xi,$$

defining $\xi = \frac{2\pi x}{a}$.

The integral has a well-known solution:

$$\int_{-\pi}^{\pi} \sin^2 \xi \, d\xi = \left[\frac{\xi}{2} - \frac{\sin 2\xi}{4} \right]_{-\pi}^{\pi} = \pi.$$

Then

$$\int_{-\frac{a}{2}}^{\frac{a}{2}} A^2 \sin^2 \left(\frac{2\pi x}{a} \right) dx = \frac{A^2 a}{2\pi} \pi = 1.$$

leading to

$$A = \sqrt{\frac{2}{a}}.$$

Therefore, the normalized wave functions are
with $n = 1$,

$$\phi_1(x) = \sqrt{\frac{2}{a}} \cos \left(\frac{\pi x}{a} \right);$$

with $n = 2$,

$$\phi_2(x) = \sqrt{\frac{2}{a}} \sin \left(\frac{2\pi x}{a} \right).$$

It can be easily shown that the normalization constants are the same for all the values of n, and therefore the normalized wave functions can be written as

$$\phi_n(x) = \sqrt{\frac{2}{a}} \cos \left(\frac{n\pi x}{a} \right), \quad \text{with } n \text{ odd integer;} \qquad (2.152)$$

$$\phi_n(x) = \sqrt{\frac{2}{a}} \sin \left(\frac{n\pi x}{a} \right), \quad \text{with } n \text{ even integer.} \qquad (2.153)$$

Moreover, from Eqs. (2.144) and (2.148) we easily get the energy eigenvalues:

$$E_n = \frac{n^2 \pi^2 \hbar^2}{2ma^2}. \qquad (2.154)$$

They are the only allowed values for the particle energy (energy quantization):

$$E_1 = \frac{\pi^2 \hbar^2}{2ma^2}, \quad E_2 = \frac{4\pi^2 \hbar^2}{2ma^2}, \quad E_3 = \frac{9\pi^2 \hbar^2}{2ma^2}, \quad \cdots,$$

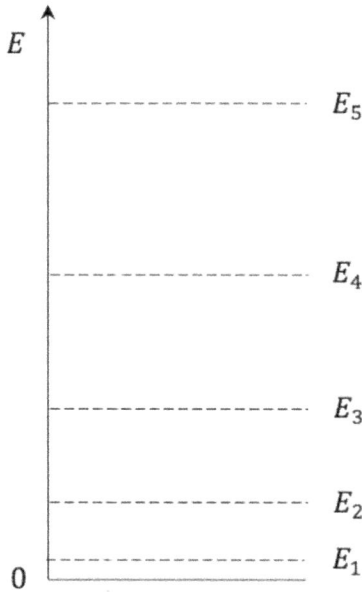

Fig. 2.13. Energy eigenvalues of an infinite potential well.

namely

$$E_2 = 4E_1, \quad E_3 = 9E_1, \quad E_4 = 16E_1, \quad E_5 = 25E_1, \quad \dots .$$

A sketch of these energy levels is given in Fig. 2.13.

It is interesting to underline that when n is an odd number we always have a maximum of the wave function at $x = 0$, while when n is an even number the wave function is equal to zero at this location. Therefore, in the first case, we have the maximum probability of finding the particle at $x = 0$, while in the second case the probability is zero.

The first five eigenfunctions ϕ_n are reported in Fig. 2.14(a) and the corresponding probability densities $|\phi_n|^2$ in Fig. 2.14(b).

A few remarks have to be made on the eigenfunctions φ_n: (i) they are alternatively even and odd functions with respect to the center of the well; (ii) starting from zero nodes at $n = 1$, the number of nodes increases with n, thus being equal to $n - 1$; (iii) it can be easily demonstrated that all the ϕ_n are a complete set of orthonormal functions (see any quantum mechanics textbook). Finally, we underline that the minimum $E_1 > 0$ is related to the uncertainty principle.

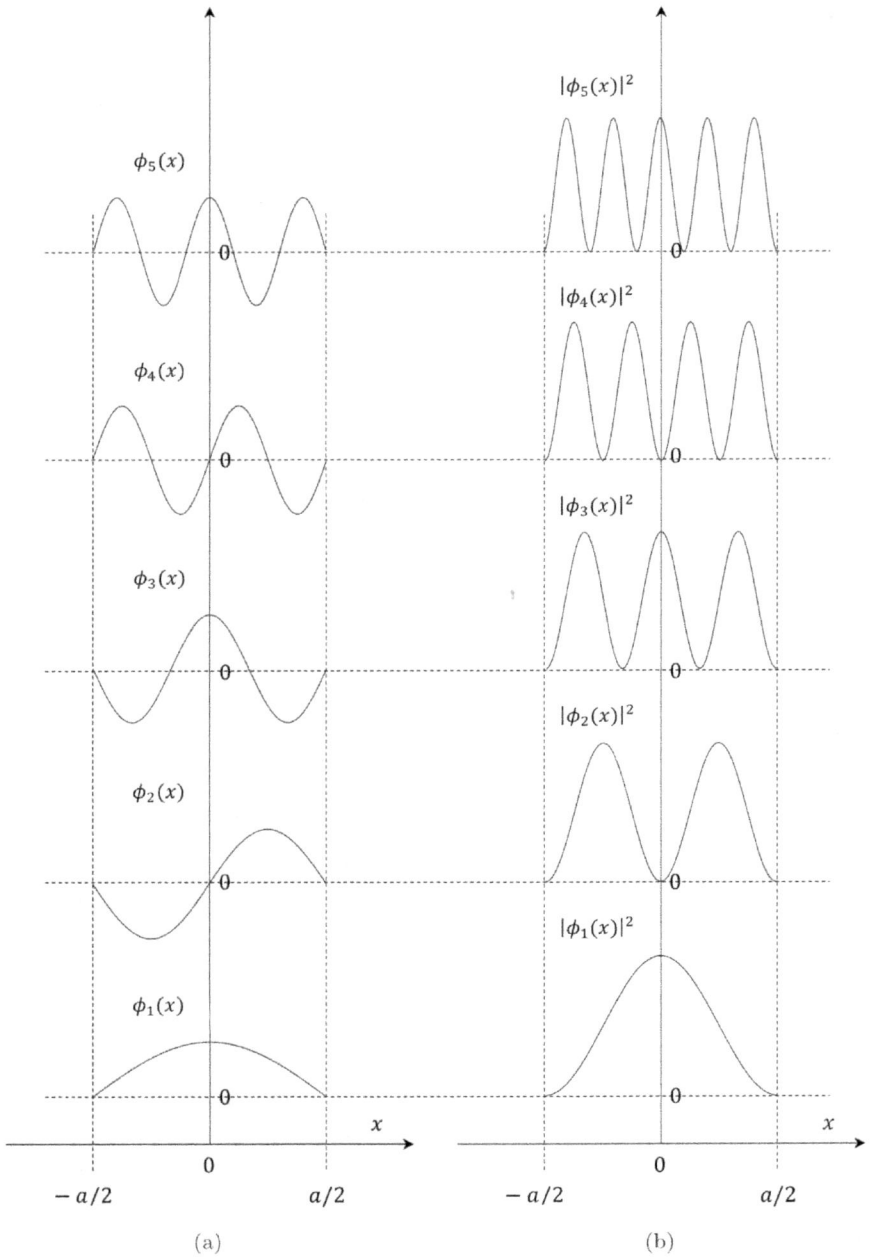

Fig. 2.14.   Infinite potential well: (a) eigenfunctions and (b) the corresponding probability density for $n = 1, 2, 3, 4, 5$.

2.10. The potential barrier and the tunnel effect

A particle not bound to a definite space volume may travel over long distances and will be described as a traveling wave. In this context, it is very important to consider the case of the potential barrier that leads to a phenomenon not possible in classical mechanics, i.e. the *tunnel effect*.

Let us first consider a *step potential barrier* made by a potential distribution defined by

$$\mathcal{V} = 0 \quad \text{for} \quad x \leq 0 \quad \text{(region I)} \quad \text{and} \quad \mathcal{V} = \mathcal{V}_0 \quad \text{for} \quad x > 0 \quad \text{(region II)}, \tag{2.155}$$

as shown in Fig. 2.15.

In region I, the Schrödinger equation is the one for the free particle:

$$\frac{d^2}{dx^2}\phi(x) + \frac{2m}{\hbar^2}E\phi(x) = 0.$$

However, in this case, we must consider the general solution with traveling waves both in the positive and in the negative **x** direction. In fact, besides the incident wave arriving from the left side of the barrier, we will have a wave reflected by the barrier itself,

$$\phi_I(x) = Ae^{ikx} + Be^{-ikx}, \tag{2.156}$$

which, as in the previous sections, leads to

$$k = \sqrt{\frac{2mE}{\hbar^2}}.$$

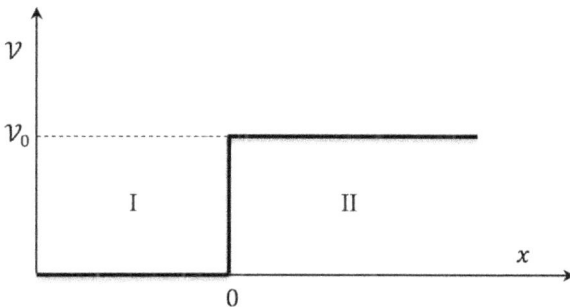

Fig. 2.15. Step potential barrier.

In region II, the Schrödinger equation includes also the potential $\mathcal{V}_0$,

$$\frac{d^2}{dx^2}\phi(x) + \frac{2m}{\hbar^2}(E - \mathcal{V}_0)\phi(x) = 0, \qquad (2.157)$$

whose solution must be given by a traveling wave:

$$\phi_{\mathrm{II}}(x) = Ce^{i\tilde{k}x}, \qquad (2.158)$$

where by substitution in Eq. (2.157) one gets

$$\tilde{k} = \sqrt{\frac{2m(E - \mathcal{V}_0)}{\hbar^2}}. \qquad (2.159)$$

Let us discuss only the case where the particle energy is lower than the potential barrier:

$$E < \mathcal{V}_0. \qquad (2.160)$$

In the classical case, this condition would simply prevent the particle from entering region II. On the contrary, the wave propagation leads to a damped but finite probability of finding the particle in this region. Under the condition (2.160), the $\tilde{k}$ parameter is a pure imaginary quantity,

$$\tilde{k} = \sqrt{\frac{2m(E - \mathcal{V}_0)}{\hbar^2}} = i\sqrt{\frac{2m(\mathcal{V}_0 - E)}{\hbar^2}} \equiv i\gamma, \qquad (2.161)$$

with γ being real. Then $i\tilde{k} = -\gamma$ and

$$\phi_{\mathrm{II}}(x) = Ce^{-\gamma x}. \qquad (2.162)$$

The wave functions can be calculated by setting the amplitude of the incident wave $A = 1$. In this way, the boundary conditions at $x = 0$ (continuity of wave functions and its derivatives) give

$$\phi_{\mathrm{I}}(0) = \phi_{\mathrm{II}}(0) \quad \rightarrow \quad 1 + B = C, \quad ik(1 - B) = -\gamma C. \qquad (2.163)$$

This system is easily solved:

$$ik(1 - B) = -\gamma(1 + B).$$

Then

$$\gamma B - ikB = -ik - \gamma, \quad \text{and then} \quad B = \frac{ik + \gamma}{ik - \gamma}, \quad C = B + 1 = \frac{2ik}{ik - \gamma}.$$

As a consequence, we have

$$\phi_{\mathrm{I}}(x) = e^{ikx} + \frac{ik + \gamma}{ik - \gamma} e^{-ikx}, \tag{2.164}$$

$$\phi_{\mathrm{II}}(x) = \frac{2ik}{ik - \gamma} e^{-\gamma x}. \tag{2.165}$$

Taking into account the complete expression of the wave function in region II, we have

$$\psi_{\mathrm{II}}(x) = \frac{2ik}{ik - \gamma} e^{-\gamma x} e^{-\frac{iE}{\hbar} t}, \tag{2.166}$$

showing that in this region we have an oscillating behavior in time with an exponential decay characterized by a penetration depth $1/\gamma$, highlighting the tunneling effect typical of wave behavior.

The second case to be considered is that of a square potential barrier as shown in Fig. 2.16. In this case, there are three regions:

$$\mathcal{V} = 0 \quad \text{for} \quad x < 0, \qquad \text{region I;}$$
$$\mathcal{V} = \mathcal{V}_0 \quad \text{for} \quad 0 \leq x \leq a, \quad \text{region II;}$$
$$\mathcal{V} = 0 \quad \text{for} \quad x < 0, \qquad \text{region III.}$$

Since in regions I and III the Schrödinger equation for the free particle holds, we write the solutions in region I as in the previous case,

$$\phi_{\mathrm{I}}(x) = Ae^{ikx} + Be^{-ikx}, \tag{2.167}$$

and in region III we keep only the forward traveling wave,

$$\phi_{\mathrm{III}}(x) = Ce^{ikx}, \tag{2.168}$$

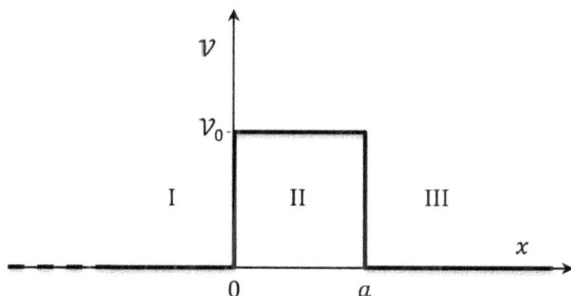

Fig. 2.16. Potential barrier.

where

$$k = \sqrt{\frac{2mE}{\hbar^2}}.$$

On the contrary, in region II the Schrödinger equation (2.157) holds taking into account the potential energy $\mathcal{V}_0$. In this case, in region II, we may have both a forward traveling wave and a reflected wave, and therefore we write the general solution as

$$\phi_{\text{II}}(x) = Fe^{i\tilde{k}x} + Ge^{-i\tilde{k}x}, \tag{2.169}$$

and, as in Eq. (2.159),

$$\tilde{k} = \sqrt{\frac{2m(E - \mathcal{V}_0)}{\hbar^2}}.$$

Let us consider first the case

(i) $E > \mathcal{V}_0$.

We use the boundary conditions at $x = 0$ and at $x = a$ (continuity of wave functions and its derivatives) to get four relations between the five coefficients:

$$\text{at } x = 0: \qquad A + B = F + G,$$

$$ik(A - B) = i\tilde{k}(F - G), \tag{2.170}$$

$$\text{at } x = a: \qquad Fe^{i\tilde{k}a} + Ge^{-i\tilde{k}a} = Ce^{ika},$$

$$i\tilde{k}Fe^{i\tilde{k}a} - i\tilde{k}Ge^{-i\tilde{k}a} = ikCe^{ika}. \tag{2.171}$$

We get

$$C' = Ce^{ika} = \frac{2k\tilde{k}}{2k\tilde{k}\cos\tilde{k}a - i(\tilde{k}^2 + k^2)\sin\tilde{k}a}. \tag{2.172}$$

By considering the amplitude of the incident wave $A = 1$, it is possible to solve the system given by Eqs. (2.170) and (2.171) to get B, C, F, G:

$$
\begin{aligned}
B - F - G &= -1, \\
-ikB - i\tilde{k}F + i\tilde{k}G &= -ik, \\
e^{i\tilde{k}a}F + e^{-i\tilde{k}a}G - C' &= 0, \\
i\tilde{k}e^{i\tilde{k}a}F - i\tilde{k}e^{-i\tilde{k}a}G - ikC' &= 0,
\end{aligned}
$$

where

$$C' = Ce^{ika},$$

necessary for calculating the transmission coefficient

$$|C'|^2 = |Ce^{ika}|^2 = |C|^2.$$

Kramer's method allows one to write

$$C' = \frac{\mathcal{N}}{\mathcal{D}},$$

where

$$\mathcal{N} = \begin{vmatrix} 1 & -1 & -1 & -1 \\ -ik & -i\tilde{k} & i\tilde{k} & -ik \\ 0 & e^{i\tilde{k}a} & e^{-i\tilde{k}a} & 0 \\ 0 & i\tilde{k}e^{i\tilde{k}a} & -i\tilde{k}e^{-i\tilde{k}a} & 0 \end{vmatrix},$$

$$\mathcal{D} = \begin{vmatrix} 1 & -1 & -1 & 0 \\ -ik & -i\tilde{k} & i\tilde{k} & 0 \\ 0 & e^{i\tilde{k}a} & e^{-i\tilde{k}a} & -1 \\ 0 & i\tilde{k}e^{i\tilde{k}a} & -i\tilde{k}e^{-i\tilde{k}a} & -ik \end{vmatrix}.$$

Then

$$\mathcal{N} = \begin{vmatrix} -i\tilde{k} & i\tilde{k} & -ik \\ e^{i\tilde{k}a} & e^{-i\tilde{k}a} & 0 \\ i\tilde{k}e^{i\tilde{k}a} & -i\tilde{k}e^{-i\tilde{k}a} & 0 \end{vmatrix} + ik \begin{vmatrix} -1 & -1 & -1 \\ e^{i\tilde{k}a} & e^{-i\tilde{k}a} & 0 \\ i\tilde{k}e^{i\tilde{k}a} & -i\tilde{k}e^{-i\tilde{k}a} & 0 \end{vmatrix}$$

$$= -ik \begin{vmatrix} e^{i\tilde{k}a} & e^{-i\tilde{k}a} \\ i\tilde{k}e^{i\tilde{k}a} & -i\tilde{k}e^{-i\tilde{k}a} \end{vmatrix} - ik \begin{vmatrix} e^{i\tilde{k}a} & e^{-i\tilde{k}a} \\ i\tilde{k}e^{i\tilde{k}a} & -i\tilde{k}e^{-i\tilde{k}a} \end{vmatrix}$$

$$= -ik(-2i\tilde{k}) - ik(-2i\tilde{k}) = -4k\tilde{k},$$

$$\mathcal{D} = \begin{vmatrix} -i\tilde{k} & i\tilde{k} & 0 \\ e^{i\tilde{k}a} & e^{-i\tilde{k}a} & -1 \\ i\tilde{k}e^{i\tilde{k}a} & -i\tilde{k}e^{-i\tilde{k}a} & -ik \end{vmatrix} + ik \begin{vmatrix} -1 & -1 & 0 \\ e^{i\tilde{k}a} & e^{-i\tilde{k}a} & -1 \\ i\tilde{k}e^{i\tilde{k}a} & -i\tilde{k}e^{-i\tilde{k}a} & -ik \end{vmatrix}$$

$$= \begin{vmatrix} -i\tilde{k} & i\tilde{k} \\ i\tilde{k}e^{i\tilde{k}a} & -i\tilde{k}e^{-i\tilde{k}a} \end{vmatrix} - ik \begin{vmatrix} -i\tilde{k} & i\tilde{k} \\ e^{i\tilde{k}a} & e^{-i\tilde{k}a} \end{vmatrix}$$

$$+ ik \begin{vmatrix} -1 & -1 \\ i\tilde{k}e^{i\tilde{k}a} & -i\tilde{k}e^{-i\tilde{k}a} \end{vmatrix} + k^2 \begin{vmatrix} -1 & -1 \\ e^{i\tilde{k}a} & e^{-i\tilde{k}a} \end{vmatrix}$$

$$= -\tilde{k}^2 e^{-i\tilde{k}a} + \tilde{k}^2 e^{i\tilde{k}a} - k\tilde{k}e^{-i\tilde{k}a} - k\tilde{k}e^{i\tilde{k}a} - k\tilde{k}e^{-i\tilde{k}a}$$

$$- k\tilde{k}e^{i\tilde{k}a} - k^2 e^{-i\tilde{k}a} + k^2 e^{i\tilde{k}a}$$

$$= -\tilde{k}^2 (e^{-i\tilde{k}a} - e^{i\tilde{k}a}) - 2k\tilde{k}(e^{-i\tilde{k}a} + e^{i\tilde{k}a}) - k^2(e^{-i\tilde{k}a} - e^{i\tilde{k}a}),$$

and then

$$\mathcal{D} = -4k\tilde{k}\cos\tilde{k}a + 2i(\tilde{k}^2 + k^2)\sin\tilde{k}a.$$

In this way, the transmission coefficient through the barrier is given by

$$T = |C|^2 = |C'|^2 = \frac{4k^2\tilde{k}^2}{4k^2\tilde{k}^2\cos^2(\tilde{k}a) + (\tilde{k}^2 + k^2)^2\sin^2(\tilde{k}a)}$$

$$= \frac{4k^2\tilde{k}^2}{4k^2\tilde{k}^2 + (\tilde{k}^2 - k^2)^2\sin^2(\tilde{k}a)}. \tag{2.173}$$

We can exploit this calculation to get the transmission coefficient in the more interesting case

(ii) $E < \mathcal{V}_0$.

We set

$$\tilde{k} = i\beta = i\sqrt{\frac{2m(\mathcal{V}_0 - E)}{\hbar^2}}. \tag{2.174}$$

In this way, the C' coefficient can be obtained:

$$C' = \frac{2ik\beta}{2ik\beta\cosh(\beta a) - (k^2 - \beta^2)\sinh(\beta a)} \tag{2.175}$$

From the previous calculation we have

$$\mathcal{N} = -4k\tilde{k} = -4ki\beta,$$

$$\mathcal{D} = -\tilde{k}^2(e^{-ika} - e^{ika}) - 2k\tilde{k}(e^{-ika} + e^{ika}) - k^2(e^{-ika} - e^{ika})$$

$$= \beta^2(e^{\beta a} - e^{-\beta a}) - 2ik\beta(e^{\beta a} + e^{-\beta a}) - k^2(e^{\beta a} - e^{-\beta a}).$$

Then

$$\mathcal{D} = -4ik\beta \cosh(\beta a) - 2(k^2 - \beta^2)\sinh(\beta a),$$

using the definition of hyperbolic functions:

$$\cosh(x) = \frac{e^x + e^{-x}}{2}, \quad \sinh(x) = \frac{e^x - e^{-x}}{2}.$$

They fulfill the identity

$$\cosh^2(x) - \sinh^2(x) = 1.$$

Then

$$C' = \frac{\mathcal{N}}{\mathcal{D}} = \frac{2ik\beta}{2ik\beta \cosh(\beta a) + (k^2 - \beta^2)\sinh(\beta a)}.$$

and the transmission coefficient through the barrier becomes

$$T = |C|^2 = |C'|^2 = \frac{4k^2\beta^2}{4k^2\beta^2 + (\beta^2 + k^2)^2\sinh^2(\beta a)}. \tag{2.176}$$

In Fig. 2.17 the transmission coefficient is plotted versus the particle energy $E/\mathcal{V}_0$ in the range $0 < E < \mathcal{V}_0$, showing that even in this case we have a probability of finding the particle on the other side of the barrier.

The result expressed by Eq. (2.176) is peculiar to quantum mechanics; in fact, in the frame of classical mechanics a particle cannot overcome a potential barrier higher than the particle energy. Of course, the effect is more relevant for a thinner barrier and the probability of finding the particle in region III decreases with the barrier thickness, as shown in Fig. 2.17. We underline that the probability is not negligible in many cases (see the curve with $a = 1$ at $E = 0.5\mathcal{V}_0$ where $T = 0.2$!), and the effect can be easily detected. We just recall that this well-known *tunnel effect* is exploited not only in small electronic devices such as optical detectors, but also in imaging apparatus such as the scanning tunnel electronic microscope.

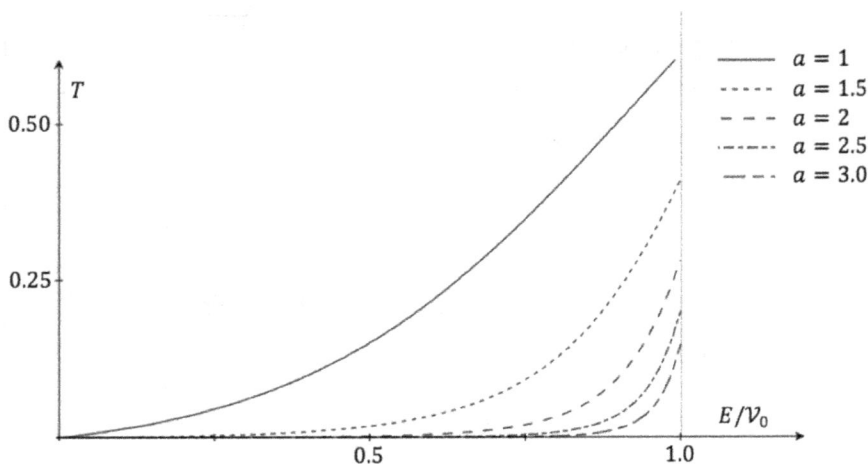

Fig. 2.17. Transmittivity of the potential barrier versus $E/\mathcal{V}_0$ when $E < \mathcal{V}_0$ for an increasing value of the barrier thickness in arbitrary units.

2.11. The harmonic oscillator

A fundamental problem of classical mechanics concerns the study of motion under the action of elastic forces, such as the motion of a point mass m linked to a spring fixed on one side. In one dimension, the force acting on the mass is proportional to the displacement and opposite to the motion direction:

$$F = -kx,$$

where k is the elastic constant. The solution of Newton's law $F = m(d^2x/dt^2)$ leads to the harmonic motion given by

$$x = x_0 \sin(\omega t + \varphi),$$

where x_0 is the maximum amplitude of the oscillation, with initial phase φ and angular frequency

$$\omega = \sqrt{\frac{k}{m}}.$$

The analysis of the elastic force shows that it is related to a potential energy:

$$\mathcal{V} = \frac{1}{2}kx^2. \qquad (2.177)$$

The study of this system from a quantum-mechanical point of view is very important for what concerns the material properties, since it can be used to describe the vibrations of atoms in molecules and in crystal lattices — vibrations which can account for several phenomena occurring in these systems. Even if in the following we will not go into a detailed description of them, which can be found in more specialized texts, in a basic treatment of quantum mechanics it is not possible to forget the main features of a quantum-mechanical harmonic oscillator.

For this aim, we start by writing the one-dimensional Schrödinger equation for a point mass under the action of a potential energy given by Eq. (2.177):

$$-\frac{\hbar^2}{2m}\frac{d^2\phi}{dx^2} + \frac{1}{2}kx^2\phi = E\phi, \quad \text{i.e.} \quad -\frac{\hbar^2}{2m}\frac{d^2\phi}{dx^2} + \frac{1}{2}m\omega x^2\phi = E\phi. \quad (2.178)$$

Taking into account the momentum operator

$$p = -i\hbar\frac{d}{dx},$$

we can write Eq. (2.178) as

$$\frac{1}{2m}[p^2 + (m\omega x)^2]\phi = E\phi. \quad (2.179)$$

Then the Hamiltonian operator can be written as

$$\mathcal{H} = \frac{1}{2m}[p^2 + (m\omega x)^2]. \quad (2.180)$$

Now we consider the following operators:

$$a_\pm \equiv \frac{1}{\sqrt{2\hbar m\omega}}[\mp ip + m\omega x]. \quad (2.181)$$

Then

$$a_-a_+ = \frac{1}{2\hbar m\omega}(ip + m\omega x)(-ip + m\omega x)$$

$$= \frac{1}{2\hbar m\omega}\{p^2 + (m\omega x)^2 - im\omega(xp - px)\}$$

$$= \frac{1}{2\hbar m\omega}\{p^2 + (m\omega x)^2\} - \frac{i}{2\hbar}[x, p]. \quad (2.182)$$

We know from Eq. (2.116) that $[x, p] = i\hbar$,

$$a_-a_+ = \frac{1}{2\hbar m\omega}\{p^2 + (m\omega x)^2\} + \frac{1}{2} = \frac{1}{\hbar\omega}\mathcal{H} + \frac{1}{2}, \quad (2.183)$$

and by changing the operators order,

$$a_+a_- = \frac{1}{\hbar\omega}\mathcal{H} - \frac{1}{2}. \quad (2.184)$$

Therefore,

$$[a_-, a_+] = 1, \quad \text{i.e. } a_- a_+ = a_+ a_- + 1, \tag{2.185}$$

and

$$\mathcal{H} = \hbar\omega \left(a_- a_+ - \frac{1}{2} \right) = \hbar\omega \left(a_+ a_- + \frac{1}{2} \right). \tag{2.186}$$

In this way, the Schrödinger equation can be written as

$$\hbar\omega \left(a_\pm a_\mp \pm \frac{1}{2} \right) \phi_n = E_n \phi_n. \tag{2.187}$$

Using Eq. (2.187), we can demonstrate that with E_n being the eigenvalue of the eigenfunction ϕ_n, if we consider the function $(a_+\phi_n)$ it is also an eigenfunction of the Hamiltonian with the eigenvalue $E_n + \hbar\omega$:

$$\begin{aligned}
\mathcal{H}(a_+\phi_n) &= \hbar\omega \left(a_+ a_- + \frac{1}{2} \right) (a_+\phi_n) \\
&= \hbar\omega \left(a_+ a_- a_+ + \frac{1}{2} a_+ \right) \phi_n \\
&= \hbar\omega a_+ \left(a_- a_+ + \frac{1}{2} \right) \phi_n = a_+ \left\{ \hbar\omega \left(a_+ a_- + 1 + \frac{1}{2} \right) \right\} \phi_n \\
&= a_+ \left\{ \hbar\omega \left(a_+ a_- + \frac{1}{2} \right) + \hbar\omega \right\} \phi_n \\
&= a_+(\mathcal{H} + \hbar\omega)\phi_n = (E_n + \hbar\omega)(a_+\phi_n). \tag{2.188}
\end{aligned}$$

In the same way, it can be calculated:

$$\mathcal{H}(a_-\phi) = (E - \hbar\omega)(a_-\phi). \tag{2.189}$$

Therefore, starting from any eigenfunction ϕ_n, the application of a_+ gives rise to the next eigenfunction, $a_+\phi_n = \phi_{n+1}$, with eigenvalue $E_{n+1} = E_n + \hbar\omega$; for this reason, it is called the *creation* operator. On the contrary, a_- is called the *destruction* operator, since it creates the next eigenfunction with lower energy $E_{n-1} = E_n - \hbar\omega$. Of course, if ϕ_0 is the minimum energy eigenfunction,

$$a_-\phi_0 = 0. \tag{2.190}$$

Then

$$\frac{1}{\sqrt{2\hbar m\omega}} [ip + m\omega x]\phi_0 = \frac{1}{\sqrt{2\hbar m\omega}} \left[\hbar \frac{d}{dx} + m\omega x \right] \phi_0 = 0. \tag{2.191}$$

It is easily integrated:

$$\frac{d\phi_0}{dx} = -\frac{m\omega}{\hbar}x\phi_0 \quad \rightarrow \quad \frac{d\phi_0}{\phi_0} = -\frac{m\omega}{\hbar}x\,dx \tag{2.192}$$

and

$$\phi_0 = Ae^{-\frac{m\omega}{2\hbar}x^2}. \tag{2.193}$$

By using ϕ_0 in the Schrödinger equation (2.187), we have
$\hbar\omega\left(a_+a_- + \frac{1}{2}\right)\phi_0 = E_0\phi_0$; since $a_-\phi_0 = 0$ it means that $\hbar\omega\frac{1}{2}\phi_0 = E_0\phi_0$
and

$$E_0 = \frac{1}{2}\hbar\omega. \tag{2.194}$$

This is the *zero-point energy* of the harmonic oscillator, i.e. its minimum energy. The whole set of eigenfunctions and eigenvalues can be found by subsequent application of a_+ to ϕ_0:

$$\phi_n = A_n(a_+)^n e^{-\frac{m\omega}{2\hbar}x^2}, \quad E_n = \left(n + \frac{1}{2}\right)\hbar\omega. \tag{2.195}$$

Analytical details and calculations of A_n and ϕ_n can be found in any text of quantum mechanics.

We just want to underline two results related to the energy eigenvalues. The first one is the equal spacing $\hbar\omega = h\nu$ of the energy levels of the harmonic oscillator, which is in agreement with Planck's postulate of the old quantum theory. The second is the peculiarity of quantum mechanics leading to the *zero-point energy* $E_0 = \frac{1}{2}\hbar\omega$, in a way similar to the result obtained for the infinite potential well (Section 1.9). This result is consistent with the uncertainty principle.

In fact, the total energy of the harmonic oscillator is

$$E = \frac{p^2}{2m} + \frac{1}{2}kx^2$$

and the uncertainty on energy can be written as

$$\Delta E_0 = \frac{(\Delta p)^2}{2m} + \frac{1}{2}k(\Delta x)^2,$$

where Δp and Δx are the uncertainties on momentum and position.

By minimization we set $\Delta E_0 = 0$. Therefore,

$$\frac{(\Delta p)^2}{2m} = -\frac{1}{2}k(\Delta x)^2 \quad \rightarrow \quad (\Delta p)^2$$

$$= -km(\Delta x)^2 \quad \rightarrow \quad (\Delta p)^2 = km\frac{\hbar^2}{(\Delta p)^2},$$

since the minimum uncertainty product is $\Delta p \Delta x = i\hbar$. Then

$$\Delta p = [km\hbar^2]^{1/4}.$$

By substituting in both terms of the expression of ΔE_0 (assumed to be of the same order of magnitude), we can estimate that

$$\Delta E_0 = \frac{[km\hbar^2]^{1/2}}{m} = \hbar\sqrt{\frac{k}{m}} = \hbar\omega.$$

This is actually an estimation of the same order of the exact value given by Eq. (2.194).

2.12. Perturbation theory: the stationary case

The perturbation theory is concerned with an approximation method to solve the Schrödinger equation when the Hamiltonian $\mathcal{H}$ can be written as the sum of two parts, one of which is to be considered as a small perturbation with respect to the main part, $\mathcal{H}_0$. This one should be simple enough to allow a solution to the Schrödinger equation, leading to corresponding energy eigenfunctions ϕ_k with eigenvalues E_k:

$$\mathcal{H}_0\phi_k = E_k\phi_k. \tag{2.196}$$

The first case to be considered is the *stationary case*, occurring when the perturbation $\mathcal{H}'$ is time-independent, with the assumption that $\mathcal{H}' \ll \mathcal{H}_0$.

Then we need to solve the new Schrödinger equation,

$$\mathcal{H}\psi = W\psi, \tag{2.197}$$

with $\mathcal{H} = \mathcal{H}_0 + \mathcal{H}'$ being, to get the new energy eigenfunctions ψ with eigenvalues W. The small perturbation condition allows expanding the new eigenfunctions and eigenvalues as power series of $\mathcal{H}'$. For this aim, we replace $\mathcal{H}'$ by $\lambda\mathcal{H}'(0 < \lambda < 1)$. In this way, the different orders of the perturbation approximation will be given by the coefficients of the corresponding power of λ. Following a common method, in the final result λ is finally replaced by 1. In this way,

$$\psi = \psi_0 + \lambda\psi_1 + \lambda^2\psi_2 + \lambda^3\psi_3 + \cdots ,$$

$$W = W_0 + \lambda W_1 + \lambda^2 W_2 + \lambda^3 W_3 + \cdots , \qquad (2.198)$$

which are substituted in the wave equation:

$$(\mathcal{H}_0 + \lambda\mathcal{H}')(\psi_0 + \lambda\psi_1 + \lambda^2\psi_2 + \cdots)$$

$$= (W_0 + \lambda W_1 + \lambda^2 W_2 + \cdots)(\psi_0 + \lambda\psi_1 + \lambda^2\psi_2 + \cdots). \qquad (2.199)$$

The identity of polyncmials is guaranteed by equating the coefficients of equal power in λ. In this way, a set of equations is obtained representing successively higher orders of perturbation:

$$\mathcal{H}_0\psi_0 = W_0\psi_0,$$

$$\mathcal{H}_0\psi_1 + \mathcal{H}'\psi_0 = W_0\psi_1 + W_1\psi_0, \qquad (2.200)$$

$$\mathcal{H}_0\psi_2 + \mathcal{H}'\psi_1 = W_0\psi_2 + W_1\psi_1 + W_2\psi_0,$$

The zero-order equation is nothing but the Schrödinger unperturbed equation. Then

$$\psi_0 = \phi_k, \quad W_0 = E_k. \qquad (2.201)$$

In order to get the first-order solution, we exploit the fact that ϕ_k are a complete set of functions fulfilling the property of orthonormality. Then we can write

$$\psi_1 = \sum_1^N a_n^{(1)}\phi_n. \qquad (2.202)$$

By substitution in the first-order approximation equation, we get

$$\mathcal{H}_0 \sum_1^N a_n^{(1)}\phi_n + \mathcal{H}'\phi_k = E_k \sum_1^N a_n^{(1)}\phi_n + W_1\phi_k. \qquad (2.203)$$

Then we multiply by ϕ_m^*,

$$\phi_m^* \mathcal{H}_0 \sum_1^N a_n^{(1)}\phi_n + \phi_m^* \mathcal{H}'\phi_k = E_k\phi_m^* \sum_1^N a_n^{(1)}\phi_n + W_1\phi_m^*\phi_k, \qquad (2.204)$$

which is to be rewritten as

$$\phi_m^* \sum_1^N a_n^{(1)} E_n\phi_n + \phi_m^* \mathcal{H}'\phi_k = E_k \sum_1^N a_n^{(1)}\phi_m^*\phi_n + W_1\phi_m^*\phi_k. \qquad (2.205)$$

The next step is integration and exploitation of orthonormality:

$$a_m^{(1)} E_m \int_{-\infty}^{\infty} \phi_m^* \phi_m dV + \int_{-\infty}^{\infty} \phi_m^* \mathcal{H}' \phi_k dV$$

$$= E_k a_m^{(1)} \int_{-\infty}^{\infty} \phi_m^* \phi_m dV + W_1 \int_{-\infty}^{\infty} \phi_m^* \phi_k dV. \qquad (2.206)$$

We divide by the normalization factor $\int_{-\infty}^{\infty} \phi_m^* \phi_m dV$ and define

$$\mathcal{H}_{km}' = \frac{\int_{-\infty}^{\infty} \phi_m^* \mathcal{H}' \phi_k dV}{\int_{-\infty}^{\infty} \phi_m^* \phi_m dV}, \qquad (2.207)$$

which becomes

$$\mathcal{H}_{km}' = \int_{-\infty}^{\infty} \phi_m^* \mathcal{H}' \phi_k dV \qquad (2.208)$$

if the normalization condition is fulfilled: $\int_{-\infty}^{\infty} \phi_m^* \phi_m dV = 1$.

Finally, we get

$$a_m^{(1)} E_m + \mathcal{H}_{km}' = E_k a_m^{(1)} + W_1 \delta_{mk}, \qquad (2.209)$$

with δ_{mk} being the Kronecker symbol.

When

$$m \neq k, \quad \delta_{mk} = 0,$$

we have

$$a_m^{(1)} = \frac{\mathcal{H}_{km}'}{E_k - E_m}. \qquad (2.210)$$

On the other hand, when $m = k$, taking into account Eq. (2.209), we have

$$\mathcal{H}_{kk}' = W_1, \qquad (2.211)$$

since $\delta_{kk} = 1$.

In this way, we have obtained the perturbed wave functions to the first order of perturbation,

$$\psi_k = \psi_0 + \psi_1 = \phi_k + \sum_1^N a_m^{(1)} \phi_m = \phi_k + \sum_1^N \frac{\mathcal{H}_{km}'}{E_k - E_m} \phi_m, \qquad (2.212)$$

whose eigenvalues are

$$W_k^{(1)} = E_k + \mathcal{H}_{kk}'. \qquad (2.213)$$

A similar procedure can be followed to get the energies and wave functions to second order in $\mathcal{H}'$. The second order correction of the energy eigenvalues is given by

$$W_k^{(2)} = E_k + \mathcal{H}'_{kk} + \sum_{k \neq n} \frac{|\mathcal{H}'_{kn}|^2}{E_k - E_n}. \tag{2.214}$$

In summary, the effect of a small stationary perturbation is to induce a modification of the wave functions and a consequent change of the corresponding eigenvalues. By considering the case of an electron in a bounded state, this means that a small perturbation to the Hamiltonian gives rise to a shift of the energy levels, which can be detected by a suitable spectroscopic technique. It occurs, for instance, when a static electric field is applied to an atomic vapor, giving rise to the well-known Stark effect.[14] By the same token, we will point out how the attractive potential of the ions in a crystal lattice may be considered a perturbation with respect to the free electrons, with important consequences for the physical behavior of the electrons in a crystal lattice.

2.13. Time-dependent perturbation theory

Several times, the perturbation on a physical system is not constant in time; then $\mathcal{H}' = \mathcal{H}'(t)$. For instance, the interaction of electromagnetic waves and the atomic system or electrons is time-dependent, owing to the oscillating character of the waves. In these cases, we need to solve the general Schrödinger equation,

$$\mathcal{H}\psi = i\hbar \frac{\partial \psi}{\partial t},$$

with $\mathcal{H}(t) = \mathcal{H}_0 + \mathcal{H}'(t)$.

Again, we develop the wave function using the eigenfunctions of the stationary case:

$$\psi = \sum_1^N a_n(t)\phi_n \exp\left(-\frac{iE_n t}{\hbar}\right). \tag{2.215}$$

By substitution in the Schrödinger equation, we get

$$[\mathcal{H}_0 + \mathcal{H}'(t)] \sum_1^N a_n(t)\phi_n \exp\left(-\frac{iE_n t}{\hbar}\right) = i\hbar \frac{\partial}{\partial t} \sum_1^N a_n(t)\phi_n \exp\left(-\frac{iE_n t}{\hbar}\right), \tag{2.216}$$

---

[14]The Stark effect consists in the splitting and shifting of electronic levels that is observed as splitting and shifting of the spectral lines of the considered element when an electric field is applied.

and therefore

$$\sum_{1}^{N} a_n(t) E_n \phi_n \exp\left(-\frac{iE_n t}{\hbar}\right) + \sum_{1}^{N} a_n(t) \mathcal{H}'(t) \phi_n \exp\left(-\frac{iE_n t}{\hbar}\right)$$

$$= i\hbar \sum_{1}^{N} \frac{da_n(t)}{dt} \phi_n \exp\left(-\frac{iE_n t}{\hbar}\right) + \sum_{1}^{N} a_n(t) E_n \phi_n \exp\left(-\frac{iE_n t}{\hbar}\right).$$

(2.217)

Then

$$\sum_{1}^{N} a_n(t) \mathcal{H}'(t) \phi_n \exp\left(-\frac{iE_n t}{\hbar}\right) = i\hbar \sum_{1}^{N} \frac{da_n(t)}{dt} \phi_n \exp\left(-\frac{iE_n t}{\hbar}\right).$$

(2.218)

Multiplication by ϕ_k^* and integration gives

$$\int_{-\infty}^{\infty} \sum_{1}^{N} a_n(t) \phi_k^* \mathcal{H}'(t) \phi_n \exp\left(-\frac{iE_n t}{\hbar}\right) dV$$

$$= i\hbar \int_{-\infty}^{\infty} \sum_{1}^{N} \frac{da_n(t)}{dt} \phi_k^* \phi_n \exp\left(-\frac{iE_n t}{\hbar}\right) dV.$$

(2.219)

By applying the orthonormality condition, we get

$$-\frac{i}{\hbar} \sum_{1}^{N} \mathcal{H}'_{kn} a_n(t) \exp\left(-\frac{iE_n t}{\hbar}\right) = \frac{da_k(t)}{dt} \exp\left(-\frac{iE_k t}{\hbar}\right)$$

(2.220)

or

$$\frac{da_k(t)}{dt} = -\frac{i}{\hbar} \sum_{1}^{N} \mathcal{H}'_{kn} a_n(t) \exp\left(-\frac{i(E_n - E_k)t}{\hbar}\right)$$

$$= -\frac{i}{\hbar} \sum_{1}^{N} \mathcal{H}'_{kn} a_n(t) \exp(i\omega_{kn} t),$$

(2.221)

where

$$\omega_{kn} = \frac{E_k - E_n}{\hbar}.$$

(2.222)

Now, we proceed as in the previous section by introducing the λ parameter $(0 < \lambda < 1)$, writing

$$\mathcal{H} = \mathcal{H}_0 + \lambda \mathcal{H}',$$

and using a power expansion for the coefficients a_n and its derivatives:

$$a_n = a_n^{(0)} + \lambda a_n^{(1)} + \lambda^2 a_n^{(2)} + \cdots,$$

$$\frac{da_k}{dt} = \frac{da_k^{(0)}}{dt} - \lambda \frac{da_k^{(1)}}{dt} + \lambda^2 \frac{da_k^{(2)}}{dt} + \cdots. \tag{2.223}$$

Then Eq. (2.221) becomes

$$\frac{da_k^{(0)}}{dt} + \lambda \frac{da_k^{(1)}}{dt} + \lambda^2 \frac{da_k^{(2)}}{dt} + \cdots$$

$$= -\frac{i}{\hbar} \sum_1^N \lambda \mathcal{H}'_{kn} [a_n^{(0)} + \lambda a_n^{(1)} + \lambda^2 a_n^{(2)} + \cdots] \exp(i\omega_{kn} t). \tag{2.224}$$

By equating the coefficients of equal power in λ we get

$$\frac{da_k^{(0)}}{dt} = 0, \tag{2.225}$$

$$\frac{da_k^{(1)}}{dt} = -\frac{i}{\hbar} \sum_1^N \mathcal{H}'_{kn} a_n^{(0)} \exp(i\omega_{kn} t), \tag{2.226}$$

$$\cdots$$

$$\frac{da_k^{(s)}}{dt} = -\frac{i}{\hbar} \sum_1^N \mathcal{H}'_{kn} a_n^{(s-1)} \exp(i\omega_{kn} t). \tag{2.227}$$

From Eq. (2.225) we have

$$a_k^{(0)} = \text{const.} \tag{2.228}$$

The zero-order values define the initial conditions specifying the system state before the application of the perturbation.

Let us assume that before perturbation the system is in a definite state m, and thus

$$a_m^{(0)} = 1, \quad a_k^{(0)} = 0 \quad \text{for} \ \ k \neq m,$$

i.e.

$$a_k^{(0)} = \delta_{mk}. \tag{2.229}$$

This means that at $t = 0$ the wave function is

$$\psi = \phi_m$$

and Eq. (2.226) becomes

$$\frac{da_k^{(1)}}{dt} = -\frac{i}{\hbar}\mathcal{H}_{km}' \exp(i\omega_{km}t),\qquad(2.230)$$

since all the terms of the summation are zero except for the one with the coefficient $a_m^{(0)} = 1$.

The physical meaning of Eq. (2.230) is a basic result of this treatment: with $da_k^{(1)}/dt \neq 0$, in the presence of an energy perturbation ($\mathcal{H}_{km}' \neq 0$), at $t > 0$, the corresponding coefficient becomes $a_k^{(1)} \neq 0$, and thus the system has now a nonzero probability of being found in the state described by the wave function ϕ_k. For this reason, $|a_k^{(1)}(t')|^2$ is the probability that in the time range $0 < t < t'$ the system undergoes a transition from the state m to the state k.

Harmonic perturbation

In order to proceed and solve Eq. (2.230), it is necessary to define the problem with more details. In particular, we need to know the time dependence of $\mathcal{H}_{km}'(t)$. In the following, we consider a harmonic perturbation at angular frequency ω, a problem that will be applied to the case of matter interaction with electromagnetic waves:

$$\mathcal{H}'(t) = \frac{H'}{2}\exp(-i\omega t) + \frac{H'^*}{2}\exp(i\omega t),\quad \text{at}\ \ t > 0,\qquad(2.231)$$

$$\mathcal{H}'(t) = 0,\quad \text{at}\ \ t \leq 0,$$

with $\mathcal{H}'$ being the time-independent amplitude of the interaction energy.

By substitution in Eq. (2.230) and integration, we get

$$a_k^{(1)}(t) = -\frac{i}{2\hbar}\int_0^t H_{km}' \exp i(\omega_{km} - \omega)t'\, dt'$$

$$-\frac{i}{2\hbar}\int_0^t (H_{km}')^* \exp i(\omega_{km} + \omega)t'\, dt'$$

$$= \frac{1}{2\hbar}\left\{H_{km}'\frac{1 - \exp i(\omega_{km} - \omega)t}{\omega_{km} - \omega} + (H_{km}')^*\frac{1 - \exp i(\omega_{km} + \omega)t}{\omega_{km} + \omega}\right\}.$$

$$(2.232)$$

If the frequency of the perturbation is close to the one associated with the transition

$$\omega \approx \omega_{km},$$

the second term of Eq. (2.232) can be neglected, and thus

$$a_k^{(1)}(t) = \frac{1}{2\hbar} H'_{km} \frac{1 - \exp i(\omega_{km} - \omega)t}{\omega_{km} - \omega} \tag{2.233}$$

and the square modulus can be easily calculated:

$$|a_k^{(1)}(t)|^2 = \frac{|H'_{km}|^2}{\hbar^2} \frac{\sin^2\left[\frac{1}{2}(\omega_{km} - \omega)t\right]}{(\omega_{km} - \omega)^2}. \tag{2.234}$$

Starting from this expression of the transition probability from a state m to a state k, we may consider different physical conditions. However, it is important to underline that up to now no assumption has been made about the energy levels E_k and E_m so that the induced transition can be upward (to higher energy) or downward (to lower energy).

Now let us suppose that $E_k > E_m$ and consider the transition from m to a continuous set of states k in a narrow range of energies around $E_k: (E_k - \varepsilon, E_k + \varepsilon)$. Additionally, we define $\rho(E_k)$ as the final states density such that $\rho(E_k)dE_k$ represents the number of final states k in the range dE_k. Then the total transition probability P_{mk} can be calculated from Eq. (2.234) by integration over all the final states:

$$P_{km}^{(1)}(t) = \frac{1}{\hbar^2} \int_{E_k - \varepsilon}^{E_k + \varepsilon} |H'_{km}|^2 \left\{ \frac{\sin^2\left[\frac{1}{2}(\omega_{km} - \omega)t\right]}{(\omega_{km} - \omega)^2} \right\} \rho(E_k)dE_k. \tag{2.235}$$

The function $F(\omega)$ in the braces shows a narrow peak around $\omega = \omega_{km}$ with a half width

$$\Delta\omega = \frac{2\pi}{t}.$$

Then, by considering t large enough, this function is nonzero in a very narrow range with respect to the one where $\mathcal{H}'_{km}$ and $\rho(E_k)$ are defined. In this way, both these functions can be considered constant in the nonzero range of $F(\omega)$, and thus Eq. (2.235) becomes

$$P_{km}^{(1)}(t) = \frac{1}{\hbar^2} |H'_{km}|^2 \rho(E_k) \int_{E_k - \varepsilon}^{E_k + \varepsilon} \frac{\sin^2\left[\frac{1}{2}(\omega_{km} - \omega)t\right]}{(\omega_{km} - \omega)^2} dE_k. \tag{2.236}$$

Now it is possible to extend the integration limits from $-\infty$ to ∞ without affecting the final result (for the very reason of the narrow peak of the $\sin^2$ function), thus allowing an analytical solution:

$$\int_{-\infty}^{\infty} \frac{\sin^2\left[\frac{1}{2}(\omega_{km}-\omega)t\right]}{(\omega_{km}-\omega)^2}\,dE_k$$

$$= \hbar \int_{-\infty}^{\infty} \frac{\sin^2\left[\frac{1}{2}(\omega_{km}-\omega)t\right]}{(\omega_{km}-\omega)^2}\,d\frac{(E_k - E_m - \hbar\omega)}{\hbar}$$

$$= \hbar \int_{-\infty}^{\infty} \frac{\sin^2\left[\frac{1}{2}(\omega_{km}-\omega)t\right]}{(\omega_{km}-\omega)^2}\,d(\omega_{km}-\omega).$$

By setting

$$\xi = (\omega_{km}-\omega)$$

and taking into account that

$$\int_{-\infty}^{\infty} \frac{\sin^2\left[\frac{1}{2}\xi t\right]}{(\xi)^2}\,d\xi = \frac{\pi}{2}t,$$

we have

$$\int_{-\infty}^{\infty} \frac{\sin^2\left[\frac{1}{2}(\omega_{km}-\omega)t\right]}{(\omega_{km}-\omega)^2}\,dE_k = \hbar\frac{\pi}{2}t.$$

Then we get

$$P_{km}^{(1)}(t) = \frac{\pi}{2\hbar}|H_{km}'|^2\rho(E_k)t. \tag{2.237}$$

From Eq. (2.237), we can write the *induced transition probability per unit time* as

$$W_{km}^{(i)} = \frac{\pi}{2\hbar}|H_{km}'|^2\rho(E_k). \tag{2.238}$$

Equation (2.238) is the well-known *golden rule* for the induced transitions.
Taking into account the basic property of the Dirac δ function,

$$\int_a^b \delta(x - x_0)f(x)dx = f(x_0) \quad \text{if } a < x_0 < b, \tag{2.239}$$

the used approximation corresponds to set into the integration

$$\frac{\sin^2\left[\frac{1}{2}(\omega_{km} - \omega)t\right]}{(\omega_{km} - \omega)^2} \approx \frac{\pi t}{2}\delta(\omega_{km} - \omega). \qquad (2.240)$$

Therefore, if we use this expression in Eq. (2.237) when dealing with the transition from m to a single k state, we can write

$$P_{km}^{(1)}(t) = \frac{\pi}{2\hbar^2}|H'_{km}|^2\delta(\omega_{km} - \omega)t$$

$$= \frac{\pi}{2\hbar}|H'_{km}|^2\delta(E_k - E_m - \hbar\omega)t, \qquad (2.241)$$

where we have taken into account the property of the δ function:

$$\delta(\alpha x) = \frac{1}{\alpha}\delta(x), \quad \text{so} \quad \delta(\omega) = \hbar\delta(\hbar\omega).$$

Finally, the *induced transition probability per unit time* between two single states becomes

$$W_{km}^{(i)} = \frac{\pi}{2\hbar}|H'_{km}|^2\delta(E_k - E_m - \hbar\omega). \qquad (2.242)$$

We can extend this expression to the case where the final energy E_k is defined by the probability distribution function $f(E_k)$ such that $f(E_k)dE_k$ is the probability of finding the energy value in the range between E_k and $E_k + dE_k$, and a suitable normalization condition holds:

$$\int_0^\infty f(E_k)dE_k = 1. \qquad (2.243)$$

In this case, the induced transition probability per unit time must be integrated over all the final states, where each term must be weighted by its own probability:

$$W_{km}^{(i)} = \frac{\pi}{2\hbar}|H'_{km}|^2\int_0^\infty \delta(E_k - E_m - \hbar\omega)f(E_k)dE_k$$

$$= \frac{\pi}{2\hbar}|H'_{km}|^2 f(E_k = E_m + \hbar\omega). \qquad (2.244)$$

Instead of the energy distribution function, we can use the so-called line shape $g(\nu)$ which fulfills the property

$$\int_0^\infty g(\nu)d\nu = 1, \qquad (2.245)$$

where ν represents the possible transition frequency, which for transitions between single states is defined by

$$\nu_{km} = 2\pi\omega_{km} = \frac{2\pi}{\hbar}(E_k - E_m). \qquad (2.246)$$

Then we can use Eq. (2.241) to obtain the transition probability:

$$P_{km}^{(1)}(t) = \frac{\pi}{2\hbar^2}|H'_{km}|^2\delta(\omega_{km} - \omega)t = \frac{1}{4\hbar^2}|H'_{km}|^2\delta(\nu_{km} - \nu)t. \qquad (2.247)$$

By integrating over all the possible final states, i.e. over all the possible transition frequencies, the induced transition probability per unit time becomes

$$W_{km}^{(i)} = \frac{1}{4\hbar^2}|H'_{km}|^2 g(\nu_{km} = \nu) = \frac{1}{4\hbar^2}|H'_{km}|^2 g(\nu). \qquad (2.248)$$

This last expression is very compact and useful since, besides the factor $1/4\hbar^2$, it reduces the induced transition probability per unit time to a product between the interaction term $|H'_{km}|^2$ between the initial and final states and the line shape $g(\nu)$ which weights the possible transition according to the distance of the perturbing frequency from the system resonance.

2.14. Two-level system

It is very useful to focus on a system with only two states, for several reasons. The first reason is that in the case of interaction between a highly monochromatic perturbation and a material system, the perturbation can be far from resonance with respect to all the possible transitions except for one connecting only two energy levels. The second reason is that even when other levels can be involved the overall effect of the interaction may possibly be approximated to the superposition of several two-level transitions.

Therefore, we consider the wave functions of the two states

$$\psi_1 = \phi_1 \exp\left(-\frac{iE_1 t}{\hbar}\right), \quad \psi_2 = \phi_2 \exp\left(-\frac{iE_2 t}{\hbar}\right). \qquad (2.249)$$

They are orthonormal functions, and a generic state of the system will be represented by the wave function

$$\psi = a_1\phi_1 \exp\left(-\frac{iE_1 t}{\hbar}\right) + a_2\phi_2 \exp\left(-\frac{iE_2 t}{\hbar}\right), \qquad (2.250)$$

which is a superposition of the two states with coefficients that must satisfy the normalization condition $|a_1|^2 + |a_2|^2 = 1$.

The probability of finding the system in the eigenstate 1 or 2 will be determined respectively by the coefficients a_i. If a time-dependent perturbation is applied to the system, this probability may change, and then these coefficients become time-dependent functions $a_i = a_i(t)$. In order to find differential equations for them, we proceed as usual by applying the Schrödinger equation to the wave function (2.250), taking into account that $\mathcal{H} = \mathcal{H}_0 + \mathcal{H}'$:

$$a_1[\mathcal{H}_0\phi_1]\exp\left(-\frac{iE_1t}{\hbar}\right) + a_2[\mathcal{H}_0\phi_2]\exp\left(-\frac{iE_2t}{\hbar}\right)$$

$$+ a_1[\mathcal{H}'\phi_1]\exp\left(-\frac{iE_1t}{\hbar}\right) + a_2[\mathcal{H}'\phi_2]\exp\left(-\frac{iE_2t}{\hbar}\right)$$

$$= i\hbar\left[\frac{da_1}{dt}\phi_1\exp\left(-\frac{iE_1t}{\hbar}\right) + \frac{da_2}{dt}\phi_2\exp\left(-\frac{iE_2t}{\hbar}\right)\right.$$

$$\left. + a_1\phi_1\left(-\frac{iE_1}{\hbar}\right)\exp\left(-\frac{iE_1t}{\hbar}\right) + a_2\phi_2\left(-\frac{iE_2}{\hbar}\right)\exp\left(-\frac{iE_2t}{\hbar}\right)\right].$$

$$(2.251)$$

Since $\mathcal{H}_0\phi_1 = E_1\phi_1$ and $\mathcal{H}_0\phi_2 = E_2\phi_2$, the first two terms on the left cancel with the last two terms on the right, and we obtain

$$a_1\left[\mathcal{H}'\phi_1\right]\exp\left(-\frac{iE_1t}{\hbar}\right) + a_2[\mathcal{H}'\phi_2]\exp\left(-\frac{iE_2t}{\hbar}\right)$$

$$= i\hbar\left[\frac{da_1}{dt}\phi_1\exp\left(-\frac{iE_1t}{\hbar}\right) + \frac{da_2}{dt}\phi_2\exp\left(-\frac{iE_2t}{\hbar}\right)\right]. \qquad (2.252)$$

Then we multiply by ϕ_1^* and integrate exploiting the orthonormality property of the eigenfunctions:

$$a_1\mathcal{H}'_{11}\exp\left(-\frac{iE_1t}{\hbar}\right) + a_2\mathcal{H}'_{12}\exp\left(-\frac{iE_2t}{\hbar}\right) = i\hbar\frac{da_1}{dt}\exp\left(-\frac{iE_1t}{\hbar}\right),$$

$$(2.253)$$

using the notation (2.207) for the matrix element $\mathcal{H}'_{ij}$. Then

$$\frac{da_1}{dt} = -\frac{i}{\hbar}[a_1\mathcal{H}'_{11} + a_2\mathcal{H}'_{12}\exp(-i\omega_0t)], \qquad (2.254)$$

where we have defined

$$\omega_0 \equiv \frac{E_2 - E_1}{\hbar}. \qquad (2.255)$$

We can follow the same steps, multiplying by ϕ_2^* and integrating Eq. (2.252), to get

$$\frac{da_2}{dt} = -\frac{i}{\hbar}[a_2 \mathcal{H}'_{22} + a_1 \mathcal{H}'_{21} \exp(-i\omega_0 t)]. \tag{2.256}$$

From symmetry it follows that $\mathcal{H}'_{11} = \mathcal{H}'_{22} = 0$, and therefore the coupled equations of the coefficients reduce to

$$\frac{da_1}{dt} = -\frac{i}{\hbar} a_2 \mathcal{H}'_{12} \exp(-i\omega_0 t), \quad \frac{da_2}{dt} = -\frac{i}{\hbar} a_1 \mathcal{H}'_{21} \exp(-i\omega_0 t). \tag{2.257}$$

These equations can be solved by iterative approximations.

Assuming that $E_2 > E_1$ and the system is initially in the low energy state, with no perturbation, we have

$$a_1(0) = 1, \quad a_2(0) = 0. \tag{2.258}$$

Then, by applying the results of the perturbation theory to this system, from Eq. (2.227) we get

$$\frac{da_1^{(s)}}{dt} = -\frac{i}{\hbar} \mathcal{H}'_{12} a_2^{(s-1)} \exp(-i\omega_0 t), \tag{2.259}$$

$$\frac{da_2^{(s)}}{dt} = -\frac{i}{\hbar} \mathcal{H}'_{21} a_1^{(s-1)} \exp(i\omega_0 t). \tag{2.260}$$

Therefore, at the first order ($s = 1$),

$$\frac{da_1^{(1)}}{dt} = 0 \quad \rightarrow \quad a_1^{(1)} = 1, \quad \frac{da_2^{(1)}}{dt} = -\frac{i}{\hbar} \mathcal{H}'_{21} \exp(i\omega_0 t). \tag{2.261}$$

Integrating Eq. (2.261), we have

$$a_2^{(1)} = -\frac{i}{\hbar} \int_0^t \mathcal{H}'_{21} \exp(i\omega_0 t') dt'. \tag{2.262}$$

Finally, we consider the harmonic perturbation given by Eq. (2.231), and therefore the first order approximation gives

$$a_2^{(1)} = -\frac{i}{2\hbar} \int_0^t \{H'_{21} \exp(i[\omega_0 - \omega]t') + (H'_{21})^* \exp(i[\omega_0 + \omega]t')\} dt'. \tag{2.263}$$

Then

$$a_2^{(1)} = \frac{1}{2\hbar} \left\{ H'_{21} \frac{1 - \exp(i[\omega_0 - \omega]t)}{\omega_0 - \omega} + (H'_{21})^* \frac{1 - \exp(i[\omega_0 + \omega]t)}{\omega_0 + \omega} \right\}. \tag{2.264}$$

When $\omega \sim \omega_0$, the second term inside the brackets is negligible with respect to the first one, and thus the square modulus is

$$|a_2^{(1)}|^2 = \frac{|H'_{21}|^2}{\hbar^2} \frac{\sin^2[(\omega_0 - \omega)t/2]}{(\omega_0 - \omega)^2}. \tag{2.265}$$

This coincides with Eq. (2.234)

We will exploit these results in Chapter 5 to obtain the atomic polarizability of an atomic system.

Chapter 3

From the Hydrogen Atom to Condensed Matter

3.1. Introduction

The aim of the present chapter is to accompany the reader through the quantum-mechanical description of matter. It is very challenging, since in only one chapter do I try to make understandable the links between the atomic model and more complex atomic aggregations following the quantum theory. The first step is to show the consequences of applying the quantum theory to a system very similar to the Bohr atom, in order to make clear the origin of the different quantum numbers describing the state of the bound electrons. This will be done by using the Schrödinger equation to study the hydrogen atom. Some issues related to atoms with a higher number of electrons will be addressed in order to describe how the structure of the atom determines its ability to give rise to molecules or condensed matter. Then a brief description of the energy quantization in molecules will be given, because they can play an important role also in determining the properties of materials in a condensed state.

Next, the origin and the different type of bonding possible between atoms and molecules will be discussed, with the aim of understanding why and when atoms like better to create a strong link among them to build molecules or more complex structures.

When we are dealing with a large number of atoms or molecules, the role of thermodynamics becomes relevant, and therefore it will be necessary to use this approach for a macroscopic description of the different phases of condensed matter. In this discussion, the link between thermodynamics and microscopic quantities will be clarified by recalling the results of statistical mechanics, which will be exploited in the following chapters devoted to the electrical and optical properties of matter.

The last part of this chapter gives a description of the main properties of the different phases shown by materials in a condensed state, with particular emphasis on crystalline lattices owing to their paramount importance in solid state physics.

3.2. Quantum-mechanical description of the hydrogen atom

The basic model for the hydrogen atom (H) sees a single electron moving around a fixed nucleus made by one proton. This assumption is justified by the mass of the proton much larger than that of the electron (about 2000 times) so that the center of mass of the atom coincides with the position of the nucleus. Then we consider the electron under the effect of the Coulomb force of the proton leading to an electron potential energy:

$$\mathcal{V} = -\frac{e^2}{4\pi\varepsilon_0 r}, \tag{3.1}$$

with r being the distance between the charges $-e$ (electron) and $+e$ (proton).

Since the proton is in a fixed position O, the Coulomb force acting on the electron is directed toward O at any time. This is the case of a *central force* $\mathbf{F}$, which in classical mechanics leads to a null torque $\mathbf{M} = \mathbf{r} \times \mathbf{F} = 0$, with $\mathbf{r}$ and $\mathbf{F}$ being parallel vectors. The consequence is conservation of angular momentum $\mathbf{L}$, since from the second equation of dynamics we have $\mathbf{M} = d\mathbf{L}/dt = 0$. The quantum-mechanical counterpart of this classical result in the presence of a central force is the commutation rule between the Hamiltonian $\mathcal{H}$ and the operator L^2 or L_z:

$$[\mathcal{H}, L^2] = [\mathcal{H}, L_z] = 0. \tag{3.2}$$

This result tells us that $\mathcal{H}, L^2, L_z$ have the same eigenfunctions. In fact, from Eq. (2.92) we know that an operator Ω not explicitly dependent on time is a constant of motion if the commutator $[\mathcal{H}, \Omega] = 0$. It means that L^2 or L_z are constant, i.e. the angular moment is constant (just like in the classical case).

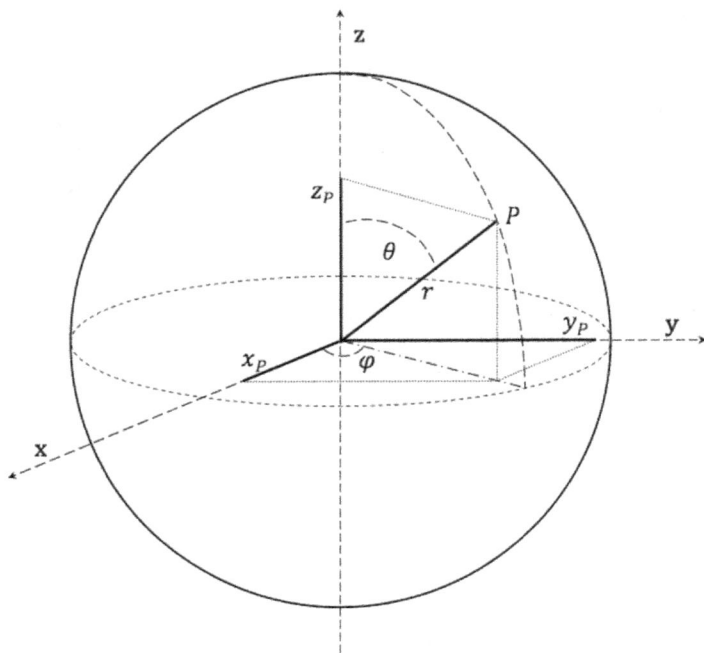

Fig. 3.1. Spherical coordinates for the position P of the electron moving around the atomic nucleus located in the origin of axes.

Equations (3.2) can be verified through transformation from the Cartesian coordinates x, y, z to spherical coordinates r, θ, φ:

$$x = r \sin\theta \cos\varphi, \quad y = r \sin\theta \sin\varphi, \quad z = r \cos\theta, \quad \text{i.e.}$$

$$r = \sqrt{x^2 + y^2 + z^2}, \quad \arccos\theta = \frac{z}{\sqrt{x^2 + y^2 + z^2}}, \quad \text{arctg } \varphi = \frac{y}{x}, \quad (3.3)$$

as shown in Fig. 3.1, where are also indicated the spherical coordinates defining the position of the electron of charge $-e$ and mass m around the nucleus of mass M located in the origin of the coordinates.

A lengthy calculation allows transformation of the operator $\mathcal{H}, L^2, L_z$ into spherical coordinates.[1]

---

[1]The method to perform the transformation of the differential operators is shown in: M.L. Boas, *Mathematical Methods in the Physical Sciences*, 2nd ed. (Wiley & Sons, 1983).

Then we have to solve the following time-independent Schrödinger equation:

$$\left[-\frac{\hbar^2}{2m}\nabla^2 - \frac{e^2}{4\pi\varepsilon_0 r} \right]\phi(\mathbf{r}) = E\phi(\mathbf{r}). \tag{3.4}$$

Owing to the symmetry of the system, it is useful to transform it into spherical coordinates; by doing that we have[2]

$$-\frac{\hbar^2}{2m}\nabla^2 = -\frac{\hbar^2}{2m}\frac{1}{r^2}\frac{\partial}{\partial r}\left(r^2\frac{\partial}{\partial r} \right) + \frac{1}{2mr^2}L^2, \tag{3.5}$$

where the square of the angular momentum operator is

$$L^2 = -\hbar^2 \left[\frac{1}{\sin\theta}\frac{\partial}{\partial\theta}\left(\sin\theta\frac{\partial}{\partial\theta} \right) + \frac{1}{\sin^2\theta}\frac{\partial^2}{\partial\varphi^2} \right]. \tag{3.6}$$

In this way, the Hamiltonian becomes

$$\mathcal{H} = -\frac{\hbar^2}{2m}\frac{1}{r^2}\frac{\partial}{\partial r}\left(r^2\frac{\partial}{\partial r} \right) + \frac{1}{2mr^2}L^2 - \frac{e^2}{4\pi\varepsilon_0 r}. \tag{3.7}$$

This is the sum of a term $\mathcal{H}_r$ operating on r and a term $\mathcal{H}_{\theta\varphi}$ operating on the angular coordinates:

$$\mathcal{H}_r = -\frac{\hbar^2}{2m}\frac{1}{r^2}\frac{\partial}{\partial r}\left(r^2\frac{\partial}{\partial r} \right) - \frac{e^2}{4\pi\varepsilon_0 r}, \tag{3.8}$$

$$\mathcal{H}_{\theta\varphi} = \frac{1}{2mr^2}L^2. \tag{3.9}$$

For this reason, we can look for a solution given by a product:

$$\phi(r, \theta, \varphi) = R(r)Y(\theta, \varphi). \tag{3.10}$$

By substitution in the Schrödinger equation (3.4), we have

$$[\mathcal{H}_r + \mathcal{H}_{\theta\varphi}]R(r)Y(\theta, \varphi) = E\,R(r)Y(\theta, \varphi). \tag{3.11}$$

That is to say,

$$Y(\theta, \varphi)\mathcal{H}_r R(r) + R(r)\mathcal{H}_{\theta\varphi}Y(\theta, \varphi) = ER(r)Y(\theta, \varphi), \tag{3.12}$$

dividing by $R(r)Y(\theta, \varphi)$:

$$\frac{1}{R(r)}\mathcal{H}_r R(r) - E + \frac{1}{Y(\theta, \varphi)}\mathcal{H}_{\theta\varphi}Y(\theta, \varphi) = 0. \tag{3.13}$$

---

[2]See I.S. Gradshteyn and I.M. Ryzhik *Table of Integrals, Series, and Products* (Academic, 1980); and D.J. Griffiths, *The Principles of Quantum Mechanics*, 2nd ed. (Prentice Hall, 2005).

Then, after multiplication by $-2mr^2/\hbar^2$,

$$\left\{ \frac{1}{R(r)} \frac{d}{dr} \left(r^2 \frac{dR(r)}{dr} \right) - \frac{2mr^2}{\hbar^2} \left[-\frac{e^2}{4\pi\varepsilon_0 r} - E \right] \right\}$$

$$+ \frac{1}{Y(\theta,\varphi)} \left\{ \frac{1}{\sin\theta} \frac{\partial}{\partial\theta} \left(\sin\theta \frac{\partial Y(\theta,\varphi)}{\partial\theta} \right) + \frac{1}{\sin^2\theta} \frac{\partial^2 Y(\theta,\varphi)}{\partial\varphi^2} \right\} = 0.$$

$$(3.14)$$

In Eq. (3.14) we have a sum of two functions of different variables; in fact, the first term in the braces is dependent only on r, while the second term is dependent only on θ and φ. For this reason, Eq. (3.14) is verified if the two functions equal the same constant with opposite signs. We set this constant[3] as $l(l+1)$:

$$\left\{ \frac{1}{R(r)} \frac{d}{dr} \left(r^2 \frac{dR(r)}{dr} \right) - \frac{2mr^2}{\hbar^2} \left[-\frac{e^2}{4\pi\varepsilon_0 r} - E \right] \right\} = l(l+1),$$

$$(3.15)$$

$$\frac{1}{Y(\theta,\varphi)} \left\{ \frac{1}{\sin\theta} \frac{\partial}{\partial\theta} \left(\sin\theta \frac{\partial Y(\theta,\varphi)}{\partial\theta} \right) + \frac{1}{\sin^2\theta} \frac{\partial^2 Y(\theta,\varphi)}{\partial\varphi^2} \right\} = -l(l+1).$$

$$(3.16)$$

Let us solve Eq. (3.16) first, writing

$$\frac{1}{\sin\theta} \frac{\partial}{\partial\theta} \left(\sin\theta \frac{\partial Y(\theta,\varphi)}{\partial\theta} \right) + \frac{1}{\sin^2\theta} \frac{\partial^2 Y(\theta,\varphi)}{\partial\varphi^2} = -l(l+1)Y(\theta,\varphi).$$

$$(3.17)$$

Also in this case, we can separate the equation into two different equations for θ and φ by defining

$$Y(\theta,\varphi) = \Theta(\theta)\Phi(\varphi). \tag{3.18}$$

Equation (3.17) becomes

$$\frac{1}{\Theta(\theta)} \sin\theta \frac{\partial}{\partial\theta} \left(\sin\theta \frac{\partial\Theta(\theta)}{\partial\theta} \right) + \frac{1}{\Phi(\varphi)} \frac{\partial^2\Phi(\varphi)}{\partial\varphi^2} = -l(l+1)\sin^2\theta.$$

$$(3.19)$$

---

[3]The reason for this choice will be clear later in this section.

Then

$$\frac{1}{\Theta(\theta)} \sin\theta \frac{\partial}{\partial\theta}\left(\sin\theta \frac{\partial\Theta(\theta)}{\partial\theta}\right) + l(l+1)\sin^2\theta + \frac{1}{\Phi(\varphi)}\frac{\partial^2\Phi(\varphi)}{\partial\varphi^2} = 0.$$

(3.20)

Again, we can choose a constant m^2 such that

$$\frac{1}{\Theta(\theta)} \sin\theta \frac{\partial}{\partial\theta}\left(\sin\theta \frac{\partial\Theta(\theta)}{\partial\theta}\right) + l(l+1)\sin^2\theta = m^2,$$ (3.21)

$$\frac{1}{\Phi(\varphi)}\frac{\partial^2\Phi(\varphi)}{\partial\varphi^2} = -m^2.$$ (3.22)

The solution of Eq. (3 22) is very easy:

$$\Phi(\varphi) = e^{im\varphi}.$$ (3.23)

Since by changing φ by 2π, without changing r and θ, we come back to the same point P (see Fig. 3.1), $\Phi(\varphi)$ must be a periodic function:

$$\Phi(\varphi) = \Phi(\varphi + 2\pi).$$ (3.24)

Therefore,

$$e^{im\varphi} = e^{im(\varphi+2\pi)} \rightarrow e^{im2\pi} = 1.$$ (3.25)

This means that

$$m = 0, \pm1, \pm2, \pm3, \ldots.$$ (3.26)

Now we are faced with solving Eq. (3.21) for $\Theta(\theta)$:

$$\sin\theta \frac{\partial}{\partial\theta}\left(\sin\theta \frac{\partial\Theta(\theta)}{\partial\theta}\right) + \left[l(l+1)\sin^2\theta - m^2\right]\Theta(\theta) = 0.$$ (3.27)

The method to solve it can be found in more specialized books. The solution is

$$\Theta(\theta) = AP_l^m(\cos\theta),$$ (3.28)

where P_l^m is the *associated Legendre function*,

$$P_l^m(x) \equiv (1 - x^2)^{|m|/2}\left(\frac{d}{dx}\right)^{|m|} P_l(x),$$ (3.29)

and $P_l(x)$ is the *l-order Legendre polynomial*,

$$P_l(x) \equiv \frac{1}{2^l l!} \left(\frac{d}{dx} \right)^l (x^2 - 1)^l. \tag{3.30}$$

The following conditions hold in (3.29) and (3.30):

$$l = 0, 1, 2, 3, \ldots, \quad |m| < l \rightarrow m = -l, -l+1, \ldots, -1, 0, 1, \ldots, l-1, l. \tag{3.31}$$

The first Legendre polynomials are

$$P_0(x) = 1, \quad P_1(x) = x, \quad P_2(x) = \frac{1}{2}(3x^2 - 1), \quad P_3(x) = \frac{1}{2}(5x^2 - 3x),$$

and the corresponding associated functions are

$$P_0^0(\theta) = 1, \quad P_1^1(\theta) = \sin\theta, \quad P_1^0(\theta) = \cos\theta, \quad P_2^2(\theta) = 3\sin^2\theta,$$

$$P_2^1(\theta) = 3\sin\theta\cos\theta, \quad P_2^0(\theta) = \frac{1}{2}(\cos^2\theta - 1).$$

Normalization must be imposed on the whole wave function $\phi(r, \theta, \varphi) = R(r)\Theta(\theta)\Phi(\varphi)$, and thus in spherical coordinates we write

$$\int_{-\infty}^{+\infty} |\phi|^2 r^2 \sin\theta \, drd\theta d\varphi = \int_0^{+\infty} |R|^2 r^2 dr \int_0^\pi \int_0^{2\pi} |Y|^2 \sin\theta \, d\theta d\varphi = 1. \tag{3.32}$$

As a consequence, it is possible to apply the normalization condition separately to the radial and to the angular part:

$$\int_0^{+\infty} |R|^2 r^2 dr = 1, \quad \int_0^\pi \int_0^{2\pi} |Y|^2 \sin\theta d\theta d\varphi = 1. \tag{3.33}$$

By performing this calculation, we obtain the final form of the normalized $Y(\theta, \varphi)$ functions, denominated as *spherical harmonics*:

$$Y_l^m(\theta, \varphi) = \kappa \sqrt{\frac{(2l+1)(l-|m|)!}{4\pi(l+|m|)!}} e^{im\varphi} P_l^m(\cos\theta), \tag{3.34}$$

where

$$\kappa = (-1)^m \quad \text{when } m \geq 0 \quad \text{and} \quad \kappa = 1 \quad \text{when } m < 0. \tag{3.35}$$

The first spherical harmonics are

$$l = 0: \quad Y_0^0 = \sqrt{\frac{1}{4\pi}};$$

$$l = 1: \quad Y_1^0 = \sqrt{\frac{3}{4\pi}} \cos\theta, \quad Y_1^{\pm 1} = \mp\sqrt{\frac{3}{8\pi}} \sin\theta e^{\pm i\varphi};$$

$$l = 2: \quad Y_2^0 = \sqrt{\frac{5}{16\pi}}(3\cos^2\theta - 1), \quad Y_2^{\pm 1} = \mp\sqrt{\frac{15}{8\pi}} \sin\theta\cos\theta e^{\pm i\varphi},$$

$$Y_2^{\pm 2} = \mp\sqrt{\frac{15}{32\pi}} \sin^2\theta e^{\pm 2i\varphi}.$$

In Fig. 3.2, $Y_0^0, Y_1^0, Y_1^{-1}, Y_1^1$ are plotted to show how the spherical harmonics characterize the shape and orientation of the electron wave function in space. We must underline that the shape depends on the number l while the orientation depends on the number m. This will obviously be reflected in the angular probability density $|Y(\theta, \varphi)|^2$.

It is worth emphasizing that the solution $Y_l^m(\theta, \varphi)$ is a consequence of the spherical symmetry of the potential energy $\mathcal{V}$, and not of its particular value, and therefore the same solution will be found in other problems with the same symmetry.

On the contrary, $\mathcal{V}$ affects the solution of Eq. (3.15) for the radial wave function $R(r)$:

$$\frac{d}{dr}\left(r^2\frac{dR(r)}{dr}\right) - \frac{2mr^2}{\hbar^2}\left[-\frac{e^2}{4\pi\varepsilon_0 r} - E\right]R(r) = l(l+1)R(r). \quad (3.36)$$

The solution is quite complex and can be written as

$$R(r) = Ce^{-\frac{\rho}{2}}\rho^l L_{n-l-1}^{2l+1}(\rho), \quad (3.37)$$

where the Laguerre polynomials are defined as[4]

$$L_n^\alpha(\rho) = \sum_{m=0}^{n}(-1)^m\binom{n+\alpha}{n-m}\frac{\rho^m}{m!}. \quad (3.38)$$

---

[4]In Eq. (3.38), the term $\binom{n+\alpha}{n-m}$ is the *binomial coefficient*. It is generally defined as: $\binom{p}{q} = \frac{p(p-1)(p-2)\ldots(p-q+1)}{q!}$, with p, q integers.

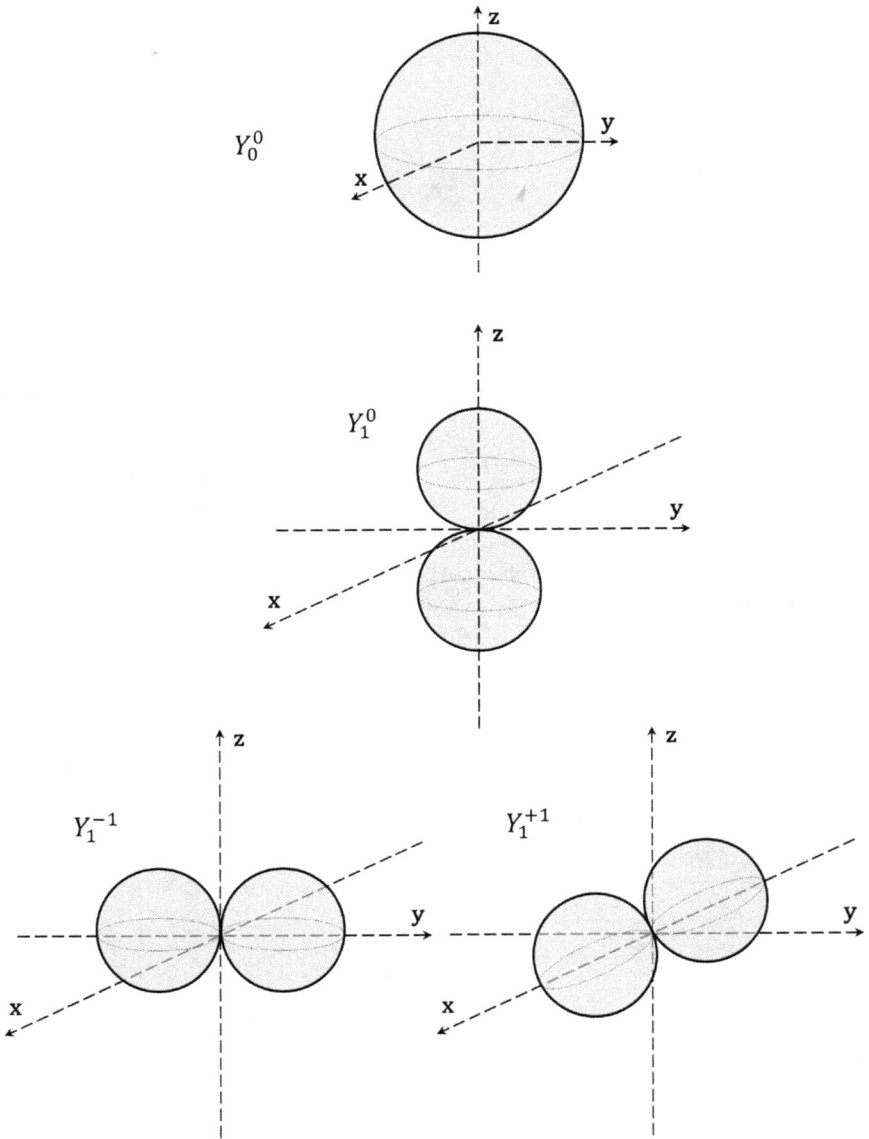

Fig. 3.2.   Plot of the spherical harmonics $Y_0^0, Y_1^0, Y_1^{-1}, Y_1^{+1}$.

C is a constant to be determined by the normalization condition,

$$C = -\sqrt{\left(\frac{2}{na}\right)^3 \frac{(n-l-1)!}{2n[(n+l)!]^3}},$$
(3.39)

and

$$\rho = \frac{2}{na}r,$$
(3.40)

where

$$a = \frac{4\pi\epsilon_0 \hbar^2}{me^2} = 0.529 \cdot 10^{-10} \text{ m},$$
(3.41)

is the Bohr radius for the electron, which corresponds to the lowest energy level as calculated by the semiclassical Bohr theory with $n = 1$.

The solution (3.37) requires that

$$n = 1, 2, 3, \ldots, \quad l = 0, 1, 2, \ldots, n-1.$$
(3.42)

According to the given solution, the lowest order radial eigenfunctions are

$$n = 1,\ l = 0: \quad R_{10} = 2a^{-\frac{3}{2}}e^{-\frac{r}{a}};$$

$$n = 2,\ l = 0, 1: \quad R_{20} = \frac{1}{\sqrt{2}}a^{-\frac{3}{2}}\left[1 - \frac{1}{2}\frac{r}{a}\right]e^{-\frac{r}{2a}},$$

$$R_{21} = \frac{1}{\sqrt{24}}a^{-\frac{3}{2}}\frac{r}{a}e^{-\frac{r}{2a}};$$

$$n = 3,\ l = 0, 1, 2: \quad R_{30} = \frac{2}{\sqrt{27}}a^{-\frac{3}{2}}\left[1 - \frac{2}{3}\frac{r}{a} + \frac{2}{27}\left(\frac{r}{a}\right)^2\right]e^{-\frac{r}{3a}};$$

$$R_{31} = \frac{8}{27\sqrt{6}}a^{-\frac{3}{2}}\left[1 - \frac{1}{6}\frac{r}{a}\right]\frac{r}{a}e^{-\frac{r}{3a}},$$

$$R_{32} = \frac{4}{81\sqrt{30}}a^{-\frac{3}{2}}\left(\frac{r}{a}\right)^2 e^{-\frac{r}{3a}}.$$

The solution of Eq. (3.36) allows finding the energy eigenvalues:

$$E_n = -\left[\frac{m}{2\hbar^2}\left(\frac{e^2}{4\pi\epsilon_0}\right)^2\right]\frac{1}{n^2}.$$
(3.43)

It can be verified that Eq. (3.43) corresponds to the expression (2.11) obtained by Bohr using the old quantum theory (!) showing the $1/n^2$ dependence, which is able to account for the atomic spectra, as discussed in Chapter 2.

In order to fully understand the meaning of $R(r)$, let us consider the radial function corresponding to the fundamental state of the hydrogen atom:

$$R_{10} = 2a^{-\frac{3}{2}}e^{-\frac{r}{a}}. \tag{3.44}$$

By defining $P(r)dr$ as the probability of finding the electron in a spherical shell with radius r' such that $r < r' < r + dr$, the normalization condition gives

$$|R(r)|^2 dV = P(r)dr. \tag{3.45}$$

Since for a spherical shell $dV = d(4/3\pi r^3) = 4\pi r^2 dr$, we have

$$|R(r)|^2 dV = 4a^{-3}e^{-\frac{2r}{a}}4\pi r^2 dr = P(r)dr. \tag{3.46}$$

Then

$$P(r) = 4a^{-3}e^{-\frac{2r}{a}}4\pi r^2. \tag{3.47}$$

The position of the electron with maximum probability can be found setting $dP(r)/dr = 0$, i.e.

$$\left|\frac{dP(r)}{dr}\right|_{r=r_{\max}} = |4a^{-3}e^{-\frac{2r}{a}}8\pi r - 8a^{-4}e^{-\frac{2r}{a}}4\pi r^2|_{r=r_{\max}} = 0, \tag{3.48}$$

and finally

$$r_{\max} = a \ ! \tag{3.49}$$

This is an amazing result of the theory: the deterministic value of the orbital Bohr radius for the fundamental state of the electron corresponds to the most probable distance of the electron from the atomic nucleus as calculated by the quantum theory.

In Fig. 3.3, the radial probability density is reported versus r in the unit of the Bohr radius a for the lowest order radial functions. It is clear how by increasing n the most probable distance for the electron increases, and therefore the orbits of the electron become wider and wider.

The complete set of normalized radial functions is

$$R_{nl}(r) = \sqrt{\left(\frac{2}{na}\right)^3 \frac{(n-l-1)!}{2n[(n+l)!]^3}} \, e^{-\frac{r}{na}} \left(\frac{2r}{na}\right)^l L_{n-l-1}^{2l+1}\left(\frac{2r}{na}\right). \tag{3.50}$$

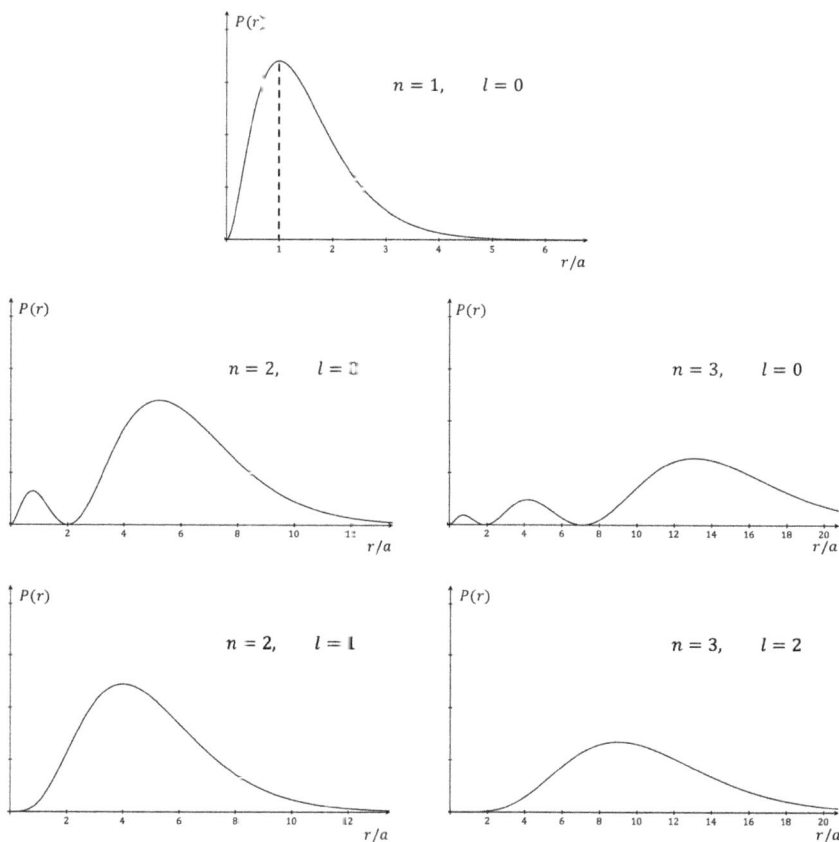

Fig. 3.3. Plots of $P(r) = 4\pi r^2 |R(r)|^2$ versus r in the unit of the Bohr radius for increasing values of the quantum numbers n and l.

Finally, the complete time-independent wave function of the electron is given by

$$\phi_{nlm}(r, \theta, \varphi) = R_{nl}(r) Y_l^m(\theta, \varphi), \tag{3.51}$$

i.e.

$$\phi_{nlm}(r, \theta, \varphi) = \kappa \sqrt{\left(\frac{2}{na}\right)^3 \frac{(n-l-1)!}{2n[(n+l)!]^3}} \sqrt{\frac{(2l+1)(l-|m|)!}{4\pi(l-|m|)!}}$$

$$\times e^{-\frac{r}{na}} e^{im\varphi} \left(\frac{2r}{na}\right)^l L_{n-l-1}^{2l+1}\left(\frac{2r}{na}\right) P_l^m(\cos\theta), \tag{3.52}$$

and the complete wave function is

$$\psi_{nlm}(r, \theta, \varphi, t) = \phi_{nlm}(r, \theta, \varphi)e^{-i\frac{E_n}{\hbar}t}. \tag{3.53}$$

The energy eigenfunctions (3.53) are identified by three integer numbers called *quantum numbers*:

Principal quantum number: $n = 1, 2, 3, \ldots$;
Azimuthal quantum number: $l = 0, 1, 2, \ldots, (n-1)$;
Magnetic quantum number: $m = -l, -l+1, \ldots, -1, 0, 1, \ldots, l-1, l$.

We have already underlined the main role of the principal quantum number: it determines the energy of the electron according to Eq. (3.43) and its average distance from the nucleus. This is clear by looking at the exponential dependence of the radial wave function $R(r)$ in Eq. (3.37),

$$e^{-\frac{\rho}{2}} = e^{-\frac{r}{na}},$$

saying that the radial distribution has a characteristic decay distance increasing with n. In fact, the exponential decay reduces to $1/e$ at a distance

$$r_{1/e} = a, 2a, 3a, \ldots \text{ as } n = 1, 2, 3, \ldots.$$

This wider and wider electron distribution is highlighted by the radial probability densities plotted in Fig. 3.3.

The meaning of the azimuthal and of the magnetic quantum numbers becomes clear by taking into account the expression of the angular momentum operator in spherical coordinates giving Eq. (3.6) for the square of the angular momentum,

$$L^2 = -\hbar^2 \left[\frac{1}{\sin\theta} \frac{\partial}{\partial\theta} \left(\sin\theta \frac{\partial}{\partial\theta} \right) + \frac{1}{\sin^2\theta} \frac{\partial^2}{\partial\varphi^2} \right],$$

and for the z component,

$$L_z = \frac{\hbar}{i} \frac{\partial}{\partial\varphi}. \tag{3.54}$$

From Eq. (3.9) we have

$$L^2 = 2mr^2 \mathcal{H}_{\theta\varphi}. \tag{3.55}$$

Then, as expected, $[\mathcal{H}_{\theta\varphi}, L^2] = 0$, and therefore the operator L^2 has the same eigenfunctions of $\mathcal{H}_{\theta\varphi}$, namely $Y_l^m(\theta, \varphi)$. From Eq. (3.17) we get

$$L^2 Y_l^m(\theta, \varphi) = \hbar^2 l(l+1) Y_l^m(\theta, \varphi). \tag{3.56}$$

Since $[L_z, L^2] = 0$ is always true, the same $Y_l^m(\theta, \varphi)$ are eigenfunctions of L_z:

$$L_z Y_l^m(\theta, \varphi) = \frac{\hbar}{i} \frac{\partial}{\partial \varphi} Y_l^m(\theta, \varphi) = \Theta(\theta) \frac{\hbar}{i} \frac{\partial}{\partial \varphi} \Phi(\varphi)$$

$$= m\hbar \Theta(\theta) \Phi(\varphi) = m\hbar Y_l^m(\theta, \varphi). \tag{3.57}$$

In this way, Eq. (3.56) gives the eigenvalues of the square of the angular momentum as $\hbar^2 l(l+1)$, and thus the modulus of the angular momentum is

$$\langle |L| \rangle = \sqrt{l(l+1)}\hbar. \tag{3.58}$$

From Eq. (3.57) the eigenvalues of the z component are obtained as

$$\langle L_z \rangle = m\hbar. \tag{3.59}$$

In conclusion: the azimuthal quantum number l determines the value of angular momentum, and the magnetic quantum number m determines its projection along the z axis. In these expressions, we have used the usual brackets to indicate the expectation value of the corresponding operator. In the following, we will drop the brackets, writing L and L_z also to indicate the values of the angular momentum and its z component.

3.3. Space quantization and spin

The results (3.58) and (3.59) allow an easy representation of the space quantization. In fact, from a classical point of view, the angular momentum vector could be oriented in any direction and therefore its z component could take any value in between $\pm |L|$, the maximum and minimum occurring when the angular momentum is oriented along the positive **z** axis or opposite to it.

On the contrary Eq. (3.59) tell us that L_z has a limited number of possible values, depending on the range of allowed integers m. This means that a limited number of different orientations θ_m are allowed for the angular momentum, taking into account that

$$L_z = m\hbar = L\cos\theta_m, \quad \text{then } L = m\frac{\hbar}{\cos\theta_m} \quad \text{and} \quad \theta_m = \arccos\frac{m}{\sqrt{l(l+1)}}. \tag{3.60}$$

This fact is shown in Fig. 3.4 for the cases $l = 1$ and $l = 2$.

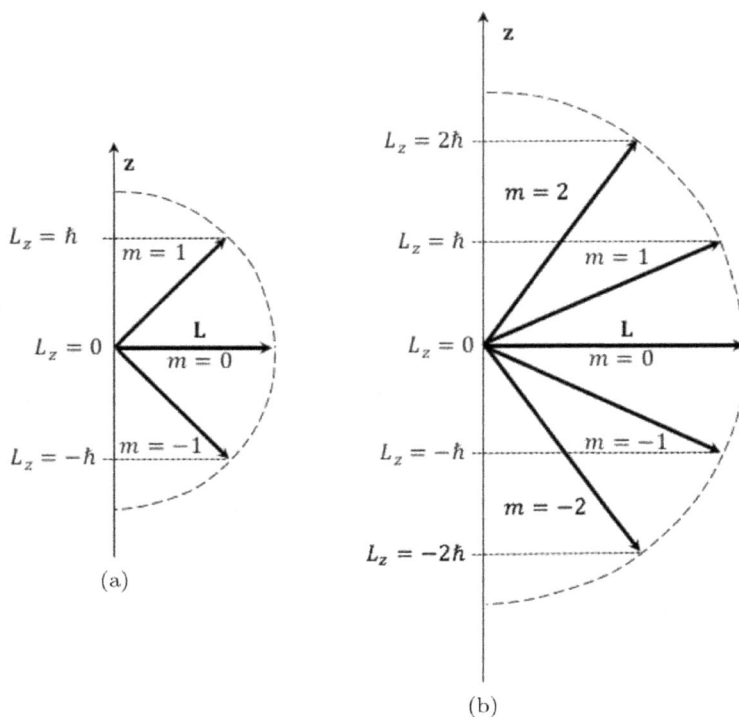

Fig. 3.4.   Space quantization for the angular momentum when $l = 1$ (a) and $l = 2$ (b).

Looking at Fig. 3.4, we have the following:

$$l = 1, \ m = 0, \pm 1; \qquad \text{then } L_z = 0, \pm \hbar, \ L = \sqrt{2}\hbar$$

$$\text{and} \quad \theta_m = \frac{\pi}{2}, \pm \frac{\pi}{4};$$

$$l = 2, \ m = 0, \pm 1, \pm 2; \quad \text{then } L_z = 0, \pm \hbar, \pm 2\hbar, \ L = \sqrt{6}\hbar$$

$$\text{and} \quad \theta_m = \frac{\pi}{2}, \pm 0.366\pi, \pm 0.196\pi.$$

It is clear from Fig. 3.4 that quantization of angular momentum concerns the orientation with respect to the z axis and does not involve the angle φ, and therefore all the directions on the conical surfaces defined by θ_m are allowed. This is actually related to the uncertainty principle, since spherical harmonics $Y_l^m(\theta, \varphi)$ are not eigenfunctions of L and therefore it has not a definite value in these states.

From the classical theory of electromagnetism, it is known that one electron of mass m_e, moving with angular momentum $\mathbf{L}$, leads to a magnetic moment given by

$$\mu = -\frac{e}{2m_e}\mathbf{L}. \tag{3.61}$$

This is easily demonstrated by recalling that a magnetic moment induced by a current closing an area A is $\mu = IA$. We write $A = \pi r^2$ for a circular loop described by a charge q with a mass m at a speed v; with T being the period, the resulting current is

$$I = \frac{q}{T} = \frac{qv}{2\pi r}.$$

Then

$$\mu = \frac{qvr}{2} = \frac{q}{2m}L, \quad \text{since } L = mvr.$$

By replacing q with $-e$ and m with m_e, we get Eq. (3.61).

Equation (3.61) is often written as

$$\boldsymbol{\mu} = \gamma_e \mathbf{L}, \tag{3.62}$$

where γ_e is defined as the *magnetogyric ratio* of the electron.

Since the interaction of a magnetic field of induction $\mathbf{B}$ with a magnetic moment gives rise to an additional potential energy that may reorient it toward the magnetic field direction, the energy of the electron submitted to a magnetic field changes by an amount

$$\Delta E_B = -\boldsymbol{\mu} \cdot \mathbf{B} = -\gamma_e \mathbf{L} \cdot \mathbf{B}. \tag{3.63}$$

Choosing the field direction along the z axis ($\mathbf{B} = B\mathbf{z}$), we get

$$\Delta E_B = -\gamma_e L_z B = m\frac{e\hbar}{2m_e}B. \tag{3.64}$$

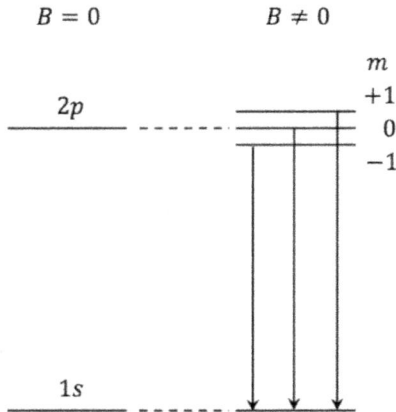

Fig. 3.5.   Spectral line splitting induced by the magnetic field owing to transitions from a state $n = 2, l = 1$ to a state $n = 1, l = 0$.

The quantity $e\hbar/2m_e$ is called *the Bohr magneton*: $\mu_B = e\hbar/2m_e = 9.274 \cdot 10^{-24}$ J/T.

In this way,

$$\Delta E_B = m\mu_B B. \tag{3.65}$$

According to Eq. (3.65), for energy levels corresponding to $l \neq 0$ (i.e. allowing different values of the quantum number m), the presence of a magnetic field makes the electron energy different for each value of m. For instance, for $l = 1$ we have three different values of the electron energy E_l^m:

$$E_1^0 \text{ unchanged}, \quad E_1^1 = E_1^0 + \mu_B B, \quad E_1^{-1} = E_1^0 - \mu_B B.$$

In this way, the observation of spectral lines of atoms under the magnetic field is able to demonstrate the space quantization as depicted in Fig. 3.5, showing the possible transitions from a state with $n = 2$, $l = 1$ to the ground state $n = 1$, $l = 0$. The single transition occurring at $B = 0$ is split in three transitions when $B \neq 0$.

As a matter of fact, a deep analysis of the lines' structure performed with high resolution methods allows detecting two lines in each of the outer lines (i.e. when $m \neq 0$). This suggests the presence of an additional property of the electron with two possible values at each energy level. Since this property is sensitive to the magnetic field, it is suitable to be treated as an intrinsic angular momentum, as if the electron were a rigid body spinning around its axis.

A definite experimental demonstration of this property was given by Stern and Gerlach[5] using an atomic beam of silver through a strongly inhomogeneous magnetic field with

$$\frac{\partial B}{\partial z} \neq 0,$$

with z being the propagation direction of the beam.

In this way, the induced potential energy is

$$U = -\boldsymbol{\mu} \cdot \mathbf{B} = -\mu_z B_z = -\mu \cos\theta B_z$$

and the force acting on the atoms is

$$F_z = -\frac{\partial U}{\partial z} = \mu \cos\theta \frac{\partial B}{\partial z},$$

where θ defines the orientation of the magnetic moment with respect to B. As a consequence, the atomic beam is deviated and its deviations can be detected on a sensitive screen. The electrons in the used atomic beam were in a fundamental state characterized by $l = 0$, which gives rise to only one value of $\mu_z = 0$; as a consequence, $F_z = 0$ and no deviation should be expected. On the contrary, two separate spots were observed (up and down with respect to the original beam direction), showing that an additional two values of the magnetic moment not related to orbital motion are present.

As a consequences of these experimental observations, the expressions (3.64) and (3.65) can be corrected using the same formalism in order to take into account this additional contribution to energy modifications:

$$\Delta E_B = m\mu_B B + m_s g\mu_B B, \tag{3.66}$$

where g is defined the *gyromagnetic ratio* and is found to be $g = 2.0023$, and m_s is the *spin magnetic quantum number*.

By analogy with Eq. (3.62), we can define a spin magnetic moment as

$$\boldsymbol{\mu}_s = \gamma_s \mathbf{S}, \tag{3.67}$$

introducing in this way the *intrinsic* or *spin angular momentum* $\mathbf{S}$ of the electron.

---

[5]W. Gerlach and O. Stern, *Z. Phys.* **9**, 349 (1922).

Equations (3.62) and (3.66) are linked by the condition $\gamma_s = g\gamma_e$, and therefore

$$\boldsymbol{\mu}_s = \gamma_s \mathbf{S} = g\gamma_e \mathbf{S} = -g\frac{\mu_B}{\hbar}\mathbf{S}. \tag{3.68}$$

The spin quantum number m_s appearing in Eq. (3.66) is defined in analogy with the quantum number m which determines the z component of the orbital angular momentum; then the z component of the spin angular momentum is written as

$$S_z = m_s\hbar \tag{3.69}$$

and

$$S^2 = s(s+1)\hbar^2, \tag{3.70}$$

with s being the *spin quantum number*. Finally, we note that in general the space quantization, described above for the angular momentum and determined by a given value of the azimuthal quantum number l, gives rise to $2l + 1$ components along the z axis. On the contrary, the experiment of Stern and Gerlach tell us that the spin magnetic moment has only two possible values, and therefore we can write

$$2s + 1 = 2 \rightarrow s = \frac{1}{2}; \quad \text{then } m_s = \pm\frac{1}{2}. \tag{3.71}$$

It is out of the scope of this book to enter into details about this subject, but we can just note that the existence of spin should be considered an additional postulate of the theory of quantum mechanics necessary for explaining the experimental observations. We recall that the result presented for the electron can be extended to any particle, the only difference being the value of the quantum number s, which in general has half-integer or integer values; this affects the particle properties because it leads to different statistical behavior.

For what concerns spin, we can represent the electron states as $\chi(\mathbf{s})$, where only two orientations are allowed for $\mathbf{s}$ ($\uparrow$ up and $\downarrow$ down). As a consequence, the state of the electron will be described by the wave function ψ_{nlm} and by its state, either $\chi(\uparrow)$ or $\chi(\downarrow)$. Namely, for each ψ_{nlm} we have two possible states of the electron. In other words, the state of the electron is identified when four quantum numbers are known: n, l, m, m_s. We will see in the next section how this affects the way electrons fill the available energy states and the consequences for the chemical–physical property of atoms.

3.4. Many-electron atoms and the periodic table

In following the method used for solving the problem of the hydrogen atom, additional difficulties arise when two or more electrons are considered dealing with heavier atoms. To solve the problem, it is necessary to adopt suitable approximations. We will not go through the calculations which can be found in many texts related to atomic physics, but we will highlight the main issues and the final consequences coming from a more complex atomic conformation. Let us consider helium (He), the two-electron atom having a positive nucleus made by two protons (charge $+2e$). In this case, not only does the electrostatic interaction occur between each electron and nucleus, but we also have a repulsive interaction between the electrons. In this way, the potential energy becomes

$$\mathcal{V}(r_1 r_2) = -\frac{2e^2}{4\pi\varepsilon_0 r_1} - \frac{2e^2}{4\pi\varepsilon_0 r_2} + \frac{e^2}{4\pi\varepsilon_0 r_{12}}, \qquad (3.72)$$

where r_1 and r_2 are the distances of the electrons from the fixed nucleus and r_{12} their mutual distance:

$$r_{12} = \sqrt{(r_{2x} - r_{1x})^2 + (r_{2y} - r_{1y})^2 + (r_{2z} - r_{1z})^2}.$$

If the electron–electron interaction is neglected, the mathematical treatment becomes very similar to those of the hydrogen atom and the final result for each electron of He will be a ground state exactly twice as much as that of the H atom and the sequence of the energy levels will follow the same dependence on the quantum number n, with degeneracy in l. However, by taking into account the third term of Eq. (3.72), one gets an additional contribution to energy, i.e. a correction of the values removing degeneracy in l, so that the electron energy depends also on the azimuthal quantum number. In general, the main variation of level energy depends on n (increasing with it), but for a fixed n the energy will be different for different values of l (increasing with it).

This result can be easily understood by recalling that the azimuthal quantum number affects the angular shape of the wave function in space. Now the electron–electron interaction can be seen as an electrostatic shielding of the positive nucleus charge on one electron; due to the other electron cloud around the nucleus. In this way, the nucleus charge is partially shielded by one electron; then its attractive force is lowered and this effect is obviously strongly dependent on the shape of the electron distribution around the nucleus. Of course, shielding becomes more and more important by increasing the number of electrons, so that some change will occur on the basic rule mentioned above about the dependence of energy distribution on the quantum numbers n and l for many-electron atoms. For instance, the level with $n = 4$, $l = 0$ corresponds to energy lower than level with $n = 3$, $l = 2$; or the level with $n = 5$, $l = 0$ corresponds to energy lower than level with $n = 4$, $l = 2$, and so on.

For what concerns the Schrödinger equation for many electrons, we must underline that it applies to wave functions dependent on the coordinates of each electron. For instance, for the He atom we have

$$\psi \equiv \psi(\mathbf{r}_1, \mathbf{r}_2), \tag{3.73}$$

and for an N-electron atom

$$\psi \equiv \psi(\mathbf{r}_1, \mathbf{r}_2, \ldots, \mathbf{r}_N). \tag{3.74}$$

Besides that, each electron can be found in either of the spin states, and therefore we could define a general eigenfunction Ψ including also the spin state determined by the quantum number m_s. For one electron it would be

$$\Psi(\mathbf{r}, m_s) \equiv \psi(\mathbf{r})\chi(m_s), \tag{3.75}$$

resulting for the many-electron atom as

$$\Psi \equiv \Psi(\mathbf{r}_1, \mathbf{r}_2, \ldots, \mathbf{r}_N, m_{s1}, m_{s2}, \ldots, m_{sN}). \tag{3.76}$$

Now we note that the quantum-mechanical point of view does not allow identifying different electrons since we cannot identify the single trajectories of the electrons, but we have only statistical information on their position; this lack of information is obviously linked to the uncertainty principle.

Let us consider a two-electron system and label each electron as #1 and #2. If they occupy two energy levels E_1 and E_2 ($E_1 < E_2$), we could have either

$$E_1(\#1)E_2(\#2) \quad \text{or} \quad E_1(\#2)E_2(\#1),$$

and the two configurations cannot be distinguished. Then, by exchanging the coordinates of the two electrons, the wave function of the system must not distinguish among them. By defining for each electron a generalized coordinate q_i including spatial coordinates and quantum numbers m_{si}, this condition can be fulfilled in only two ways[6]:

$$\Psi(q_1, q_2) = \pm\Psi(q_2, q_1). \tag{3.77}$$

---

[6]In fact, $\Psi(q_2, q_1)$ and $-\Psi(q_2, q_1)$ give the same probability density, $|\Psi(q_2, q_1)|^2$.

That is to say, the wave function can be either symmetric or antisymmetric for electron exchange:

$$\Psi_\pm(q_1, q_2) = A\{\Psi_1(q_1)\Psi_2(q_2) \pm \Psi_2(q_1)\Psi_1(q_2)\}, \qquad (3.78)$$

where Ψ_1 and Ψ_2 are different states of the single particle and A is a normalization constant. It is an experimental result that for electrons (like for all particles where the spin quantum number is half-integer) the wave function (3.78) must be antisymmetric, and thus *for electrons*,

$$\text{if } \Psi_1 = \Psi_2, \quad \text{then } \Psi_\pm(q_1, q_2) = A\{\Psi_1(q_1)\Psi_1(q_2) - \Psi_1(q_1)\Psi_1(q_2)\} = 0,$$
$$(3.79)$$

which means that it must be $\Psi_1 \neq \Psi_2$.

This condition represents the well-known *Pauli exclusion principle*, which states that *in an atomic system it is not possible that two electrons are represented by the same wave function, i.e. they cannot have the same four quantum numbers.* In other words, the same spatial wave function ψ_{nlm} can describe the state of two electrons at most, each in a spin state opposite to the other.

It is now possible to understand how the electrons fill the available energy levels in different atoms. First of all, let us recall the traditional denomination assigning a letter to each l state at a given value of n:

$$\begin{aligned} l &= 0 \rightarrow s, \\ l &= 1 \rightarrow p, \\ l &= 2 \rightarrow d, \\ l &= 3 \rightarrow f, \\ l &= 4 \rightarrow g, \\ &\cdots \end{aligned} \qquad (3.80)$$

Then energy levels are defined with a number indicating the principal quantum number n and a letter indicating l according to Eqs. (3.80); for instance, a level with $n = 1$, $i = 0$ is ($1s$) and a level with $n = 3$, $l = 2$ is ($3d$). The number of electrons filling a specific configuration is given as the apex on the brackets; for example, if four electrons are in the $3d$ state, we write $(3d)^4$.

According to the Pauli exclusion principle, there are a limited number of available states for each value of n: since for each value of n we have[7]

---

[7]Since for each value of n there are $2l + 1$ different wave functions, the total number of wave functions from the ground state to the nth shell is $\sum_{l=0}^{n-1}(2l + 1) = n^2$, the well-known arithmetic progression.

n^2 possible spatial wave functions, at most $2n^2$ electrons can fill the energy *shell* (as it is often defined as the set of levels under the same value of n). The different wave functions and electron states are shown below for $n = 1, 2, 3$.

$$
\begin{aligned}
n = 1: \quad & l = 0, \quad m = 0, \quad \text{wave function} \quad \psi_{100} \quad (1s); \\
n = 2: \quad & l = 0, \quad m = 0, \quad && \psi_{200} \quad (2s); \\
& l = 1, \quad m = -1, \quad && \psi_{21-1} \quad (2p); \\
& \quad\quad\; m = 0, \quad && \psi_{210}; \\
& \quad\quad\; m = +1, \quad && \psi_{211}; \\
n = 3: \quad & l = 0, \quad m = 0, \quad && \psi_{300} \quad (3s); \\
& l = 1, \quad m = -1, \quad && \psi_{31-1} \quad (3p); \\
& \quad\quad\; m = 0, \quad && \psi_{310}; \\
& \quad\quad\; m = +1, \quad && \psi_{311}; \\
& l = 2, \quad m = -2, \quad && \psi_{32-2} \quad (3d); \\
& \quad\quad\; m = -1, \quad && \psi_{32-1}; \\
& \quad\quad\; m = 0, \quad && \psi_{320}; \\
& \quad\quad\; m = +1, \quad && \psi_{321}; \\
& \quad\quad\; m = +2, \quad && \psi_{322},
\end{aligned}
\tag{3.81}
$$

where each state can be filled by two electrons. Then, for each l value, we have the following numbers of available states:

Azimuthal quantum number	Denomination	Number of electronic states
$l = 0$	s	2
$l = 1$	p	6
$l = 2$	d	10
$l = 3$	f	14

According to this and taking into account that in steady state conditions the electrons fill the lowest energy levels, it is easy to find which are the occupied energy levels depending on the number of electrons in the atom. We show this in Table 3.1 for the first ten elements of the periodic table, where in the last column we represent each wave function with a box that can include two electrons shown as arrows pointing up or down, depending on the spin. In this sketch, we took into account the first Hund rule, which says that the configuration with the highest value of total spin

Table 3.1.   Occupation of energy levels in the first ten elements of the periodic table.

Element	Number of electrons	Configuration	State occupation and spin				
H	1	$(1s)$	↑				
He	2	$(1s)^2$	↑↓				
Li	3	$(1s)^2(2s)$	↑↓	↑			
Be	4	$(1s)^2(2s)^2$	↑↓	↑↓			
B	5	$(1s)^2(2s)^2(2p)$	↑↓	↑↓	↑		
C	6	$(1s)^2(2s)^2(2p)^2$	↑↓	↑↓	↑	↑	
N	7	$(1s)^2(2s)^2(2p)^3$	↑↓	↑↓	↑	↑	↑
O	8	$(1s)^2(2s)^2(2p)^4$	↑↓	↑↓	↑↓	↑	↑
F	9	$(1s)^2(2s)^2(2p)^5$	↑↓	↑↓	↑↓	↑↓	↑
Ne	10	$(1s)^2(2s)^2(2p)^6$	↑↓	↑↓	↑↓	↑↓	↑↓

is more stable and the stability is decreasing with decreasing total spin (*principle of maximum spin multiplicity*).

Now we will present a few examples in order to clarify how the electronic configuration around the nucleus affects the physical–chemical properties of atoms determining if and under which conditions atoms can link to each other to give rise to molecules and/or to condensed states of matter. By doing that, we will introduce the concept of the *valence electron*, which is of paramount importance, in the following chapters of this book.

Let us consider sodium (Na), a very common element having 11 electrons. In this case, the electronic configuration must be $(1s)^2(2s)^2(2p)^6(3s)$. Then two shells ($n = 1$ and $n = 2$) are full while only one electron is in the shell $n = 3$ at a much higher energy. This makes sodium highly reactive, trying to lend the external electron to another atom in order to keep a stable electronic configuration given by $(1s)^2(2s)^2(2p)^6$, where the external shell has eight electrons.

The tendency of each atom to go toward an external configuration of the complete shell filled by eight electrons is actually an empirical law called the *octet rule*. This concerns the outermost electrons of the atom, called *valence electrons*, which are the ones most important for our purposes since they are responsible for bonding between atoms and are the main ones responsible for the electrical and optical properties of matter. Elements like Na belong to the first group of the periodic table; they have one valence electron and are called *alkali metals*. They include Li, K, Rb, Cs and Fr.

The opposite behavior is shown, for instance, by Cl (chlorine), with 17 electrons. Following the described filling methods, the electronic configuration is found to be $(1s)^2(2s)^2(2p)^6(3s)^2(3p)^5$, i.e. only one electron is missing, to get the stable configuration with eight outermost valence electrons according to the octet rule. Therefore, this atom likes to borrow one electron from another atom (for instance Na, which likes to lend one electron) to get the more stable external configuration $(3s)^2(3p)^6$. This tendency of atoms to gain electrons is called *electronegativity* and plays a very important role in chemical bonding. Atoms like Cl belong to the VII group of the periodic table and are highly reactive as F, Br, I and At, called *halogens*.

According to the remarks made on the octet rule, we expect that all the atoms completing such a configuration in the steady state are very stable, having very low chemical reactivity. They are the *noble gases*, such as He, Ne, Kr, Xe and Rn. In fact, as an example, we have already seen in Table 3.1 that the electronic configuration of Ne is $(1s)^2(2s)^2(2p)^6$.

In conclusion, we note that all the chemical elements are included in the periodic table of elements in a tabular arrangement (Table 3.2), based on the electronic configuration and recurring chemical properties, presented in order of increasing atomic number. Details can be found in any text on basic chemistry.

A very important property that we need to remember is the *first ionization potential*, since it plays an important role in chemical bonds. It is defined as the energy necessary for removing the outermost electron from the atom in the steady state. It is dependent on the atomic number Z and generally decreases with increasing Z owing to the increasing shielding effect of the inner electrons; however, it presents peaks when the electronic configuration is stable, and therefore maxima occur in correspondence with the noble gas, as depicted in Fig. 3.6.

3.5. Chemical bonds

The next step toward the description of the basic properties of matter is the analysis of the possible interaction between close atoms determining when it can lead to bonding between two or more atoms.

We should recognize three different cases:

(i) Bonding leading to the build-up of *molecules*, namely a stable combination of two or more atoms;
(ii) Bonding leading to a condensed state without previous formation of molecules;
(iii) Bonding leading to a condensed state as a consequence of formation of molecules through process (i).

Table 3.2.

Periodic Table of Elements

| Atomic Number |
| Symbol |
| Atomic Weight |

1	2	3	4	5	6	7	8	9	10	11	12	13	14	15	16	17	18
1 H 1.008																	2 He 4.003
3 Li 6.941	4 Be 9.012											5 B 10.811	6 C 12.011	7 N 14.007	8 O 15.999	9 F 18.998	10 Ne 20.180
11 Na 22.990	12 Mg 24.305											13 Al 26.982	14 Si 28.086	15 P 30.974	16 S 32.066	17 Cl 35.453	18 Ar 39.948
19 K 39.098	20 Ca 40.078	21 Sc 44.956	22 Ti 47.88	23 V 50.942	24 Cr 51.996	25 Mn 54.938	26 Fe 55.933	27 Co 58.933	28 Ni 58.693	29 Cu 63.546	30 Zn 65.39	31 Ga 69.732	32 Ge 72.61	33 As 74.922	34 Se 78.972	35 Br 79.904	36 Kr 81.80
37 Rb 84.468	38 Sr 87.62	39 Y 88.906	40 Zr 91.224	41 Nb 92.906	42 Mo 95.95	43 Tc 98.907	44 Ru 101.07	45 Rh 102.906	46 Pd 106.42	47 Ag 107.868	48 Cd 112.411	49 In 114.818	50 Sn 118.71	51 Sb 121.760	52 Te 127.6	53 I 126.904	54 Xe 131.29
55 Cs 132.905	56 Ba 137.327	57-71	72 Hf 178.49	73 Ta 180.948	74 W 183.85	75 Re 186.207	76 Os 190.23	77 Ir 192.22	78 Pt 195.08	79 Au 196.967	80 Hg 200.59	81 Tl 204.383	82 Pb 207.2	83 Bi 208.980	84 Po [208.982]	85 At 209.987	86 Rn 222.018
87 Fr 223.020	88 Ra 226.025	89-103	104 Rf [261]	105 Db [262]	106 Sg [266]	107 Bh [264]	108 Hs [269]	109 Mt [268]	110 Ds [269]	111 Rg [272]	112 Cn [277]	113 Nh [286]	114 Fl [289]	115 Mc [289]	116 Lv [298]	117 Ts [294]	118 O [294]

57 La 138.906	58 Ce 140.115	59 Pr 140.906	60 Nd 144.24	61 Pm 144.913	62 Sm 150.36	63 Eu 151.966	64 Gd 157.25	65 Tb 158.925	66 Dy 162.50	67 Ho 164.930	68 Er 167.26	69 Tm 168.934	70 Yb 173.04	71 Lu 174.967
89 Ac 227.028	90 Th 232.038	91 Pa 231.036	92 U 238.029	93 Np 237.048	94 Pu 244.064	95 Am 243.061	96 Cm 247.070	97 Bk 247.070	98 Cf 251.080	99 Es [254]	100 Fm 257.095	101 Md 258.1	102 No 259.101	103 Lr [262]

Legend: Alkali Metal, Alkaline Earth, Transition Metal, Basic Metal, Semimetal, Nonmetal, Halogen, Noble Gas, Lanthanide, Actinide

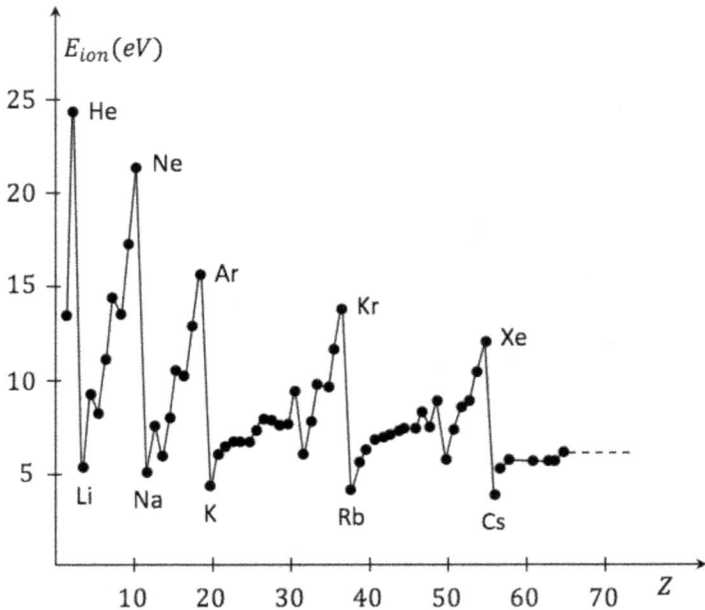

Fig. 3.6.   The first ionization potential of elements versus the atomic number Z.

We should note that electronic configuration of interacting atoms is the most important factor determining the kind of bonding occurring, the possible condensed state formation and its physical properties.

Since it is not possible to build a logic hierarchy like the atom→ molecule→condensed state, we describe in the following the different kinds of interactions, mentioning typical cases where they play a major role.

It is well known that a group of atoms occupying a volume V much bigger than the actual volume of all the atoms (*covolume*) and interacting only through elastic collisions (interaction potential energy $V = 0$) is called an *ideal gas*. As will be discussed later in this chapter, by varying the thermodynamical parameters of the system (pressure P, volume V, temperature T) the conditions for an ideal gas may not be fulfilled anymore and both attractive and repulsive forces may appear.

Let us take into account two interacting atoms: if the interaction potential has a negative minimum for a specific finite distance between them, this bonding state will be more probable than the system made by two free atoms. This means that bonding minimizes the total energy and a molecule will be formed. The same argument can be applied to N atoms interacting to form a condensed state: in order to give rise to the new structure, the energy of the N-atom system must be lower in the condensed state than that of N free atoms. The difference between these two values of energy is just the cohesion energy — the one to be provided to the system in order to break bonding.

Of course, bonding will occur if there is a range of distances between atoms such that the long range attractive forces (for example due to electrostatic interaction between the valence electrons and positive nuclei) overcome the short range repulsive ones (for example due to overlapping of the full electronic shells, which according to the Pauli principle promote electrons to higher energy levels).

For the sake of clarity, let us consider the interaction between two atoms, since the same arguments can be applied to a higher number of atoms. Their interaction depends on two factors: (i) the charge state of the atoms (i.e. if they are neutral or ionized) and (ii) their electronic configuration, in particular that of the valence electrons, distinguished as "closed shell" and "open shell" according to the octet rule. Therefore, it is possible to make the following classifications related to different types of chemical bonds:

(1) Interaction between a closed shell positive ion and a closed shell negative ion;
(2) Interaction between two closed shell neutral atoms;
(3) Interaction between a closed shell ion and an open shell ion;
(4) Interaction between two open shell neutral atoms.

In general, the potential energy of the interaction can be written as

$$V(r) = -\frac{\alpha}{r^p} + \frac{\beta}{r^q}, \tag{3.82}$$

with r being the distance between the atomic nuclei and α, β, p, q being parameters taking into account the nature of the interaction and the features of the atoms. A typical plot of Eq. (3.82) is shown in Fig. 3.7.

Repulsive potential

The increase in a repulsive potential with reducing distances between atoms is due to: (i) the electrostatic interaction between charges of the same sign (positive nuclei and electronic clouds); (ii) the Pauli principle

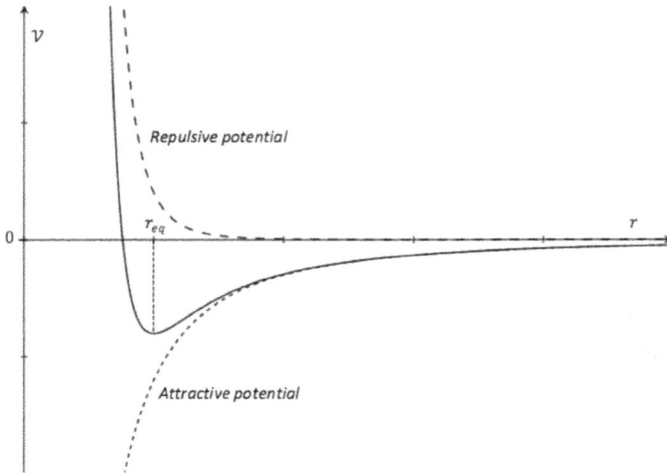

Fig. 3.7.   Typical shape of the interaction potential $\mathcal{V}(r)$. The attractive and repulsive contributions are also shown.

if the approaching atoms have closed shells. In fact, when the electronic distributions of two atoms try to overlap, electrons of atom A will occupy states of atom B and vice versa. Since this overlap is allowed only if empty states are available according to the Pauli principle, it will be forbidden in the case of closed shells leading to an increase in energy for the electrons, thus resulting in a repulsive force. This explains the meaning of the classification reported above as the existence of closed shells will play a strong antibonding role (this effect is responsible for incompressibility of liquids and solids).

The most-used empirical expression for the repulsive potential is the *Lennard–Jones potential,*

$$\mathcal{V}(r) = \frac{B}{r^{12}}, \tag{3.83}$$

or the *Born–Mayer potential,*

$$\mathcal{V}(r) = Ae^{-r/\rho}, \tag{3.84}$$

with A, B and ρ being parameters to be determined through experiments.

When bonding takes place, the attractive potential will overcome the repulsive one with a minimum for a specific distance between atoms. We briefly describe below the main types of bonding between atoms.

Ionic bond

This occurs between atoms with open shells. One of them usually misses one or two electrons to fill the shell, while the other has one or two electrons in excess of the closed shell. As a consequence, when the atoms are close to each other the excess electrons of one will fill the open shell of the other, thus creating positive and negative ions that interact strongly through electrostatic forces. Then, the ionic bond occurs between atoms with a positive oxidation number and atoms with a negative oxidation number, with the *oxidation number* being the number of electrons that are given (positive) or acquired (negative) by the atom. Using the term *electronegativity* for the tendency of an atom to attract electrons, we can say that the ionic bond occurs between pairs of atoms with a large electronegativity difference. In general, if Z_1 is the number of electrons removed from one atom and Z_2 the number of electrons added to the other atom, the attractive potential will be

$$\mathcal{V}(r) = -\frac{(Z_1 e)(Z_2 e)}{4\pi\epsilon_0 r} \tag{3.85}$$

when only two types of atoms are involved, $Z_1 = Z_2 \equiv Z$; then

$$\mathcal{V}(r) = -\frac{(Ze)^2}{4\pi\epsilon_0 r}. \tag{3.86}$$

The spherical symmetry of the interaction potential produces a nondirectional bonding. Nevertheless, a slight modification from the spherical symmetry is caused by the dipole moment that each ion induces on the other. Atoms realizing this strong bonding usually give rise to crystal phases without first passing through a molecular arrangement. This is typical of halogens of alkyl metals such as NaCl or LiF and occurs between atoms of group I and group VII of the periodic table. The halogen atom has seven electrons in the sp shell, missing only one electron to close the shell; on the other hand, alkyl metals have only one electron in the outermost shell which will be removed and added to the halogen atom generating the two ions (e.g. Na^+ and Cl^-) which form the strong electrostatic bonding. It is worth noting that the positive ion having an excess positive charge will attract its inner electrons in a stronger way to produce electronic clouds with lower average distances from the nucleus. The opposite will occur in the negative ion, where electrons will be characterized by higher average distances.

In these systems, ions form an ordered structure where each of them has six ions of opposite sign close to it (*nearest neighbors*) and a face-centered cubic crystal lattice (see Section 3.9). The structure is determined by ion arrangement maximizing the attractive force that balances the short range repulsive one. Actually, in such a structure, the interaction takes place among several ions, even if the most relevant effect is due to nearest neighbors. In this case, the calculation of the interaction potential must take into account a large number of pairs of interacting atoms (attractive between opposite sign ions; repulsive between same sign ions).

Considering $Z = 1$, the interaction potential between ion i and ion j at distance r_{ij} becomes

$$\mathcal{V}_{ij}(r) = Ae^{-r_{ij}/\rho} \pm \frac{e^2}{4\pi\epsilon_0 r_{ij}}, \tag{3.87}$$

using the *Born–Mayer* expression for the repulsive potential; the choice of the sign in the second term depends on the sign of interacting ions. Therefore, the total potential energy of each ion is

$$\mathcal{V}_i(r) = \sum_j \mathcal{V}_{ij}(r). \tag{3.88}$$

By neglecting the surface effects where ions are surrounded by a lower number of different ions, the total potential energy of the crystal can be written as

$$\mathcal{V}_{\text{tot}}(r) = N\mathcal{V}_i(r), \tag{3.89}$$

with N being the number of pairs of ions of opposite sign. It corresponds to the energy which is to be provided to "melt" the lattice.

Including Eqs. (3.87) and (3.88) in Eq. (3.89), one gets

$$\mathcal{V}_{\text{tot}}(R_0) = -\left[N\alpha\frac{e^2}{4\pi\epsilon_0 R_0}\right]\left(1 - \frac{\rho}{R_0}\right), \tag{3.90}$$

where R_0 is the equilibrium distance defined by $(\frac{\partial\mathcal{V}_{\text{tot}}}{\partial R})_{R=R_0} = 0$ and R is the nearest neighbor distance. The number α is dependent on the crystalline structure and determines the main contribution to $\mathcal{V}_{\text{tot}}$ given by the square brackets term called the *Madelung energy*; in fact usually $\rho \ll R_0$, and thus the second term on the right-hand side of Eq. (3.90) is ≈ 1.

In order to get Eq. (3.90), we define R as the distance between the nearest neighbors, and therefore the distances between the interacting ions can be written as $r_{ij} = p_{ij}R$, with p_{ij} being integers characteristic of the crystal lattice. By taking into account the repulsive term only for the nearest neighbors, we have

$$\mathcal{V}_{\text{tot}}(r) = N \sum_j \left\{ Ae^{-r_{ij}/\rho} \pm \frac{e^2}{4\pi\epsilon_0 r_{ij}} \right\}$$

$$= N \left\{ Ane^{-R/\rho} \pm \sum_j \frac{e^2}{4\pi\epsilon_0 p_{ij} R} \right\}$$

$$= N \left\{ Ane^{-R/\rho} - \alpha \frac{e^2}{4\pi\epsilon_0 R} \right\},$$

where n is the number of nearest neighbors and

$$\alpha = \sum_j \left(\pm \frac{1}{p_{ij}} \right)$$

is the *Madelung constant* of the crystal lattice. The equilibrium distance R_0 is found by setting

$$\left(\frac{\partial \mathcal{V}_{\text{tot}}}{\partial R} \right)_{R=R_0} = 0.$$

Then

$$\frac{d}{dR} \left\{ Ane^{-R/\rho} - \alpha \frac{e^2}{4\pi\epsilon_0 R} \right\} = 0 \rightarrow -\frac{An}{\rho} e^{-R_0/\rho} + \alpha \frac{e^2}{4\pi\epsilon_0 R_0^2} = 0$$

$$\rightarrow Ane^{-R_0/\rho} = \alpha\rho \frac{e^2}{4\pi\epsilon_0 R_0^2},$$

and then

$$\mathcal{V}_{\text{tot}}(R_0) = N \left\{ \alpha\rho \frac{e^2}{4\pi\epsilon_0 R_0^2} - \alpha \frac{e^2}{4\pi\epsilon_0 R_0} \right\} = - \left[N\alpha \frac{e^2}{4\pi\epsilon_0 R_0} \right] \left(1 - \frac{\rho}{R_0} \right).$$

The *cohesion energy* of each pair of ions is defined as the total energy divided by the number (N) of pairs:

$$E_{\text{coh}} = \frac{\mathcal{V}_{\text{tot}}(\bar{R}_0)}{N} = - \left[\alpha \frac{e^2}{4\pi\epsilon_0 R_0} \right] \left(1 - \frac{\rho}{R_0} \right) \qquad (3.91)$$

Of course, the crystalline structure is more and more stable for increasing values of E_{coh}, which will produce increasing values of the fusion temperature. Accordingly, the ionic bonding is a strong and long-range bond giving rise to structures stable from the mechanical and chemical points of view.

Covalent bond

There is another type of strong bonding between atoms occurring when the electronegativity difference between the interacting atoms is low or even zero. It involves neutral atoms instead of ions and has a preferred direction corresponding to the overlapping direction of the atoms wave functions; nevertheless, it gives rise to cohesion energies of the same order of the ionic bond. It occurs when the atomic shells are open; then the Pauli principle does not give rise to repulsive forces because electronic wave functions can be overlapped without promotion of electrons to higher energy levels. In this way, electron clouds are delocalized on all the atoms participating in the bonding. The typical case is that of elements of the IV group with four valence electrons (such as C, Si or Ge) then with the external (sp) shell half-filled: they miss just four electrons to close the shell according to the octet rule. Bonding is realized by delocalization of the valence electrons around the participating atoms. Again, we have an electrostatic interaction between electrons and nuclei, but its description is more complex with respect to the ionic bonding case.

In order to better understand the nature of the covalent bond, we consider the simple case concerning the formation of the H_2 molecule. In this case, each atom has one electron in the $1s$ state. When the electronic clouds overlap, each electron may occupy the $1s$ state of the other atom without violation of the Pauli principle, if its spin is opposite to that of the other electron. Then each electron feels a stronger electrostatic interaction, since it is due to the two positive charges of the nuclei; as a consequence, this electron sharing leads to lowering of the system's total energy. On the contrary, if electrons have the same spin they cannot share the same state and the overlapping must lead to an increase in the total energy with a repulsive effect.

Let us focus on the effect on each electron by considering the H_2^+ ion where one electron lies in the field of two single proton positive nuclei. This example helps in understanding the property of the electron in the case of a possible bonding between two atoms. We follow the Hund–Mulliken methods of *molecular orbitals*. The system Hamiltonian is

$$\mathcal{H} = -\frac{\hbar^2}{2m}\nabla^2 - \frac{e^2}{4\pi\epsilon_0 r_A} - \frac{e^2}{4\pi\epsilon_0 r_B} + \frac{e^2}{4\pi\epsilon_0 R}, \tag{3.92}$$

where r_A and r_B are the distances of the electron from each nucleus and R their distance, and the last term represents the repulsive electrostatic interaction between them. Of course, the problem must be solved through the Schrödinger equation, which allows finding the system wave functions and the corresponding energy levels; this treatment can be found in specialized texts on molecular physics. Here we will discuss some main features that allow understanding the formation of a binding state, following a method that can be applied also to more complex molecular systems.

The electron wave functions in H_2^+ are expected to have a large amplitude around the nuclei, where the electrostatic interaction is bigger, in the same way as for a single atom; therefore, they must give rise to a probability of finding the electron near the positive charges higher than in between them. The lowest level of approximation (*LCAO* — *linear combination of atomic orbitals*) leads to trying an electron wave function as a linear combination of the atomic wave functions ψ_A and ψ_B, either symmetric,

$$\psi_s = \frac{1}{\sqrt{2}}\{\psi_A + \psi_B\}, \tag{3.93}$$

or antisymmetric,

$$\psi_c = \frac{1}{\sqrt{2}}\{\psi_A - \psi_B\}, \tag{3.94}$$

with respect to exchange of ψ_A and ψ_B, where $1/\sqrt{2}$ is the normalization factor.

The electron wave function ψ_A and ψ_B are sketched in Fig. 3.8.

The corresponding probability densities are shown in Fig. 3.9; they are similar, but with a strong qualitative difference. In fact, comparing $|\psi_s|^2$ to $|\psi_a|^2$, we observe that for the symmetric case the probability of finding the electron in between the nuclei is higher than for the antisymmetric case, where there is a minimum in between them. Since the electron between the nuclei gives rise to shielding of their repulsive interactions, we expect this configuration to help the bonding effect with respect to the other. Therefore, we expect a lower energy associated with ψ_s with respect to the

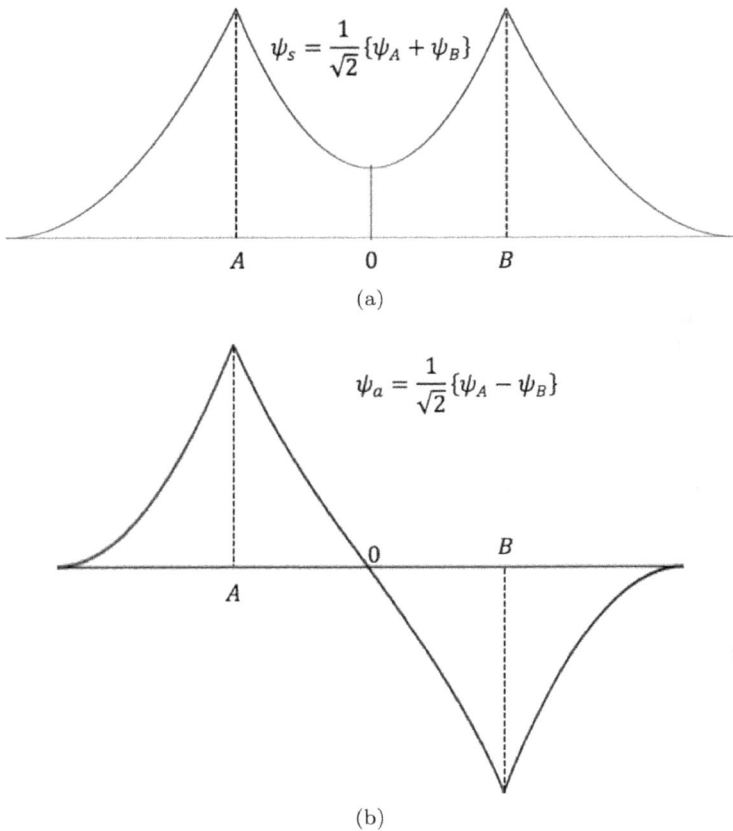

$$\psi_s = \frac{1}{\sqrt{2}}\{\psi_A + \psi_B\}$$

A   0   B

(a)

$$\psi_a = \frac{1}{\sqrt{2}}\{\psi_A - \psi_B\}$$

0   B

A

(b)

Fig. 3.8.   Electron wave functions (a) symmetric and (b) antisymmetric in the molecules H_2^+, sketched versus the position of the atomic nuclei.

one associated with ψ_a. This can be easily checked, by comparing these energy values to those of the single atom.

For the symmetric wave function, we can write the expectation value as

$$E_s = \int_{-\infty}^{\infty} \psi_s^* \mathcal{H} \psi_s dV$$

$$= \frac{1}{2}\left\{ \int_{-\infty}^{\infty} \psi_A^* \mathcal{H} \psi_A dV + \int_{-\infty}^{\infty} \psi_B^* \mathcal{H} \psi_B dV \right.$$

$$\left. + \int_{-\infty}^{\infty} \psi_A^* \mathcal{H} \psi_B dV \int_{-\infty}^{\infty} \psi_B^* \mathcal{H} \psi_A dV \right\}. \qquad (3.95)$$

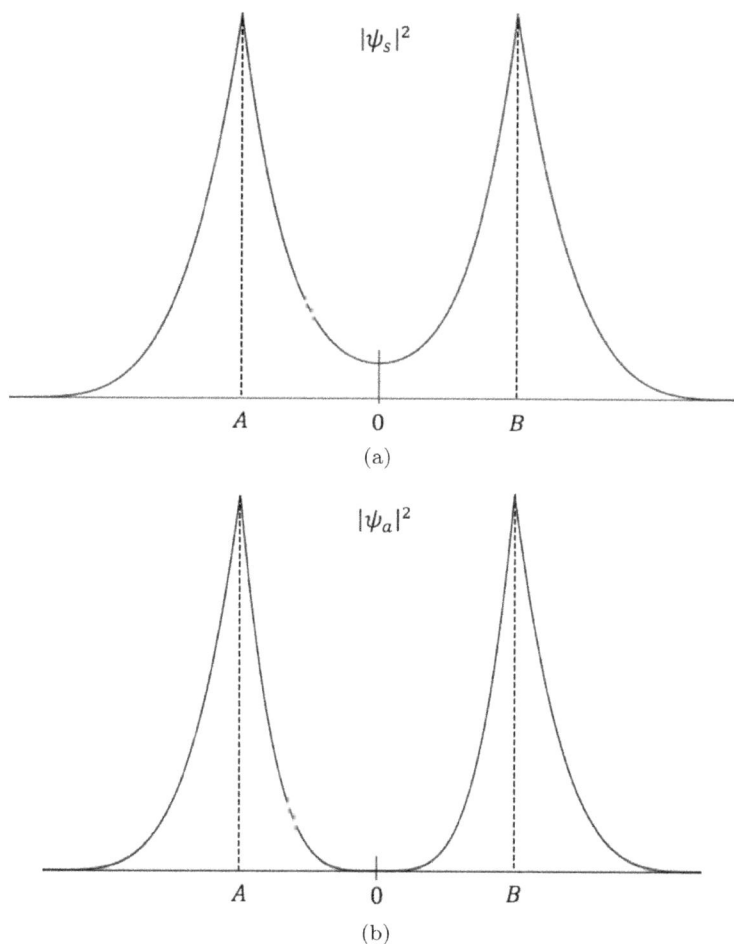

Fig. 3.9. Probability density for wave functions (a) symmetric and (b) antisymmetric in the molecules H_2^+, sketched versus the position of the atomic nuclei.

Since ψ_A and ψ_B are equivalent, the first term is equal to the second one and the third term is equal to the fourth one. Then

$$E_s = \int_{-\infty}^{\infty} \psi_A^* \mathcal{H} \psi_A dV + \int_{-\infty}^{\infty} \psi_A^* \mathcal{H} \psi_B dV. \qquad (3.96)$$

The first term is known as the *atomic or Coulomb integral* and corresponds to the energy of the single atom, while the second term is known as the *resonance integral* and is due to the overlapping of the orbitals of the two atoms.

Let us write Eq. (3.96) as

$$E_s = \alpha + \beta. \tag{3.97}$$

By repeating the calculation for the antisymmetric wave function, we will get

$$E_a = \alpha - \beta. \tag{3.98}$$

Therefore, the difference in energy between the molecular and atomic states is just

$$\Delta E_s = E_s - \alpha = \beta \tag{3.99}$$

for the symmetric wave function and

$$\Delta E_a = E_a - \alpha = -\beta \tag{3.100}$$

for the antisymmetric wave function.

The expressions (3.99) and (3.100) are the result of strong approximations. In fact, by writing Eqs. (3.93) and (3.94) with a normalization factor $1/\sqrt{2}$, we have neglected the wave function overlapping. A more general definition could be written as

$$\psi_{s,a} = \{c_A \psi_A \pm c_B \psi_B\},$$

with the expectation value of energy given by

$$E_{s,a} = \frac{\int \psi_{s,a}^* \mathcal{H} \psi_{s,a}}{\int \psi_{s,a}^* \psi_{s,a}}.$$

The straightforward calculation can be found in any textbook of molecular physics. It leads to

$$\psi_{s,a} = \frac{1}{\sqrt{2(1 \pm S)}} \{\psi_A \pm \psi_B\},$$

with the corresponding energies

$$E_{s,a} = \frac{\alpha \pm \beta}{1 \pm S},$$

where

$$S = \int_{-\infty}^{\infty} \psi_A \psi_B dV.$$

Anyway, since $S < 1$ the qualitative discussion that follows keeps its validity.

The quantum-mechanical calculation can show that $\beta < 0$. Therefore, the symmetric state corresponds to an energy lower than in the single atom ($E_s < \alpha$) and gives rise to a *bonding* effect, while the antisymmetric state leads to an energy higher than in the single atom ($E_a > \alpha$) and gives rise to an *antibonding* state. In accordance with that, ψ_s and ψ_a are called bonding and antibonding orbitals. Of course, other bonding and antibonding orbitals originate from the overlapping of excited atomic orbitals. The energy diagram for the discussed case is given in Fig. 3.10.

We must underline the quantum origin of the bond, since it comes out from the wave function overlapping represented by β. Another aspect to be noted is that the molecular bonding of two equal atoms having the same energy levels leads to splitting these levels into two different states of lower and higher energy, respectively.

The presented approach is very approximate, but it makes understandable the origin of the covalent bond and it can be generalized to more complex systems, even if obvious theoretical refinements will be necessary. Moreover, the representation of the molecular orbitals in space is useful for figuring out the corresponding symmetries. Considering the example

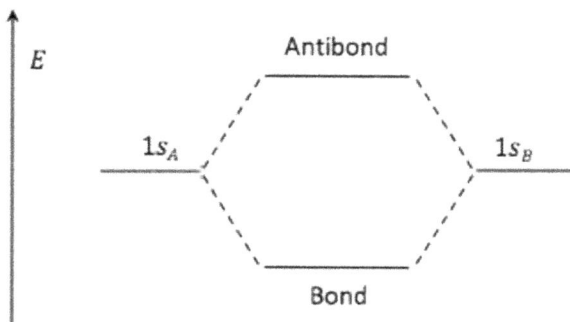

Fig. 3.10. Energy diagram showing the formation of molecular states from the $1s$ state of the atomic hydrogen.

Table   3.3.   Correspondence   between
atomic and molecular orbitals.

Atomic orbitals	s	p	d	f
Molecular orbitals	σ	π	δ	ϕ

discussed above which is related to the $1s$ state of the hydrogen atom, we observe the molecular orbitals to have a cylindrical symmetry along the line connecting the two atoms. These molecular counterparts of the s atomic orbitals are called σ orbitals. The symmetric bonding state is indicated by a subscript g, while the antisymmetric antibonding state is indicated by a subscript u. Then the two molecular orbitals are σ_g and σ_u.

Following similar arguments, we get: molecular orbitals π with a nodal plane along the line connecting the two nuclei originated by atomic p states; molecular orbitals δ with two nodal planes along the line connecting the two nuclei originated by atomic d states; etc.

The denomination correspondence is shown in Table 3.3.

It is a general feature of bonding that one state of the single atom in the molecule corresponds to a number of states equal to the number of interacting atoms in the molecule, some of them having a bonding character and the others an antibonding one. In the case of a large number of shared electrons, we get the energy bands, which will be described in Chapter 4.

Obviously, several bonds among atoms have both ionic and covalent features. For instance, atoms of the III and V groups (such as Ga and As) have three and five valence electrons, respectively. Then the occurrence of a covalent bonding necessary for realizing a closed shell around each atom leads to an asymmetric charge distribution around the nuclei: the average charge on the atom of valence 3 will be negative with respect to that on the atom with valence 5, thus inducing an electrostatic interaction between them.

An approach similar to the one discussed above for describing the covalent bond can be used for polyatomic molecules with a degree of complexity that is higher and higher by increasing the number of interacting atoms. The starting point will be a linear combination of atomic wave functions to get the molecular orbitals through resolution of a number of equations increasing with the number of atoms.

We must underline that the covalent bond leads to bonding energy of the same order as that of ionic bonding. It is worth noting that the covalent bond leads to molecules able to form condensed phases under suitable thermodynamic conditions. In this case, one finds an interatomic strong covalent bond and an intermolecular bond weaker than the former one, and therefore the phase transition from the condensed to the "gas" phase does not lead to molecular dissociation. On the other hand, it is also possible to have the formation of crystals from atoms via covalent bonding without the preliminary formation of molecules; this happens for semiconductors like Si, Ge and GaAs.

Metallic bond

This can be considered a special case of the covalent bond where a few links between one atom and its neighbors are missing and the corresponding electrons are delocalized along the macroscopic dimension of the material. Therefore, in a crystal, we may have fixed positive ions located in the lattice sites and free electrons in the crystal volume. As a consequence, we have electrostatic interaction between the positive ions and the cloud of delocalized electrons. The metallic bond occurs in elements with 1, 2 or 3 valence electrons that are not sufficient to fill all the atomic orbitals. Of course, the directional character typical of the covalent bond is lost in this case, and therefore metallic bonding has spherical symmetry and leads to condensed phases rather than molecules.

The cohesion energy (and the fusion temperature) increases with the number of valence electrons of the interacting atoms.

van der Waals bond

When atoms have spherical symmetry and closed external shells (full octect), neither an ionic nor a covalent bond is possible. However, the dynamical character of the spherical symmetry due to the electronic distribution around the atomic nucleus leads to local fluctuations of the charge density that can give rise to a weak electric dipole. An electric dipole p_1 will be able to induce on the close atom an electric dipole proportional to the originated electric field:

$$p_2 \propto \frac{p_1}{r^3}. \tag{3.101}$$

Since the interaction energy between two dipoles is proportional to $p_1 p_2 / r^3$, we get an attractive potential energy:

$$\mathcal{V} = -\frac{A}{r^6}. \tag{3.102}$$

When this weak bonding is able to give rise to a condensed state (e.g. liquid helium), the potential energy of the system can be written as

$$\mathcal{V} = C \sum_{ij} \left(\frac{B}{r_{ij}^{12}} - \frac{A}{r_{ij}^{6}} \right), \tag{3.103}$$

where the Lennard–Jones repulsive potential has been used and summation is performed over all the couples of interacting atoms. This potential is responsible for the formation of the so-called *molecular crystals*, characterized to be soft and with a low melting point.

Hydrogen bond

An electronic distribution around atoms, which are part of a molecule, without spherical symmetry gives rise to a permanent dipole that is obviously able to induce an interaction with neighbor molecules either by inducing another dipole with a mechanism similar to the one described above (*dipole-induced dipole interaction*) or by direct coupling with the dipole of the other molecule (*dipole–dipole interaction*). These kinds of interactions have an important role in the formation of the condensed phase of several materials. A very important case is the dipolar bonding involving hydrogen, called the *hydrogen bond*. It is consequent to the formation of a covalent bonding between different atoms, like the one occurring in the molecule H_2O. In fact, in this case, each atom of hydrogen gives one electron to the oxygen atom in order to allow it to close the external shell. In this way, two dipoles appear owing to the uneven electronic distribution over the molecules. The dipoles are oriented from the O atom (negative) to each H atom (positive). The hydrogen bonding is the one linking one H atom to the O atom of the next molecule and it is responsible for the formation of the condensed phase, thus connecting all the H_2O molecules, as shown in Fig. 3.11.

The hydrogen bond is much stronger than the van der Waals bond, since it is originated by permanent dipoles. It plays a fundamental role in the structure of DNA and other biomolecules.

3.6. Energy levels in molecules

The example of the ionized hydrogen molecule presented in the previous section to explain the nature of the covalent bond has shown how the energy levels of the valence electron are modified when the molecule is formed. Similar concepts are valid in all cases.

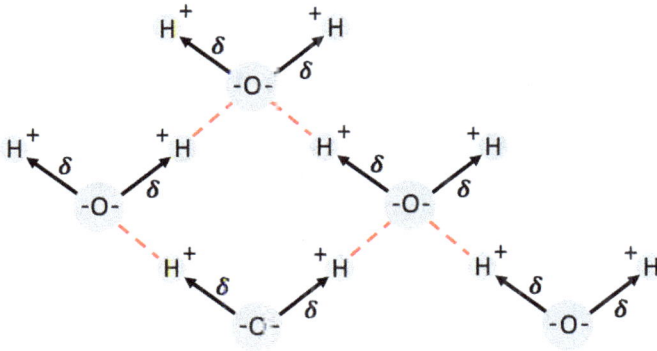

Fig. 3.11.   Hydrogen bonding in the condensed phase of H_2O.

However, we have taken into account only the electronic energy, and not that of the whole system, so we must also consider the motion of the positive nuclei. This motion is not negligible and produces a correction to the energy even if the main contribution comes from the electronic levels. The problem is generally simplified owing to the different time scale of the motion of electrons and nuclei, which allows a separate treatment. Simple classical arguments can be used to support this assumption, known in quantum mechanics as the Born–Oppenheimer approximation. In the following, we consider a diatomic molecule where the average distance between the two nuclei is a. This is also the indetermination in the electron localization, and therefore its momentum will be of the order of $p \approx \hbar/a$; then the kinetic energy of the electron is $E_e = \hbar^2/2m_ea^2$. Using $a \approx 1$ Å and the values $\hbar = 1.06 \cdot 10^{-34}$ J·s and $m_e = 9.1 \cdot 10^{-31}$ Kg, we find the order of magnitude of the electronic energy to be in the range of one or a few electron volts. This confirms what has already been pointed out in the discussion of the hydrogen atom, i.e. transitions between states with different principal quantum numbers correspond to photons in the visible or near-UV region of the electromagnetic spectrum.

Besides the translation corresponding to a macroscopic motion of the material that includes the molecule, we are interested in possible vibrations and rotations of nuclei with respect to the center of mass. Taking into account that the elastic force linking nuclei to electrons has the same value and treating electrons and nuclei as harmonic oscillators, we can write the angular frequency of vibration for the electron as $\omega_e = \sqrt{k/m_e}$ and the one for nucleus as $\omega_N = \sqrt{k/M_N}$, and thus the corresponding ratio between the vibrational energies is $\hbar\omega_N/\hbar\omega_e = \sqrt{m_e/M_N}$. Since the ratio between the masses is in the range of 10^{-3}–10^{-5}, the ratio between the vibrational frequencies and energies is in the range of 0.03–0.003, i.e. a nuclear vibration is about two orders of magnitude slower than that of electrons. Therefore, the energy of a nuclear vibration is in the range of 10^{-2} eV, corresponding to transitions in the IR region of the electromagnetic spectrum.

In order to estimate the rotational motion of a diatomic molecule, we just consider the case of two equal masses rotating around the center of mass located at a distance $a/2$ from each nucleus. In this case, the moment of inertia of the system is[8] $I = M_N a^2/8 + M_N a^2/8 = M_N a^2/4$, while the rotational energy of the rigid rotator can be written as $E_R = I\omega^2/2 = L^2/2I$, with L being the angular momentum. Thus, taking $\hbar$ as the order of magnitude of the angular momentum, we have $E_R \approx 2\hbar^2/M_N a^2 \approx 4(m_e/M_N)E_e$. This results in the range of 10^{-3}–10^{-5} eV corresponding to transitions in the far-IR or microwave region. Since the period of motion is of the order of $\hbar$ divided by the energy, this rotation is much slower than the electronic motion.

Based on these arguments, the Born–Oppenheimer approximation allows solving the Schrödinger equation for the wave function ψ of the molecule by separating electronic and nuclei wave functions in such a way that

$$\psi(\mathbf{R}; r_1, \ldots, r_N) = \sum_q F_q(\mathbf{R})\phi_q(\mathbf{R}; r_1, \ldots, r_N), \qquad (3.104)$$

where ϕ_q are the electronic wave functions parametrically dependent on the intermolecular coordinate $\mathbf{R}$ (held fixed during the calculation) and $F_q(\mathbf{R})$ are the wave functions representing the nuclear response when the electronic system is in the state q.

For a diatomic molecule, the complete Schrödinger equation would be

$$\left[-\frac{\hbar^2}{2\mu}\nabla^2 - \frac{\hbar^2}{2m_e}\sum_i \nabla_i^2 + \mathcal{V} \right]\psi = E\psi. \qquad (3.105)$$

The first term is the kinetic energy operator of nuclei where $\mu = M_A M_B/(M_A + M_B)$ is the reduced mass of the two nuclei; the second term is the kinetic energy operator for all the valence electrons and the potential energy $\mathcal{V}$ includes three terms: the Coulomb attraction between electrons and nuclei, the Coulomb repulsion between electrons, and the Coulomb repulsion between nuclei.

---

[8]Each mass contributes to the moment of inertia with a term: $(M_N/2)a^2/4$.

The detailed solution of this problem can be found in textbooks dealing with molecular physics and molecular spectroscopy; it is obviously of increasing complexity when polyatomic molecules are considered. We just summarize here the main features of the resulting energy states for a diatomic molecule.

The total energy should be written as the sum of the electronic energy $E_e(R_0)$ calculated at the equilibrium distance between the two nuclei, the vibrational energy E_v and the rotational energy E_r of the nuclei:

$$E = E_e(R_0) + E_v + E_r. \tag{3.106}$$

A good representation of the electron energy versus R is given by

$$E_e(R) = E_e(\infty) + \mathcal{V}(R), \tag{3.107}$$

where

$$\mathcal{V}(R) = [E_e(\infty) - E_e(R_0)][e^{-2\alpha(R-R_0)} - 2e^{-\alpha(R-R_0)}],$$

the Morse potential, and $E_e(\infty)$ corresponds to free atoms.

The vibrational energy is the one coming from the solution to the harmonic oscillator problem sketched in Chapter 2, and therefore

$$E_v = \left(v + \frac{1}{2}\right)\hbar\omega_0, \quad \text{where } \omega_0 = \sqrt{\frac{k}{\mu}} \quad \text{or} \quad \nu_0 = \frac{1}{2\pi}\sqrt{\frac{k}{\mu}}, \tag{3.108}$$

where $v = 0, 1, 2, \ldots$. The quantity

$$D_e = E_e(\infty) - E_e(R_0) \tag{3.109}$$

is the difference between the energy at infinite distance and the energy at the minimum of the Morse curve. Since the minimum energy of the molecule is $E_e(R_0) + (1/2)h\nu_0$, the quantity

$$D_0 = D_e - \frac{1}{2}h\nu_0 \tag{3.110}$$

is the *dissociation energy* of the molecule, i.e. the energy required to break the bonding between the linked atoms. Both ν_0 and D_e are very important parameters characteristic of each molecular bonding and need to be taken into account in several applications related to interaction of electromagnetic waves and matter.

The rotational energy can be calculated considering the molecule as a rigid rotator around the axis of symmetry, and therefore

$$E_r = \frac{L^2}{2I} = \frac{\hbar^2 J(J+1)}{2\mu R_0^2} \equiv BJ(J+1), \quad J = 1, 2, 3, \dots, \qquad (3.111)$$

where we have used the eigenvalues of L^2 calculated when dealing with the hydrogen atom, but we employ a different symbol to represent the related quantum number since now it applies to the whole molecule rather than the single atom. We have also defined the rotational constant of the molecule as

$$B = \frac{\hbar^2}{2\mu R_0^2} = \frac{\hbar^2}{2I_0}. \qquad (3.112)$$

The energy of the molecule versus R is reported in Fig. 3.12, where $E(R)$ is given for two values of the principal quantum number of the electronic atomic level, indicated as the "ground state" and the "excited state."

We observe that in the excited state the equilibrium distance R_0 is bigger because the atoms are more loosely bound. In general, by increasing the electronic excitation this curve becomes shallower and broader until the value $E_e(\infty)$ is reached, leading to molecule dissociation.

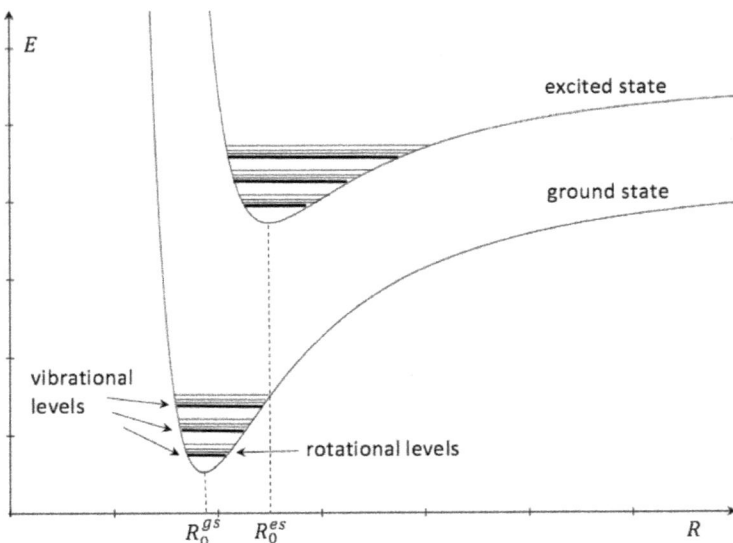

Fig. 3.12.   Energy levels of a diatomic molecule versus the intermolecular distance R. For each value of $E(R)$, a set of vibrational levels is found and a set of rotational levels for each vibrational level. R_0^{gs} is the equilibrium distance in the ground state and R_0^{es} the one in the excited state.

We have already pointed out that the energy range related to pure electronic, vibrational and rotational transitions is very different, and therefore they involve very different parts of the electromagnetic spectrum. Now we are able to make some additional remarks on the information that can be retrieved by a molecular spectrum.

Let us consider the change in energy ΔE_r between two successive rotational states:

$$\Delta E_r = E_r^{J+1} - E_r^J$$
$$= B\{(J+1)(J+2) - J(J+1)\} = 2B(J+1). \qquad (3.113)$$

Thus, the quantity $2B$ determines the minimum change in rotational energy (occurring when $J = 0$):

$$\Delta E_r^{\min} = 2B = \frac{\hbar^2}{\mu R_0^2} = \frac{\hbar^2}{I_0}. \qquad (3.114)$$

Then the analysis of rotational spectra gives a measurement of the momentum of inertia $I_0 = \mu R_0^2$ of the molecule and an estimation of the equilibrium distance R_0 if the ion masses are known.

The expression (3.114) allows evaluating the frequency and the wavelength of photons connected with transitions between pure rotational states. Such transitions are possible only for molecules with a permanent dipole moment, such as HCl, HF or LiH. The frequency of photons exciting the molecules to a higher energy rotational state or generated through decay from a higher to a lower energy rotational state is

$$h\nu_r = \Delta E_r^{\min} = \frac{\hbar^2}{\mu R_0^2} \rightarrow \nu_r = \frac{\hbar}{2\pi\mu R_0^2} \rightarrow \lambda_r = \frac{2\pi c\mu R_0^2}{\hbar}. \qquad (3.115)$$

As an example, let us consider hydrogen chloride (HCl) with $R_0 = 1.27$ Å and $\mu = 1.63 \cdot 10^{-27}$ Kg, where $c = 3.0 \cdot 10^8$ m/s and $\hbar = 1.06 \cdot 10^{-34}$ J $\cdot$ s. We get $\nu_r \approx 6.4 \cdot 10^{11}$ s^{-1} and $\lambda_r \approx 0.047$ m $= 4.7$ cm. The corresponding photon energy is $h\nu_r \approx 4.2 \cdot 10^{-22}$ J $\approx 2.6 \cdot 10^{-3}$ eV.

On the other hand, from Eq. (3.108) we get the minimum energy change between two vibrational states as

$$E_v^{v+1} - E_v^v = \hbar\omega_0 = h\nu_0 = \hbar\sqrt{\frac{k}{\mu}}. \qquad (3.116)$$

Taking the elastic constant of the HCl molecule as $k \approx 480$ N/m, we have $\nu_0 \approx 8.6 \cdot 10^{13}$ s^{-1}, with the corresponding wavelength $\lambda_0 \approx 3.5$ μm and photon energy $h\nu_0 = 3.6 \cdot 10^{-1}$ eV.

Wave number

In molecular spectroscopy, the quantity commonly used to measure the frequency of the rotovibrational transitions is the *wave number*, defined as the inverse of the corresponding wavelength,

$$\tilde{\nu} \equiv \frac{1}{\lambda} = \frac{\nu}{c},$$

usually measured in the unit of cm^{-1}. Referring to the estimations made for the molecule of HCl, one gets the vibrational frequency of the molecule $\tilde{\nu}_0 = \nu_0/c \approx 2.9 \cdot 10^3$ cm^{-1} and the change of rotational frequency $\tilde{\nu}_r = \nu_r/c \approx 2.1 \cdot 10^1$ cm^{-1}.

As a matter of fact, the usual way of proceeding is to get information on the molecule's properties (μ, k, R_0) using absorption spectroscopy in the infrared. We will discuss in the next section the statistical distribution of molecular energy at a given temperature; we just anticipate here that at room temperature we have molecules only in the lowest energy vibrational state, while different rotational states are generally filled. For a diatomic molecule, the allowed transitions need to correspond to changes of 1 for the vibrational number v and ± 1 for the rotational number J. If there is no change of the electronic state (i.e. keeping the same principal quantum number n), the photon absorption may involve two different transitions, starting from the energy

$$E_0^J = \frac{1}{2}h\nu_0 + BJ(J+1), \tag{3.117}$$

either to

$$E_1^{J+1} = \left(\frac{1}{2}+1\right)h\nu_0 + B(J+1)(J+2) \tag{3.118}$$

or to

$$E_1^{J-1} = \left(\frac{1}{2}+1\right)h\nu_0 + B(J-1)J. \tag{3.119}$$

In this way, we may have transitions for the photon energies

$$E_1^{J+1} - E_0^J = h\nu_0 + 2B(J+1), \tag{3.120}$$

$$E_1^{J-1} - E_0^J = h\nu_0 - 2BJ, \tag{3.121}$$

with the corresponding photon frequencies

$$\nu_{J \to J+1} = \nu_0 + \frac{2B}{h}(J+1), \tag{3.122}$$

$$\nu_{J \to J-1} = \nu_0 - \frac{2B}{h}J. \tag{3.123}$$

As a consequence, the rotovibrational absorption spectrum consists of two groups of lines: the *P branch* for transitions with $\Delta J = -1$ and the *R branch* for transitions with $\Delta J = +1$ (when transitions with $\Delta J = 0$ are possible, we speak of the *Q branch*). The lines are separated by the quantity $2B/h$ except in the central part of the spectrum, where the separation is $4B/h$; in fact, from Eqs. (3.122) and (3.123) we have

$$\nu_{J=0 \to J=1} = \nu_0 + \frac{2B}{h}, \quad \nu_{J=1 \to J=0} = \nu_0 - \frac{2B}{h}. \tag{3.124}$$

Therefore, the central frequency between these two lines is the vibrational frequency ν_0.

A typical rotovibrational absorption spectrum of a molecule[9] is shown in Fig. 3.13, and rotovibrational constants of some molecules are reported in Table 3.4. In the figure, we notice that the peak separation is not exactly constant, increasing slightly at higher frequencies. This happens because an increase in the rotational number J leads to an increase in

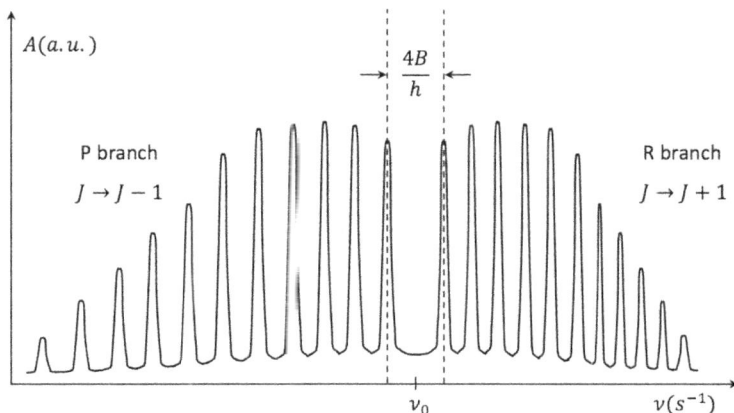

Fig. 3.13.   Typical absorption spectrum due to rotovibrational transitions of a diatomic molecule. ν_0 is the central vibrational frequency.

---

[9]See for instance: G. Hertzberg, *Molecular Spectra and Molecular Structure* — I. Spectra of Diatomic Molecules (D. Van Nostrand, 1950).

Table 3.4.   Rotovibrational constants of some diatomic molecules.

Molecule	R_0 (Å)	D_0 (eV)	ν_0 (cm^{-1})	B (eV)
H_2	0.74	4.48	4395	$7.6 \cdot 10^{-3}$
N_2	1.09	9.75	2360	$2.5 \cdot 10^{-4}$
O_2	1.21	5.08	1580	$1.8 \cdot 10^{-4}$
HCl	1.27	4.43	2990	$1.3 \cdot 10^{-3}$
NaCl	2.51	4.22	380	$2.4 \cdot 10^{-5}$

the atom separation, thus increasing the molecular moment of inertia. The consequence is a decrease in the factor B.

3.7. Thermodynamics and phase transitions

The physical properties of matter on a macroscopic scale involve a large number of "particles" (namely electrons or atoms or molecules). This fact makes necessary a statistical approach that allows evaluating the average mechanical properties of the particle, linking them to the thermodynamic parameters describing the state of the system as temperature T, pressure P and volume V. Therefore, the equilibrium conditions of the system of many particles must fulfill the laws of thermodynamics based on the definitions of the extensive[10] state functions U (the internal energy) and S (the entropy). Actually, in addition, combinations of U and S are often used, depending on the specific system transformations under investigation; these are:

$$
\begin{array}{lll}
\text{the enthalpy:} & H = U + PV; & \\
\text{the free energy:} & F = U - TS; & (3.125) \\
\text{the Gibbs free enthalpy:} & G = U - TS + PV. &
\end{array}
$$

The benefit of defining these quantities is evident by considering infinitesimal (reversible) transformations such as

$$dH = dU + PdV + VdP = dQ + VdP = TdS + VdP,$$

$$dF = dU - TdS - SdT = -PdV - SdT, \qquad (3.126)$$

$$dG = dU - TdS - SdT + PdV + VdP = -SdT + VdP,$$

---

[10]The thermodynamic parameters are categorized as intensive and extensive quantities. Intensive parameters (such as temperature T, pressure P, density ρ) characterize the local properties of the system and do not depend on the size of the system. Extensive parameters (such as mass M, volume V, internal energy U) are additive quantities; their value is determined by the sum of the values they have in different parts of the system.

where we have used the condition of the *first principle of thermodynamics*:

$$dU = dQ - dL = TdS - PdV, \tag{3.127}$$

with L being the mechanical work and Q the exchanged heat.

Indeed, depending on the kind of transformation analyzed, a suitable *thermodynamic potential* can be minimized. It is quite useful to recall that this denomination comes from the property of finding the thermodynamic parameters by differentiating the thermodynamic potentials with respect to a specific variable, keeping the other constant. Then we have

$$T = \left(\frac{\partial U}{\partial S}\right)_V, \quad P = -\left(\frac{\partial U}{\partial V}\right)_S \tag{3.128}$$

and

$$T = \left(\frac{\partial H}{\partial S}\right)_P, \quad V = \left(\frac{\partial H}{\partial P}\right)_S,$$

$$P = -\left(\frac{\partial F}{\partial V}\right)_T, \quad S = -\left(\frac{\partial F}{\partial T}\right)_V, \tag{3.129}$$

$$V = \left(\frac{\partial G}{\partial P}\right)_T, \quad S = -\left(\frac{\partial G}{\partial T}\right)_P,$$

in the same way as the mechanical force is found by differentiating the potential energy of a mechanical system. As mentioned, the equilibrium is easily found by minimization of a suitable thermodynamic potential. For instance, for a system with constant volume and constant entropy ($dV = 0$; $dS = 0$), the equilibrium condition according to Eq. (3.127) is $dU = 0$, while if the equilibrium occurs at constant temperature and volume ($dV = 0$; $dT = 0$), from Eqs. (3.126) we have $dF = 0$.

The expressions given above can be generalized to multicomponent systems, where the concentration of each component is allowed to change. In this case, we have an additional term in the mechanical work proportional to the variations of the number of individual components dn_i of the system; in this way, Eq. (3.127) becomes

$$dU = TdS - PdV + \sum_i \mu_i dn_i, \tag{3.130}$$

where μ_i is the *chemical potentials* of the i component. Therefore,

$$\mu_i = \frac{\partial U}{\partial n_i}, \tag{3.131}$$

and similar expressions hold for the other thermodynamic potentials.

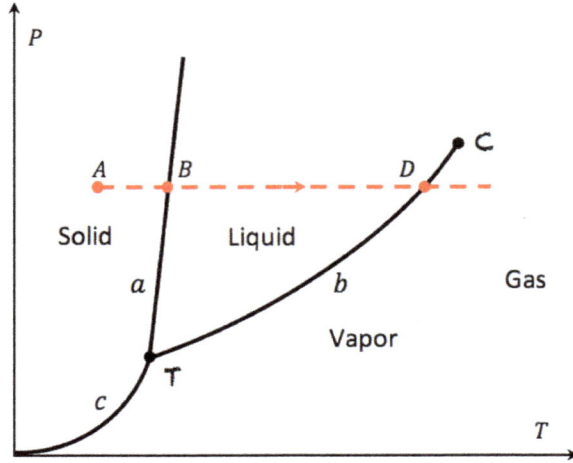

Fig. 3.14.   Phase diagram using the thermodynamic parameters P and T at constant volume V. The horizontal dashed red line represents a transformation at constant pressure.

The values of the thermodynamic parameters affect the stability of a macroscopic system with a specific aggregation state, and thus their changes determine the *phase transitions*, such as liquefaction from the solid to the liquid state, or evaporation from the liquid to the vapor state. The values of the thermodynamic parameters allowing stability of a specific phase and the ones corresponding to phase transitions are reported in the phase diagram characteristic of each material. Often, a phase diagram is built by using P and T as variable parameters at constant volume showing the stability regions for the solid, liquid and vapor phases. Of course, more complex phase diagrams can be drawn where multicomponents are present or other phases occur, such as liquid-crystalline states or glass states.

A simple case is sketched in Fig. 3.14. The lines separate regions where a specific aggregation state is stable: the correspondent values T and P are the ones allowing coexistence of the two phases. The meaning of this diagram is easily understood by following a possible transformation at constant pressure starting from a point (A) corresponding to the solid state and moving along a line parallel to the positive direction of the T axis. The crossing point (B) with line a corresponds to fusion at the specific pressure of the transformation. By continuing to provide heat to the system, the temperature will not change until all the volume of the material is in the liquid state, since the heat is used to break the cohesion. After that, the temperature again starts to increase in the region of stability of the liquid phase. The temperature rise will stop again at the crossing point (D) with line b corresponding to vaporization, and it will start to increase again only after complete vaporization of the system.

Looking at Fig. 3.14 it is clear that, by following the same process at higher pressure, the temperature of the phase transition increases as well. It is worth underlining that line a is almost parallel to the P axis; the positive slope with respect to the abscissa is due to the higher density of the material in the solid state with respect to the liquid state; as a consequence, a rise in pressure makes the solid state more stable, requiring a higher temperature for fusion. This property is highlighted by moving along a line parallel to the P axis (at constant temperature) around the point (B) corresponding to the phase transition: at higher pressure we find a stable solid state; at lower pressure we find a stable liquid state. Similar arguments can be used for any point of curve b separating the liquid from the vapor phase. The *triple point* T is the one corresponding to equilibrium of the three phases. At temperatures and pressures below the triple point, the liquid phase cannot exist and along line c we have the equilibrium between the solid and the vapor phase; then, by heating the material in the solid state under these thermodynamic conditions, we get sublimation to the vapor state. Another interesting point of the diagram is the *critical point* C, which defines the *critical temperature* T_c, and the *critical pressure* P_c. When $T > T_c$ and $P > P_c$, the material is in the state of a *supercritical fluid*. Under these conditions, there is no discontinuity between the gaseous and the liquid state. The physical properties of the material are mixed between those of a liquid and a gas; for instance, density is typical of liquids while viscosity is typical of gases.

We usually speak of *gas* when the state of a material at room temperature is over the critical point C but below the critical pressure $(T > T_c$ and $P < P_c)$. In other words, the difference between vapor and gas is that vapor can liquefy by increasing pressure, while gas cannot, being transformed into a supercritical fluid at $P > P_c$.

In summary, the three lines a, b, c represent equilibrium states between the phases: solid–liquid, liquid–vapor, solid–vapor. We recall that the fusion and evaporation points taken at a pressure $P = 1$ atm $(= 1.01325 \cdot 10^5$ Pa$)$ are called "normal". We should also recall that the actual phase diagram of a specific material might be much more complex since by varying the thermodynamic parameters other condensed phases could appear, such as liquid-crystalline states or glassy phases.

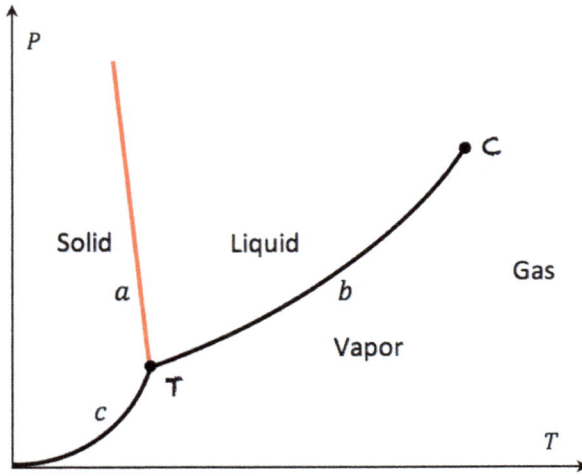

Fig. 3.15.   Phase diagram of H_2O.

In dealing with the phase diagram, we must mention the peculiarities of water, or, as is better to say, H_2O. In fact, as can be observed in Fig. 3.15, line a shows a negative slope.

This means that the rise in pressure decreases the phase transition temperature from the solid to the liquid state, in opposition to what generally happens. This occurs because in this case the mass density of the liquid state is higher than that of the solid state. This property is a consequence of the important role of the hydrogen bonding in the liquid state. It is interesting to consider the behavior for a temperature below the triple point: increasing the pressure, we favor the phase transition solid–liquid (this is the phenomenon allowing an ice-skater to skate!).

Data related to H_2O at $P = 1.01 \cdot 10^5$ Pa (=1 atm) are reported below:

Mass of molecule:	$m_{H_2O} = 2.992 \cdot 10^{-26}$ kg
Mass density of water:	$\rho_{water} = 1$ kg/m^3
Mass density of ice:	$\rho_{ice} = 917$ kg/m^3
Latent heat of evaporation:	$\lambda_{l-v} = 2.3 \cdot 10^6$ J/kg $= 6.9 \cdot 10^{-20}$ J/molecule
Latent heat of fusion:	$\lambda_{s-l} = 3.3 \cdot 10^5$ J/kg $= 1.0 \cdot 10^{-20}$ J/molecule

Latent heat of sublimation:	$\lambda_{s-v} = 2.8 \cdot 10^6$ J/kg
	$= 8.5 \cdot 10^{-20}$ J/molecule
Specific heat of ice:	$c_{\text{ice}} = 2.1 \cdot 10^3$ J/kg$\cdot$K
Specific heat of water:	$c_{\text{water}} = 4.2 \cdot 10^3$ J/kg$\cdot$K
Electric dipole moment of water molecule:	$\mu_0 = 6.17 \cdot 10^{-30}$ C$\cdot$m = 1.85 D

For H_2O, the *fusion normal point* ($P = 1$ atm) corresponds to $T = 0°C$ and the *evaporation normal point* falls at $T = 100°C$. The triple point is given by $T_T = 0.0098°C$ and $P_T = 611$ Pa ($= 0.006$ atm), while the critical point is at $T_c = 374°C$ and $P_c = 2.21 \cdot 10^7$ Pa ($= 218$ atm).

It is well known that the link between the thermodynamic parameters and microscopic properties is described by statistical mechanics, the most important result being the *equipartition theorem*, which relates the temperature to the average translational kinetic energy K of a system. According to it, we write

$$K = \frac{3}{2}Nk_BT, \tag{3.132}$$

with N being the number of molecules[11] and $k_B = 1.38066 \cdot 10^{-23}$ J/K the Boltzmann constant. Since each free molecule has three translational degrees of freedom,[12] the quantity $3N$ represents the degrees of freedom of the system, and therefore the contribution of each degree of freedom to the average kinetic energy is $(1/2)k_BT$.

We know that in the vapor state this kinetic energy is the most important contribution to the total energy, with the potential energy $\mathcal{V}$ between the molecules being negligible, since the molecules are far apart and not interacting. Then we can state that in the vapor phase

$$\mathcal{V} \ll K \tag{3.133}$$

or

$$\mathcal{V} \ll \frac{3}{2}Nk_BT, \tag{3.134}$$

---

[11] Here (and in the following discussion), "molecule" can also stand for "atom," depending on the material's composition.

[12] We have already discussed the other internal degrees of freedom represented by the rotational and vibrational states of the molecule.

or, for each molecule,

$$\mathcal{V}_{\text{mol}} \ll \frac{3}{2}k_B T. \tag{3.135}$$

It is clear that by decreasing the temperature (at constant pressure) the former inequalities become weaker and weaker until we reach a condition where

$$\mathcal{V} \approx \frac{3}{2}Nk_B T, \tag{3.136}$$

i.e.

$$\mathcal{V} \approx K. \tag{3.137}$$

When approaching this condition, we find the transition of the material from the vapor to the liquid state, i.e. a phase transition to a condensed state where the volume becomes a property of the aggregation state rather than a property of the container of the vapor. Now, the nearest molecules interact with each other with *short-range* forces. This means that in this state there is no long-range positional and orientational correlation among the molecules; in other words, the knowledge of the orientation and location of a molecule does not give any information on that of a distant molecule. This is the isotropic liquid phase showing a higher mass density with respect to the vapor state and low compressibility (zero in the ideal case).

By a further decrease in temperature, it is possible to reverse the (3.133) or (3.134) condition, i.e. to get

$$\mathcal{V} \gg K. \tag{3.138}$$

In this case, the molecular interaction is strong, producing effects on macroscopic lengths; as a consequence, the distances between molecules are fixed. The system becomes a solid "rigid body," differently from the liquid and vapor phases, which can be described as fluids. In this case, the location and orientation of a molecule is correlated with that of all the other molecules of the material.

As an example, it is possible to check the fulfillment of (3.138) in the case of a crystalline ionic solid. With the interatomic distance being of the order of $a \sim 1$ Å $= 10^{-8}$ cm, we can evaluate the potential energy of the ionic bonding between two ions with charges $+e$ and $-e$, respectively: $\mathcal{V} = (1/4\pi\epsilon_0)(e^2/a) \sim 2.3 \cdot 10^{-11}$ erg $= 14$ eV. We can compare this value to the estimated kinetic energy of electrons, since the kinetic energy of ions fixed on the lattice sites is negligible with respect to it. For an electron localized on the atomic site of size a, the maximum value of the wave vector is $k \sim 2\pi/a$; so $K = \hbar^2 k^2/2m \sim \hbar^2 (2\pi/a)^2/2m = 3.8\,\mathrm{eV}$. Then $K < V$, as expected for a solid phase. We may also notice that this phase is stable at room temperature (300 K) since in this case the additional thermal kinetic energy is just of the order of $k_B T \sim 0.026\,\mathrm{eV} \ll V$.

3.8. Statistical concepts

The thermodynamic arguments discussed above, as mentioned, have their basis in statistical considerations that become necessary when one is dealing with a large amount of particles, as we can recognize taking into account that one cubic centimeter of a vapor phase consists of more than 10^{19} molecules. Additionally, a statistical analysis of such a system requires being able to describe the dynamical state of each particle starting from some basic assumptions. This leads to formulating a probability distribution of the particles with respect to the different possible dynamical states. To know the probability distribution means to know how the particles $n_1, n_2, \ldots, n_i, \ldots$ of the system are distributed among the possible energy states $E_1, E_2, \ldots, E_i, \ldots$, to produce a total energy given by

$$\mathcal{U} = n_1 E_1 + n_2 E_2 + \cdots + n_i E_i + \cdots = \sum_j n_j E_j. \qquad (3.139)$$

This expression defines the *partition* of the ensemble of the n_j particles. Of course, for an isolated system the total energy is constant and therefore any change in the distribution increasing the energy of a number of particles must correspond to another change in the occupation numbers leading to an equal decrease in energy. What is interesting to know is the partition given by Eq. (3.139) that is most probable for each macroscopic (thermodynamic) state. We describe as the state of *statistical equilibrium* the one corresponding to the mentioned partition, when the physical conditions of the system are defined. This state of statistical equilibrium can be modified only by an external effect leading to a specific change in the physical system. The theory of statistical mechanics allows finding the most probable partition for a system with a large number of particles.

The classical approach considers *identical* but *distinguishable* particles. Following thermodynamic arguments, assuming that the probability of a particular partition is proportional to the number of different particle distributions producing the same partition, we find that the total number of particles can be written as

$$N = e^{-\alpha} \sum_j g_j e^{-\beta E_j}, \tag{3.140}$$

where α is a parameter referring to the physical properties of the system and g_j accounts for the degeneracy of the energy level E_j (i.e. it counts the number of energy levels with the same energy, E_j). The factor β is obtained as

$$\beta = \frac{1}{k_B T}, \tag{3.141}$$

with T being the absolute temperature (K) and k_B the Boltzmann constant.

From Eq. (3.140), the *partition function* Z is defined as

$$Z = \sum_j g_j e^{-\beta E_j}. \tag{3.142}$$

After additional steps, the partition with the highest probability is obtained and written as

$$n_i = \frac{N}{Z} g_i e^{-\frac{E_i}{k_B T}}. \tag{3.143}$$

This is the classical *Boltzmann distribution*. Since the exponential part is a decreasing function of $E_i/k_B T$, the larger this ratio is the smaller will be the occupation number n_i. Therefore, at a given temperature it decreases at increasing values of E_i. On the other hand, by increasing the temperature T the occupation number n_i at a given E_i increases as well.

According to it, the ratio between the occupation numbers of two different energy states E_k and E_m is given by

$$\frac{n_k}{n_m} = \frac{g_k}{g_m} e^{-\frac{E_k - E_m}{k_B T}} = \frac{g_k}{g_m} e^{-\frac{\Delta E}{k_B T}}, \tag{3.144}$$

where $\Delta E = E_k - E_m$.

In order to understand the important meaning of Eq. (3.144), we can calculate the ratio n_k/n_m for different values of ΔE and T.

In Table 3.5, three values of ΔE are considered, corresponding respectively to transitions between different rotational, vibrational and electronic states. In the first case, we see that rotational states have approximately the

Table 3.5. Energy differences for transitions between rotational, vibrational and electronic states, and the ratio of the occupation numbers of the corresponding energy levels. Calculations have been performed for $g_k = g_m$.

ΔE (eV)	n_k/n_m ($T = 100$ K)	n_k/n_m ($T = 300$ K)	n_k/n_m ($T = 1000$ K)
10^{-4} (rotational states)	0.988	0.996	0.999
$5 \cdot 10^{-2}$ (vibrational states)	$3 \cdot 10^{-3}$	0.146	0.560
1.0 (electronic states)	$4.15 \cdot 10^{-51}$	$1.98 \cdot 10^{-17}$	$8.9 \cdot 10^{-6}$

same occupation number at any temperature. In contrast, the vibrational states appear more sensitive to temperature changes since the ratio between the occupation numbers increases significantly with temperature. Concerning different electronic states, the ratio is extremely small at any temperature, i.e. at statistical equilibrium the upper level can be considered empty with respect to the lower level.

The result (3.143) has been obtained ignoring additional restrictions that may apply to the distributions of particles among the different states associated with the energy levels. These restrictions may originate from the quantum-mechanical properties of the considered particles; as a consequence, they affect the probability distribution among the energy states, thus giving rise to a *quantum statistic*.

First of all, *undistinguishable* particles must be considered. Additionally, when the particles, such as electrons in atoms, must be described by antisymmetric wave functions as discussed in Section 3.4, they obey the Pauli exclusion principle. In this case, they follow the Fermi–Dirac statistical distribution and are called *fermions*. The additional restriction is that two particles cannot be in the same dynamical state, and this strongly affects their statistical distribution. On the contrary, when this restriction does not apply, the particles follow the Bose–Einstein statistical distribution and are called *bosons*. At high temperature, the two quantum-statistical distributions merge into the Boltzmann distribution.

The application of these concepts to statistics leads to the quantum partition function for fermions, given by

$$n_j = \frac{g_j}{e^{\frac{E_j - \mu}{k_B T}} + 1}, \tag{3.145}$$

where n_j is the number of particles occupying the state E_j of degeneracy g_j at a given temperature T and μ is the chemical potential. From Eq. (3.145) it is possible to get the well-known *Fermi–Dirac* probability distribution,

$$f(E, T) = \frac{1}{e^{\frac{E-\mu}{k_B T}} + 1}, \tag{3.146}$$

which is the probability that a state at energy E be occupied at temperature T.

Taking into account that

$$\lim_{T \to 0} e^{-\frac{E-\mu}{k_B T}} = \begin{cases} 0 & \text{if } E < \mu \\ \infty & \text{if } E > \mu \end{cases}, \tag{3.147}$$

it follows that at $T = 0$ all the lowest energy levels up to the value μ are occupied, differently from the classical Boltzmann distribution law giving all the particles in the ground state in the limit of $T = 0$. The difference is due to the exclusion principle preventing all the particles from staying in the same energy state. At $T = 0$, the energy of the highest occupied level is called the *Fermi energy*, E_F, i.e. $E_F = \mu$, the *chemical potential*.

Therefore, as shown in Fig. 3.16, at $T = 0$ the distribution function is steplike (all available states filled up to E_F; all states empty for $E > E_F$), while at higher temperature the occupation probability for states at $E > E_F$ increases with a corresponding decrease for $E < E_F$, being always

$$f(\mu = E_F, T) = \frac{1}{2}. \tag{3.148}$$

The range of energies where the distribution changes with temperature is of the order of $k_B T$ around μ, and usually at room temperature

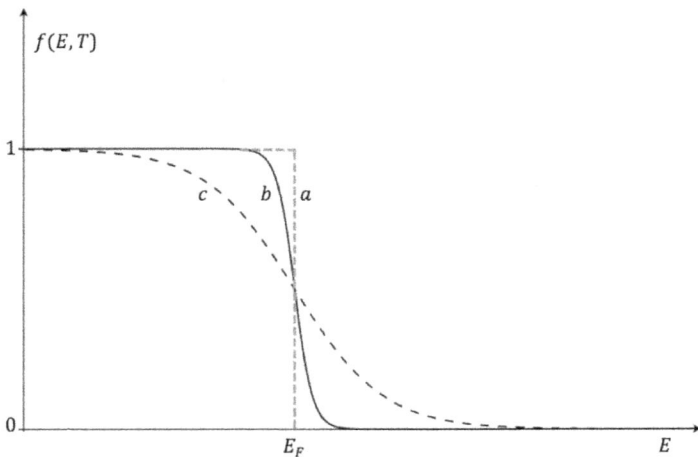

Fig. 3.16.   *Fermi–Dirac distribution: (a) $T_a = 0$ K, (b) $T_b > T_a$, (c) $T_c > T_b$.*

the condition $k_B T \ll \mu$ holds and thus modifications in the particle distribution are relevant only around the initial Fermi energy. This is why usually one sets $\mu \approx E_F$, thus writing the *Fermi–Dirac* distribution as

$$f(E,T) = \frac{1}{e^{\frac{E-E_F}{k_B T}} + 1}. \tag{3.149}$$

When $E - E_F \gg k_B T$, we have

$$f(E,T) = \frac{1}{e^{\frac{E-E_F}{k_B T}} + 1} \approx e^{-\frac{E}{k_B T}}, \tag{3.150}$$

corresponding to the Boltzmann distribution.

We will see in the next chapter the fundamental role played by this statistical distribution in the electronic properties of crystalline structures.

For the sake of completeness in the discussion of the results of the statistical (classical and quantum) mechanics, we must mention the distribution found for quantum particles not obeying Pauli's exclusion principle. It is called *Bose–Einstein* distribution and is formally similar to the Fermi–Dirac distribution even if it describes a very different behavior for *bosons*. The occupation number in this case becomes

$$n_j = \frac{g_j}{e^{\alpha + \frac{E_j}{k_B T}} - 1}, \tag{3.151}$$

where $\alpha > 0$. The dependence of the occupation number on energy becomes clear by considering the corresponding continuous function, setting $g_j = 1$:

$$n(E,T) = \frac{1}{e^{\alpha + \frac{E}{k_B T}} - 1}. \tag{3.152}$$

This function is shown in Fig. 3.17 where the trend to enhance the occupation number at the lowest energy by decreasing the temperature T is evident. This is the well-known effect of *Bose condensation*.

An example of quantum particles fulfilling the Bose–Einstein statistical distribution is given by photons: simple arguments allow obtaining the already-discussed blackbody radiation spectrum starting from this statistical distribution of photons.

3.9. The crystalline solid state

3.9.1. *Basics of crystallography*

The most-ordered condensed state of matter is the one where atoms are located in a regular way in the bulk of the material having a long-range

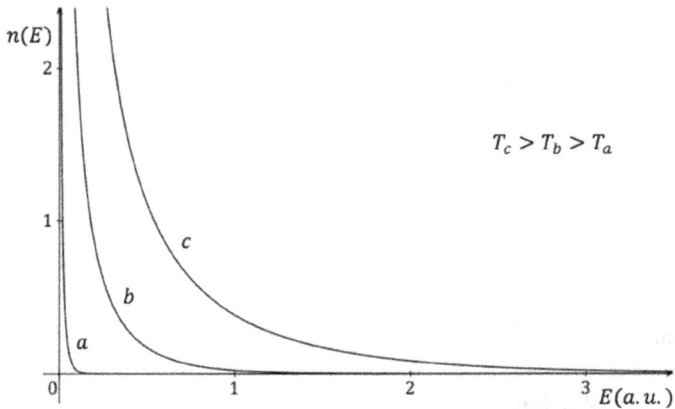

Fig. 3.17.   *Bose–Einstein* distribution.

positional order. In this case, the ordered structure reaches a macroscopic scale with possible inclusion of defects of different types and origins. When the macroscopic ordered structure does not include any defect, it is denominated a "single crystal".

A basic concept for describing the structure of a crystalline solid is that of a *lattice*, which defines the periodic array of atoms. Then a lattice is a geometrical concept defined by all the points whose position is given by the vector

$$\mathbf{R} = n_1\mathbf{a} + n_2\mathbf{b} + n_3\mathbf{c}, \tag{3.153}$$

where $\mathbf{a}$, $\mathbf{b}$, $\mathbf{c}$ are three vectors (not all in the same plane) called *primitive vectors* and n_1, n_2, n_3 are three integer numbers. That is to say, the arrangement and orientation appear "*exactly the same, from whichever of the points the array is viewed.*"[13] This is actually the definition of the *Bravais lattice*.

Starting from a lattice point, any translation vector

$$\mathbf{T} = m_1\mathbf{a} + m_2\mathbf{b} + m_3\mathbf{c} \tag{3.154}$$

with m_1, m_2, m_3 integer numbers leads to another lattice point from which the structure appears the same. According to the definition (3.153), $\mathbf{T}$ is a

---

[13]N.W. Ashcroft and N.D. Mermin, *Solid State Physics* (Harcourt College Publishers, 1976), p. 64.

lattice vector and could be expressed by the difference between two lattice vectors:

$$T = R_2 - R_1. \tag{3.155}$$

This means that translation is the basic *symmetry operation* fulfilled by any crystal structure. The concept of symmetry is of paramount importance in condensed matter physics, since from existing symmetries some physical properties can be deduced. Depending on the arrangement of the periodic array of points, different lattices may fulfill different symmetry operations. From the mathematical point of view, this means that any space function associated with the lattice does not change if the symmetry operation is applied to the space coordinates:

$$f(\mathbf{r}) = f(\hat{\mathbf{S}}\mathbf{r}), \tag{3.156}$$

where $\hat{\mathbf{S}}$ represents a specific transformation of the coordinate $\mathbf{r}$. The symmetry transformations can be translations, rotations around an axis, reflections with respect to a plane, or inversions with respect to a point. All together they make a *point group* and the crystal lattice is said to be invariant with respect to a specific point group, defined by a number of symmetry operations.

Of course, the lattice definition given above (and the consequent symmetry operations that can be fulfilled) applies to an infinite lattice. In the real world, we have a finite volume and interfaces between the crystal and the "outside", and therefore we must take into account that these concepts can be properly applied far from the crystal boundary surfaces. It is out of the scope of this text to discuss in detail the concept of symmetry in crystals; in the box below is a simple example of a triangular two-dimensional lattice with only the aim of clarifying the meaning of symmetries.

The concept of symmetry can be easily understood by looking at the regular two-dimensional array of points in Fig. IN3.1 (triangular lattice). The figure shows that there are six rotation angles that map the lattice into itself around any axis orthogonal to the lattice plane passing through a lattice point: rotations of angle m $(2\pi/6)$, where $m = 0, 1, \ldots, 5$. In the figure is shown the $2\pi/6$ rotation of point A around the axis passing through point O, leading the point A in the lattice point B. These are six fold axes of symmetry. After any of these rotations, the lattice looks exactly the same as before the rotation. Besides that, this lattice allows additional symmetry operations not shown in the figure. Any reflection around a surface (orthogonal to the lattice plane) including the line connecting an infinite number of lattice points is a symmetry operation. Six symmetry planes of this kind can be found in this triangular structure.

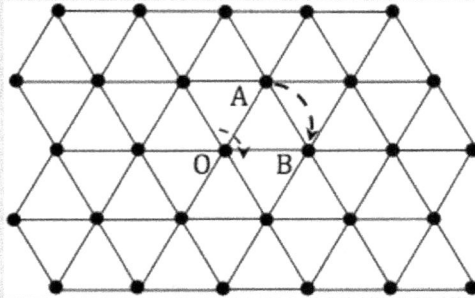

Fig. IN3.1. A two-dimensional triangular lattice and the symmetry operation corresponding to the lattice rotation around an axis orthogonal to the lattice plane passing through lattice point O.

Examples of lattice structures are shown in Fig. 3.18.

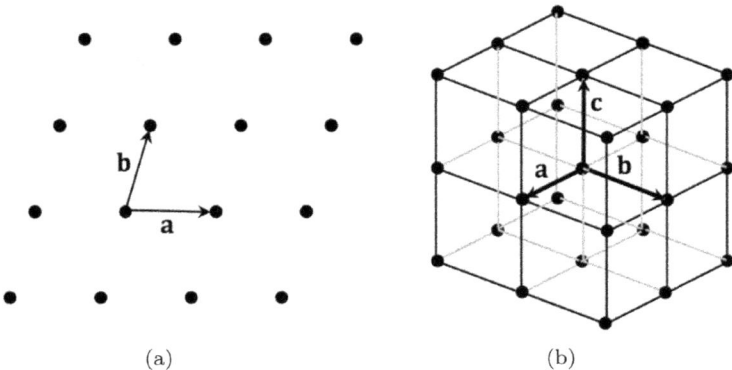

(a)   (b)

Fig. 3.18. A two-dimensional lattice (a) and a three-dimensional lattice (b) with indication of the primitive vectors.

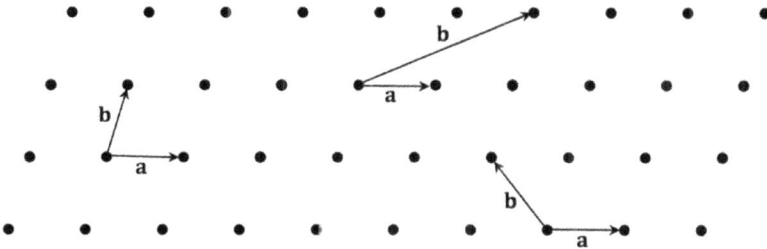

Fig. 3.19. Different possible choices of primitive vectors in a two-dimensional lattice.

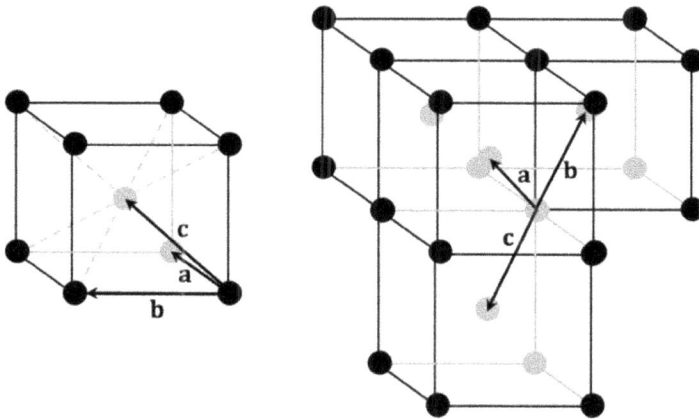

Fig. 3.20. A body-centered cubic lattice with different possible choices of primitive vectors.

We recall that the choice of the primitive vectors is not unique; namely, different choices allow fulfilling the lattice definition (3.153). This is clear by looking at Figs. 3.19 and 3.20.

In three dimensions there are only 14 different geometrical periodic arrays of points fulfilling the condition (3.153), these are the space Bravais lattices. They are represented in a more convenient way by the unit cells. The *unit cell* of a lattice is a volume including one or more lattice points that can fill the whole space when suitable translations following the lattice symmetry are applied. Therefore, the primitive cell is the unit cell of the smallest size. However, it is often useful to represent the lattice through the unit cell rather than through the primitive cell, because the former is able to reproduce the lattice symmetries, while often this is not the case with

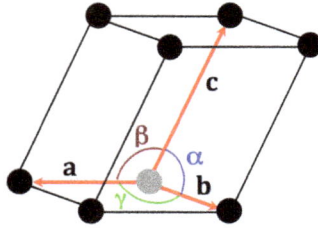

Fig. 3.21.   The definition of the unit vectors and their mutual angles.

the primitive cell. The unit cell is defined by the mutual length of the unit vectors a, b, c and the angles α, β, γ between them, as shown in Fig. 3.21.

The unit cells of the 14 Bravais lattices are shown in Fig. 3.22. In accordance with the definition of Fig. 3.21, Table 3.6 summarizes their characteristics.

Looking at Fig. 3.22, we see that the possible different configurations in each lattice system depend on the different arrangement of the lattice points. For instance, in the three cubic lattices we have: point lattices only on the eight corners of the cubic cell for the simple cubic, one additional lattice point in the diagonal crossing point in the body-centered cubic, and additional points in the center of each face in the face-centered cubic.

For each type of lattice, the number of points closest to a specific lattice point is different; these are *nearest neighbors* to each point. This number is a property of the lattice and is called the *coordination number*. For instance, a simple cubic lattice has a coordination number equal to 6.

Another concept useful for the description of crystalline solids is that of the lattice plane, since the orientation of the lattice planes with respect to the specific excitation source (such as electromagnetic waves) affects the observations. A lattice plane must include at least three lattice points but, owing to the translational symmetry, it contains an infinite number of lattice points, and therefore it gives rise to a two-dimensional lattice. Of course, there are infinite parallel lattice planes with the same orientation. The orientation of a lattice plane is usually given by three numbers known as *Miller indices*. They are defined in the following way: the plane including the lattice points intersects the axis determined by the orientation of the unit vectors a, b, c; the inverse of such intercepts defines the Miller indices, as shown in Fig. 3.23.

A lattice is a geometrical concept useful for describing the ordered arrangements of atoms in space, and therefore the *crystal structure* can be considered the association of a single atom or a group of atoms (called a

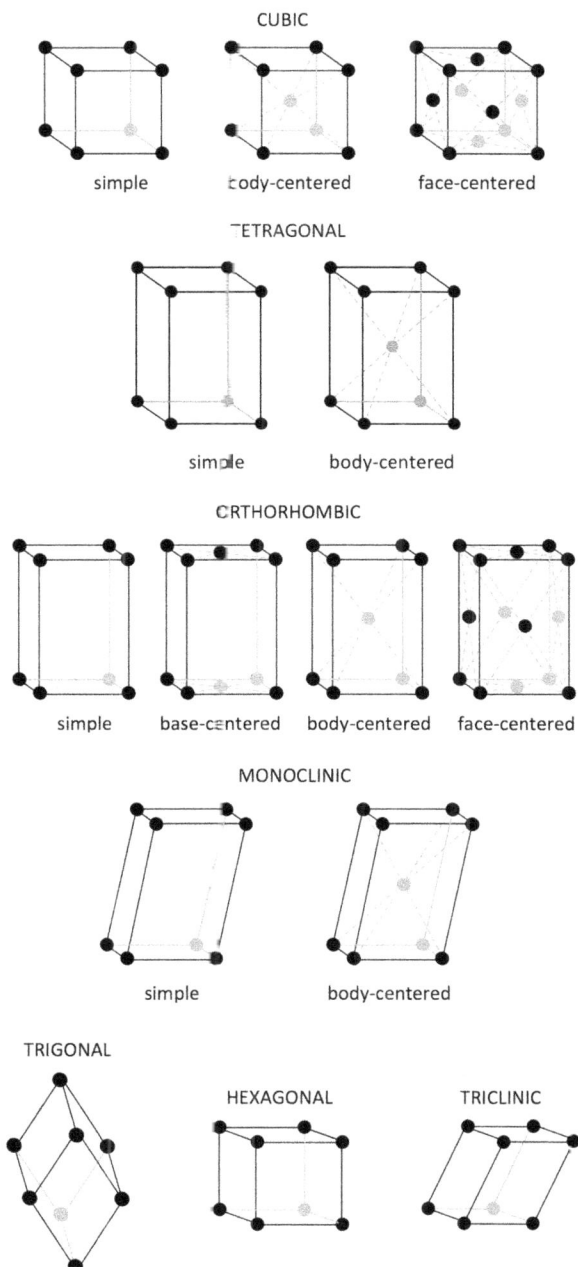

Fig. 3.22. The 14 Bravais lattices represented through their unit cells.

Table 3.6.   Characteristics of the Bravais lattice.

Lattice system	Number of lattice systems	Lattice symbols	Characteristics of the unit cell
Cubic	3	P or SC I or BCC F or FCC	$a = b = c$ $\alpha = \beta = \gamma = 90°$
Tetragonal	2	P, I	$a = b \neq c$ $\alpha = \beta = \gamma = 90°$
Orthorhombic	4	P, C, I, F	$a \neq b \neq c$ $\alpha = \beta = \gamma = 90°$
Monoclinic	2	P, C	$a \neq b \neq c$ $\alpha = \gamma = 90° \neq \beta$
Trigonal	1	R	$a = b = c$ $\alpha = \beta = \gamma < 120° \neq 90°$
Hexagonal	1	P	$a = b \neq c$ $\alpha = \beta = 90°$ $\gamma = 120°$
Triclinic	1	P	$a \neq b \neq c$ $\alpha \neq \beta \neq \gamma$

base) to a lattice:

$$\text{Crystal structure} = \text{lattice} + \text{base}.$$

This means that around or close to lattice points we have a mass distribution, which can be described by a density function $\rho(\mathbf{r})$ such that

$$\rho(\mathbf{r}) = \sum_{i,n} m_i \delta(\mathbf{r} - \mathbf{R}_n - \mathbf{c}_i), \qquad (3.157)$$

where m_i is the mass of the atom i located in the position given by the vector $\mathbf{c}_i$ in the unit cell and δ represents the Dirac delta function. The vector $\mathbf{R}_n$ identifies each lattice point. Of course, in the perfect crystal the density of matter is periodic, i.e. it is invariant under a lattice translation:

$$\rho(\mathbf{r}) = \rho(\mathbf{r} + \mathbf{T}). \qquad (3.158)$$

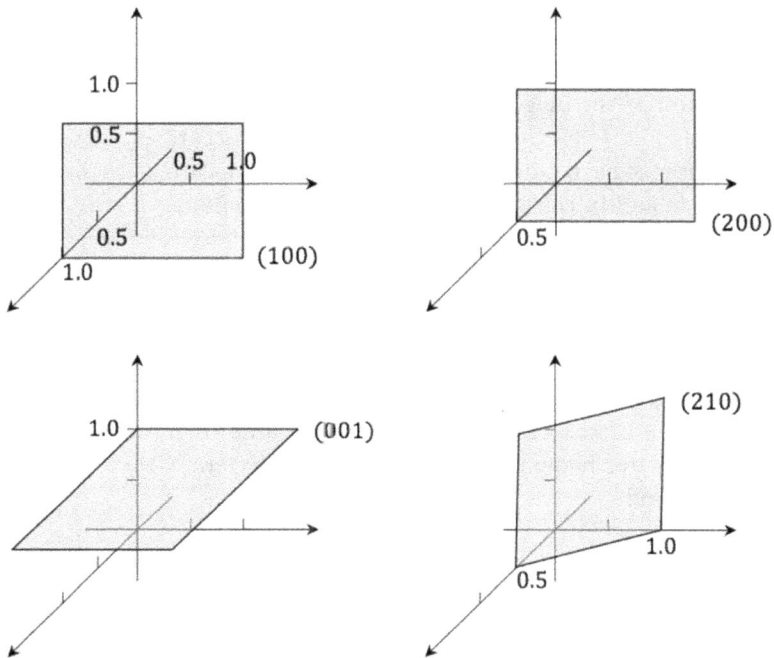

Fig. 3.23. Planes identified by Miller indices in a cubic lattice (identified by three orthogonal axes). The Miller indices of each plane are the ones in brackets.

A very common example is given by sodium cloride (NaCl), whose unit cell of the face-centered cubic lattice is shown in Fig. IN3.2.

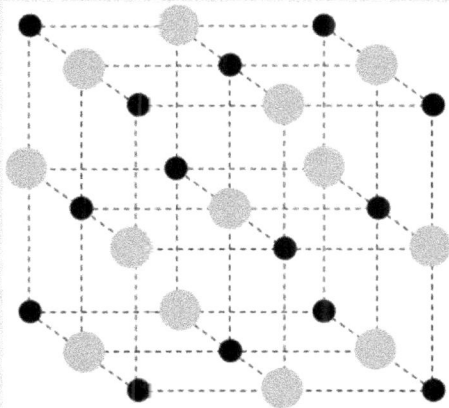

Fig. IN3.2. Crystal structure of NaCl. The small black spheres represent Na atoms, and the larger spheres Cl atoms.

In this case, the base is made up of one atom of Na located on the corner of the cube at coordinates $(0,0,0)$ and one atom of Cl located in the center of the diagonal of the cube at coordinates $(1/2, 1/2, 1/2)$. Then in this unit cell the Na atoms are on the corners of the cube and in the center of each cubic face, while the Cl atoms are at the midpoint of each side and in the center of the cube diagonals.

3.9.2. *X-ray diffraction and the reciprocal lattice*

Typical interatomic distances in crystalline solids are in the range of a few angstroms (1 Å $= 10^{-8}$ cm); then this is the order of magnitude of the wavelength of the electromagnetic waves able to interact with these periodic structures being diffracted by them. They are X-rays with photon energy of around

$$h\nu = \frac{hc}{\lambda} \approx 1.2 \cdot 10^4 \text{ eV}.$$

The same arguments used in Section 1.7 to discuss diffraction by a thick phase grating can be applied here for the X-ray diffraction by a crystal. In this case, the diffracting planes are lattice planes including the regular pattern of lattice points, fulfilling the condition (1.201), which defines the Bragg grating. Following a standard denomination, here we call d the period of the structure,[14] and therefore by looking at Fig. 3.24 we can again find Bragg's law of diffraction, connecting the diffraction angle θ to the period d of the structure and the impinging wavelength:

$$2d \, \sin\theta = m\lambda, \quad \text{with } m = 1, 2, 3, \ldots. \tag{3.159}$$

In fact, $2d \sin\theta$ is the optical path difference between two rays reflected by neighboring planes that must be equal to an integer multiple of the light wavelength to get interference maxima. Of course, the diffraction pattern depends on the lattice characteristics since they define the period d of the structure. Actually, for any crystal lattice, different periodic structures can be found, as different lattice planes can be identified. Therefore, for each wavelength λ, by varying the incidence angle one can find diffraction patterns corresponding to different periodicities $d_1, d_2, d_3, \ldots$.

---

[14] It corresponds to Λ in Eq. (1.201), where d was the thickness of the grating.

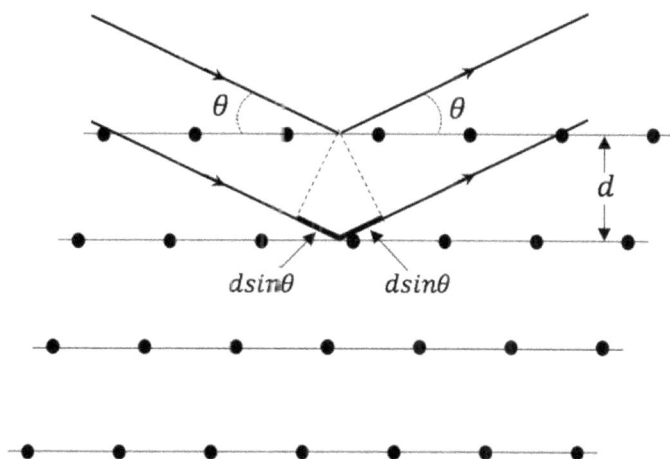

Fig. 3.24. A simple geometrical demonstration of Bragg's law of diffraction.

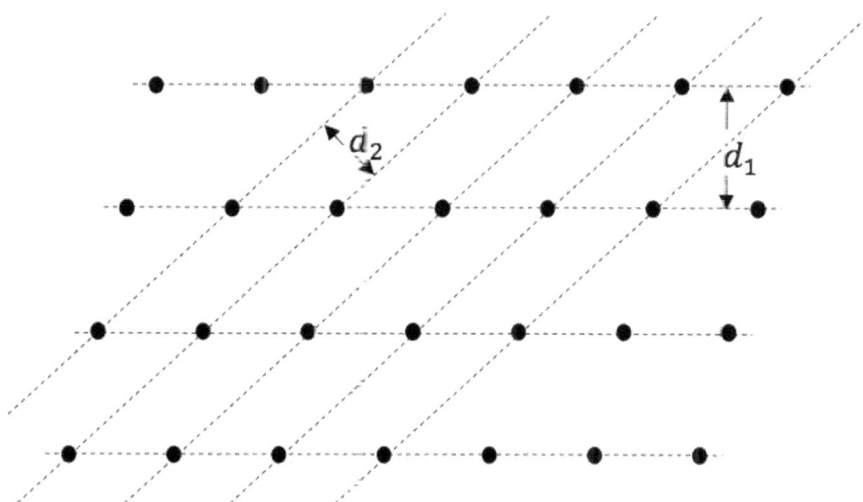

Fig. 3.25. Two different periods in the same crystal structure.

As an example, Fig. 3.25 shows two different series of lattice planes in the same periodic structure of Fig. 3.24.

Laue's method is very convenient for getting quickly the crystal orientation. In this case, the X-rays impinging on the crystal are highly collimated, keeping the fixed incident angle, but cover a broad spectrum.

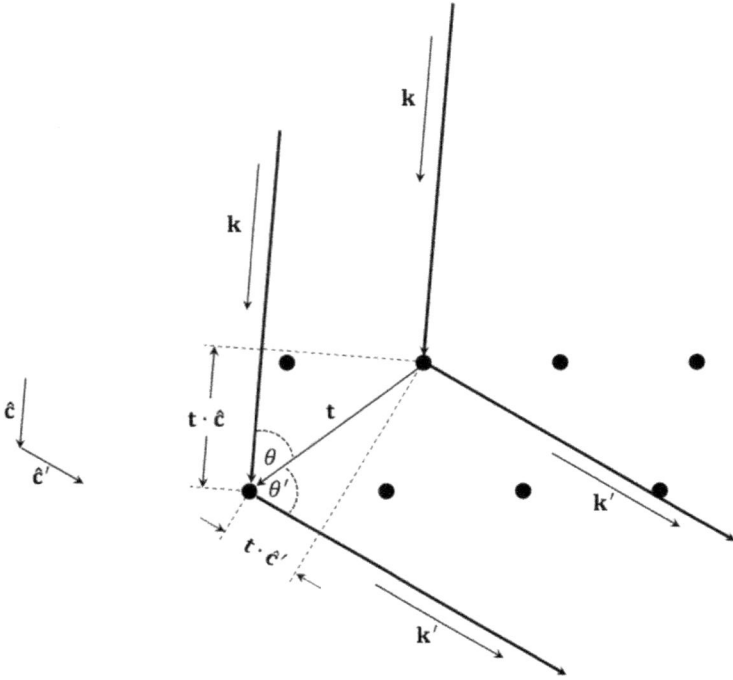

Fig. 3.26.   Calculation of the diffraction condition with Laue's method.

Fulfillment of Eq. (3.159) occurs for each wavelength for each possible d and gives rise to the diffraction pattern made by bright spots on a dark field. The analysis is performed finding the directions of the diffracted beams in the following way. We define $\hat{\mathbf{c}}$ and $\hat{\mathbf{c}}'$ as the unit vector corresponding to the direction of the incident wave and of the diffracted wave originated by scattering by atoms located at the lattice sites. The corresponding wave vectors will be $\mathbf{k} = (2\pi/\lambda)\hat{\mathbf{c}}$ and $\mathbf{k}' = (2\pi/\lambda)\hat{\mathbf{c}}'$. Looking at Fig. 3.26, the optical path difference between two scattered rays with wave vector $\mathbf{k}'$ is

$$t\cos\theta + t\cos\theta' = \mathbf{t} \cdot (\hat{\mathbf{c}} - \hat{\mathbf{c}}'), \qquad (3.160)$$

where $\mathbf{t}$ is a lattice vector connecting two lattice points. Therefore, the interference maxima occur when

$$\mathbf{t} \cdot (\hat{\mathbf{c}} - \hat{\mathbf{c}}') = m\lambda, \quad \text{with } m = 1, 2, 3\ldots, \qquad (3.161)$$

or

$$\mathbf{t} \cdot (\mathbf{k} - \mathbf{k}') = m2\pi. \qquad (3.162)$$

This condition must hold for any lattice vector $\mathbf{R}$, and therefore we can write

$$\mathbf{R} \cdot (\mathbf{k} - \mathbf{k}') = m2\pi, \tag{3.163}$$

which can be written as

$$e^{i\mathbf{R} \cdot \Delta\mathbf{k}} = 1, \tag{3.164}$$

defining $\Delta\mathbf{k} = \mathbf{k}' - \mathbf{k}$.

Equation (3.164) shows the relationship between the lattice structure that determines the set of vectors $\mathbf{R}$ and the diffracted pattern created by all the directions $\mathbf{k}'$ that fulfill Eq. (3.164) for a given incident direction $\mathbf{k}$. It is useful to define another set of vectors $\mathbf{G}$ as

$$e^{i\mathbf{R} \cdot \mathbf{G}} = 1. \tag{3.165}$$

The vectors $\mathbf{G}$ define in the k-space the *reciprocal lattice* made by the points connected by vectors:

$$\mathbf{G} = g_1\mathbf{A} + g_2\mathbf{B} + g_3\mathbf{C}, \tag{3.166}$$

where g_i are integer numbers and $\mathbf{A}$, $\mathbf{B}$, $\mathbf{C}$ define the unit cell of the reciprocal lattice in the k-space.

Taking into account the scalar product $\mathbf{R} \cdot \mathbf{G}$,[15] the condition (3.163) is fulfilled when

$$\begin{array}{lll} \mathbf{a} \cdot \mathbf{A} = 2\pi, & \mathbf{a} \cdot \mathbf{B} = 0, & \mathbf{a} \cdot \mathbf{C} = 0; \\ \mathbf{b} \cdot \mathbf{A} = 0, & \mathbf{b} \cdot \mathbf{B} = 2\pi, & \mathbf{b} \cdot \mathbf{C} = 0; \\ \mathbf{c} \cdot \mathbf{A} = 0, & \mathbf{c} \cdot \mathbf{B} = 0, & \mathbf{c} \cdot \mathbf{C} = 2\pi. \end{array} \tag{3.167}$$

Therefore, we can write

$$\mathbf{A} = 2\pi \frac{\mathbf{b} \times \mathbf{c}}{\mathbf{a} \cdot \mathbf{b} \times \mathbf{c}}, \quad \mathbf{B} = 2\pi \frac{\mathbf{c} \times \mathbf{a}}{\mathbf{a} \cdot \mathbf{b} \times \mathbf{c}}, \quad \mathbf{C} = 2\pi \frac{\mathbf{a} \times \mathbf{b}}{\mathbf{a} \cdot \mathbf{b} \times \mathbf{c}}. \tag{3.168}$$

We notice that $v_L = \mathbf{a} \cdot \mathbf{b} \times \mathbf{c}$ is the volume of the lattice unit cell and $v_R = \mathbf{A} \cdot \mathbf{B} \times \mathbf{C} = 8\pi^3/v_L$ is the volume of the corresponding unit cell of the reciprocal lattice. In Eq. (3.165), one can find that $\mathbf{A}$ is perpendicular to $\mathbf{b}$ and $\mathbf{c}$, and similar relationships for the other unit vectors. This is actually a particular case of a general property that can be easily demonstrated: any reciprocal lattice vector is perpendicular to a crystal lattice plane.

---

[15]$\mathbf{R} \cdot \mathbf{G} = (n_1\mathbf{a} + n_2\mathbf{b} + n_3\mathbf{c}) \cdot (g_1\mathbf{A} + g_2\mathbf{B} + g_3\mathbf{C}) = n_1g_1\mathbf{a} \cdot \mathbf{A} + n_1g_2\mathbf{a} \cdot \mathbf{B} + n_1g_3\mathbf{a} \cdot \mathbf{C}$
$+ n_2g_1\mathbf{b} \cdot \mathbf{A} + n_2g_2\mathbf{b} \cdot \mathbf{B} + n_2g_3\mathbf{b} \cdot \mathbf{C} + n_3g_1\mathbf{c} \cdot \mathbf{A} + n_3g_2\mathbf{c} \cdot \mathbf{B} + n_3g_3\mathbf{c} \cdot \mathbf{C}.$

It can be easily shown that the reciprocal of the reciprocal lattice is the original crystal lattice, and therefore we can also write

$$\mathbf{a} = 2\pi \frac{\mathbf{B} \times \mathbf{C}}{\mathbf{A} \cdot \mathbf{B} \times \mathbf{C}}, \quad \mathbf{b} = 2\pi \frac{\mathbf{C} \times \mathbf{A}}{\mathbf{A} \cdot \mathbf{B} \times \mathbf{C}}, \quad \mathbf{c} = 2\pi \frac{\mathbf{A} \times \mathbf{B}}{\mathbf{A} \cdot \mathbf{B} \times \mathbf{C}}. \quad (3.169)$$

These expressions are quite useful, since the observation of the diffraction pattern gives a direct measurement of the reciprocal lattice parameters, and therefore using Eqs. (3.169) it is possible to get quickly the information on the crystal lattice.

Through Eq. (3.165) we have defined the reciprocal lattice vector as

$$\mathbf{G} = \mathbf{\Delta k} = \mathbf{k}' - \mathbf{k}. \quad (3.170)$$

This means that maxima of the diffraction pattern are connected by a vector $\mathbf{G}$. In other words, they correspond to the reciprocal lattice points, giving an image of the reciprocal lattice, as shown in Fig. 3.27. This allows a measurement of $\mathbf{A}, \mathbf{B}$ and $\mathbf{C}$ and calculation of the lattice unit vectors through Eqs. (3.169).

Of course, a more detailed analysis of the diffraction pattern that also takes into account the intensity distribution gives information too on the mass distribution around the lattice sites. Additional details can be found in specialized texts of solid state physics.

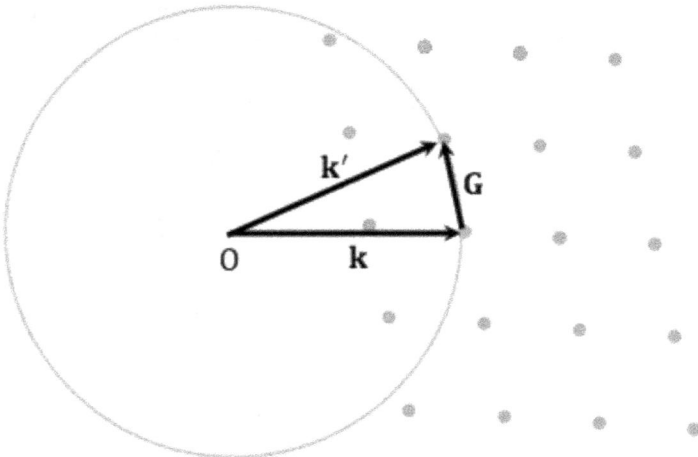

Fig. 3.27. The vector $\mathbf{G} = \mathbf{k}' - \mathbf{k}$, connecting two points of the reciprocal lattice. The first one corresponds to the end of the incident wave vector $\mathbf{k}$, and the second one to the end of the diffracted wave vector $\mathbf{k}'$.

3.10. The liquid and the glassy state

The liquid state is macroscopically defined as the aggregation phase of matter with a given volume, but without a fixed shape, which is that of the container of the material.

This reflects the nonrigid character of the liquid (the distance between each pair of atoms or molecules changes with time), but the volume is approximately constant (the ideal liquid is incompressible). The observation that phase transition from the gas state to the liquid state at a temperature above the critical temperature T_c occurs in a continuous way confirms the disordered arrangement of molecules in a liquid similar to the vapor phase. However, the much higher mass density close to that of the solid phase points out the much stronger interaction and packing between the molecules. Table 3.7 reports the density of a few materials appearing at room temperature in the different phases.

From the microscopic point of view, we have already outlined that the molecules (or atoms)[16] in the liquid state have a kinetic energy of the same order of magnitude as the potential energy owing to the interaction with the surrounding molecules; this property leads to correlation on a microscopic scale, while no long-range correlation between the molecules is present.

For this reason, the conventional approach to studying the liquid state properties is based on a statistical description. Starting from a density distribution function that describes the molecule distribution in

Table 3.7. The mass density of a few elements appearing at room temperature in different phases.

	Material	Density (g/cm^3)
Gas	Hydrogen	$0.09 \cdot 10^{-3}$
	Oxygen	$1.43 \cdot 10^{-3}$
	Helium	$0.18 \cdot 10^{-3}$
Liquids	Water	1.00
	Ethyl alcohol	0.80
	Mercury	13.6
Solids	Aluminum	2.70
	Gold	19.3
	Diamond	3.51

---

[16] In the following, we mention only molecules, but the discussion applies to atoms as well.

the volume of the material, the most commonly used approach exploits a pair distribution function that gives the probability of finding a molecule in a small region around another molecule. Considering isotropic liquids and defining R as the distance between the molecules, the pair distribution function can be written as

$$g(R) = \frac{1}{\langle n \rangle_V} \frac{dn(R, R + dR)}{dv(R, R + dR)}, \tag{3.171}$$

where $\langle n \rangle_V$ is the average molecular density of the material, and dn/dv is the density of molecules at a distance between R and $R + dR$ from the reference molecule. This is also called the radial distribution function. By considering a sphere with radius R around one molecule of the liquid, the average number of particles included in a shell with the radius ranging from R to $R + dR$ is given by

$$dn(R) = g(R)\langle n \rangle_V \cdot 4\pi R^2 dR. \tag{3.172}$$

By looking at Eq. (3.171), the meaning of $g(R)$ becomes clear. In fact, it gives an evaluation of how different the local density is from the average density. If R_0 is the molecule radius, $g(R) = 0$ if $R \leq R_0$, then it increases up to a first peak occurring at $R = R_1$, as shown in Fig. 3.28, with smoother and smoother oscillations around the value $g(R) = 1$ for increasing values of R. The peaks correspond to the distances with the highest probability of finding molecules; in the usual solid state denomination, the first peak is due to the nearest neighbor molecules. The typical shape of the function

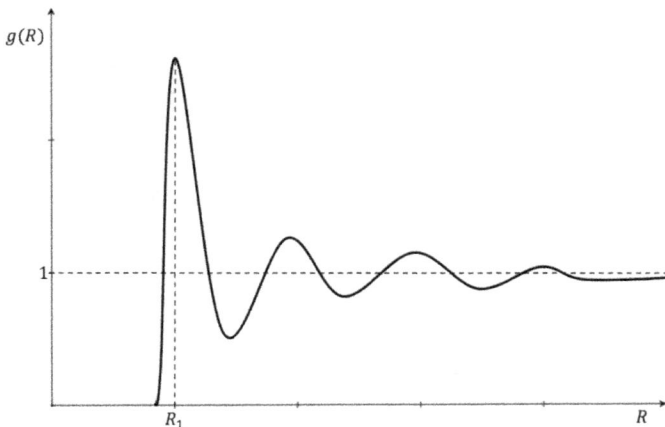

Fig. 3.28. Pair distribution function versus R for an isotropic liquid.

$g(R)$ for an isotropic liquid given in Fig. 3.28 shows that the density around one molecule of a liquid is not uniform, with values higher than the average density (positive peaks) owing to the interaction with neighbor molecules and approaching the asymptotic value $g(R) = 1$ at a distance corresponding to a few molecular diameters, where any correlation between the molecules is lost, so that their density equals the average one.

It is quite interesting to compare this function for a liquid compound to the one occurring in the case of a vapor system and of a crystalline solid. In the case of vapor, we expect a step function, being zero for $R < l$ and $g(R) = 1$ if $R \geq l$, if l is the mean free path of the particle. On the contrary, for a crystalline solid we expect a number of sharp peaks at fixed values of R corresponding to the distances between the considered atom and the other components of the crystal lattices.

The nonrigid character of a liquid allows continuous displacement of the molecule's position in a kind of random walk similar to the one performed by a particle dispersed in a fluid. This is the classical Brownian motion, named by the botanist Robert Brown, who in 1827 observed with an optical microscope the motion of pollen grains suspended in water. An example of the trajectory of a particle dispersed in a liquid is shown in Fig. 3.29.

Of course, a molecule is not moving alone, owing to the strong interaction with the neighbor molecules, and therefore we should consider the mass of a group of molecules as being equivalent to the particle performing the Brownian motion. In this way, we can use the Langevin equation based on Einstein's explanation of this phenomenon. The motion is due to the random collisions with surrounding molecules possessing a kinetic energy proportional to temperature $K = (3/2)k_BT$, causing a continuous transfer of energy and momentum between the molecules. The second law of dynamics ($\mathbf{F} = m\mathbf{a}$) in this case is written as

$$\mathbf{F}_e(t) - \gamma \frac{d\mathbf{r}}{dt} = M \frac{d^2\mathbf{r}}{dt^2}, \qquad (3.173)$$

where $\mathbf{F}_e(t)$ is the stochastic force due to random collisions fluctuating very rapidly so that its time average is $\langle \mathbf{F}_e(t) \rangle = 0$; $-\gamma(d\mathbf{r}/dt)$ is the friction and M is the mass of the group of molecules. Equation (3.173) is easily solved to get the average square displacement $\langle \mathbf{r}^2(t) \rangle$.[17]

---

[17]This is done since the significant information is not linked to the motion direction; in fact, $\langle \mathbf{r}(t) \rangle = 0$. The notation $\mathbf{r}^2$ reads $\mathbf{r} \cdot \mathbf{r} = r^2$.

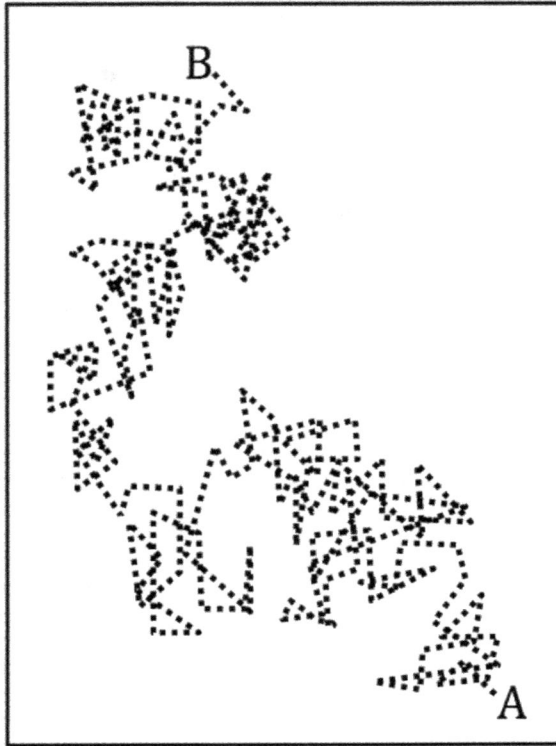

Fig. 3.29.   The typical trajectory of a particle suspended in a liquid.

With this aim we perform the scalar product by $\mathbf{r}$ of both members of the equation:

$$M\mathbf{r} \cdot \frac{d^2\mathbf{r}}{dt^2} = -\gamma\mathbf{r} \cdot \frac{d\mathbf{r}}{dt} + \mathbf{r} \cdot \mathbf{F}_e(t)$$

Then we remark that

$$\frac{d\mathbf{r}^2}{dt} = 2\mathbf{r} \cdot \frac{d\mathbf{r}}{dt}, \quad \frac{d^2\mathbf{r}^2}{dt^2} = 2\mathbf{r} \cdot \frac{d^2\mathbf{r}}{dt^2} + 2\left(\frac{d\mathbf{r}}{dt}\right)^2,$$

and perform the time average by defining $\langle \mathbf{r}^2(t) \rangle = s$. We get

$$M\frac{d^2s}{dt^2} + \gamma\frac{ds}{dt} - \left\langle 2M\left(\frac{d\mathbf{r}}{dt}\right)^2 \right\rangle = \langle \mathbf{r} \cdot \mathbf{F}_e(t) \rangle = 0.$$

Since the last term on the left side of the equation is proportional to the kinetic energy K, we write

$$M\frac{d^2 s}{dt^2} + \gamma\frac{ds}{dt} = 6k_B T.$$

This differential equation can be solved by setting the initial conditions $s = 0$ and $ds/dt = 0$.

The solution is

$$s(t) = \langle \mathbf{r}^2(t) \rangle = \frac{6k_B T}{\gamma}\left[t + \frac{M}{\gamma}(e^{-\frac{\gamma}{M}t} - 1)\right]. \qquad (3.174)$$

It shows the presence of a characteristic time M/γ such that when $t \gg M/\gamma$ the average square displacement from the initial position becomes a linear function of time:

$$s(t) = \langle \mathbf{r}^2(t) \rangle = \frac{6k_B T}{\gamma}t. \qquad (3.175)$$

In Fig. 3.30 the typical behavior of $s(t)$ is plotted for a microparticle suspended in a liquid, showing the linear asymptotic behavior usually reached in a fraction of microseconds. It is remarkable that the mass M is affecting the characteristic time, but does not affect the slope of $s(t)$ where the material properties are present only in the friction constant γ, a parameter determining the viscosity experienced by a body flowing in the considered liquid.

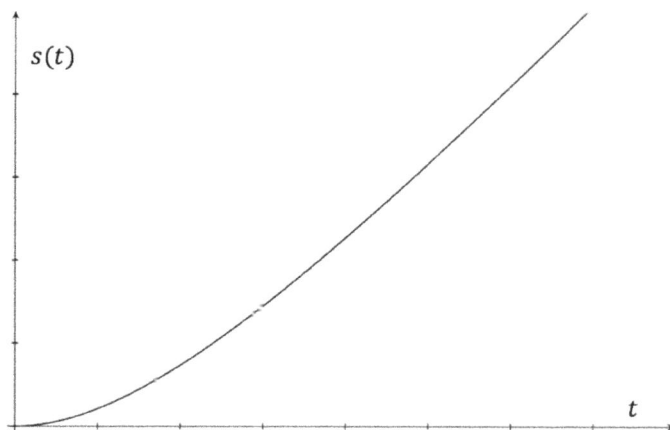

Fig. 3.30.   Plot of the average square displacement $s(t)$ in the Brownian motion.

The slope of the asymptotic linear part of $s(t)$ is rewritten using a single *diffusion constant D*,

$$D = \frac{k_B T}{\gamma},\qquad(3.176)$$

including the material properties (γ) and the temperature T. According to Eq. (3.175), we can write the time dependence of the modulus of displacement from the initial position as

$$|\mathbf{r}(t)| = \sqrt{6Dt},\qquad(3.177)$$

as reported by Einstein in his famous paper[18] on Brownian motion of particles suspended in a liquid.

Even if this phenomenon has been described for microparticles dispersed in fluids, it should also apply to a microscopic volume of the same liquid which spontaneously goes farther and farther from an initial position.

Using the Stokes expression for the friction coefficient,

$$\gamma = 6\pi R\eta,\qquad(3.178)$$

where R is the radius of the diffusing particle and η the viscosity of the medium, from Eq. (3.178) we get

$$D = \frac{k_B T}{6\pi R\eta},\qquad(3.179)$$

showing the diffusion constant dependence on the inverse of the particle size. When applying these concepts to uniform liquids, we should consider R as the molecular size. In this case, we get for simple liquids $D \approx 10^{-6} \div 10^{-5}$ cm^2/s.

A direct consequence of the described motion of the liquid molecules is the continuous transition occurring at the liquid surface from the liquid to the vapor state, causing the phenomenon called *evaporation*. We have already pointed out that liquids are disordered structures very similar to gas and therefore their molecules follow a very similar statistical distribution of velocities at a given temperature T of the medium. In this distribution, we have some molecules with a kinetic energy higher than the cohesion energy responsible for the aggregation of molecules in the liquid phase, and therefore, if these molecules are close to the open surface of the liquid, they can escape, thus passing to the vapor phase even if the temperature of the

---

[18]A. Einstein, *Annalen der Physik* **17**, 549 (1905).

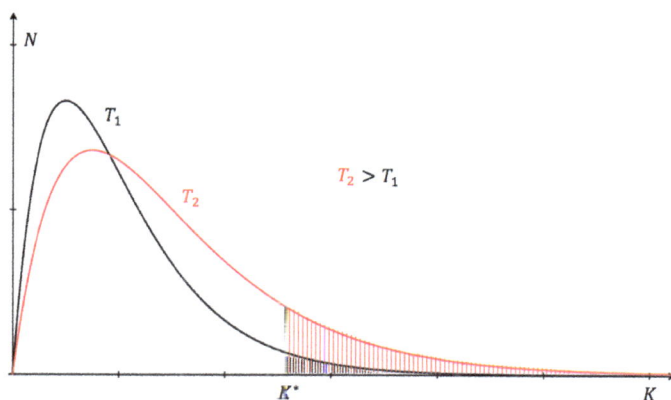

Fig. 3.31. Plot of the fraction of molecules in the liquid state versus the corresponding kinetic energy at two different temperatures. The fraction of molecules with kinetic energy higher than K^* can escape from the liquid; the shaded areas under the curves show the increasing number of molecules able to escape by increasing the temperature.

liquid is much below the thermodynamical phase transition temperature at the given pressure. This concept is illustrated in Fig. 3.31, where the fractional distribution of kinetic energies in a liquid is shown, highlighting the group of molecules which have enough energy to escape.

If the liquid surface is open the evaporation process is continuous and reduces the total volume in the condensed state, while if it is limited in a closed volume the process becomes balanced by the opposite occurring from the vapor to the liquid state; the equilibrium occurs when the vapor pressure is achieved in the closed volume, such that the number of molecules which evaporate corresponds to the condensation in the liquid phase of the same number of molecules.

The glassy state

When a liquid is cooled at a given pressure, it can crystallize if the process is slow enough to keep the material always close to the thermodynamic equilibrium. This process is associated with a sudden change of some physical parameters; among them the specific heat at the melting temperature T_m is usually considered to identify the phase transition. Crystallization leads to a spatially ordered structure where atoms are in definite locations forming a geometrical lattice, as discussed in Section 3.9. On the contrary, when the cooling is quick enough the liquid phase is still observed at temperatures $T < T_m$, and this is called *undercooled* or *supercooled liquid*; by further

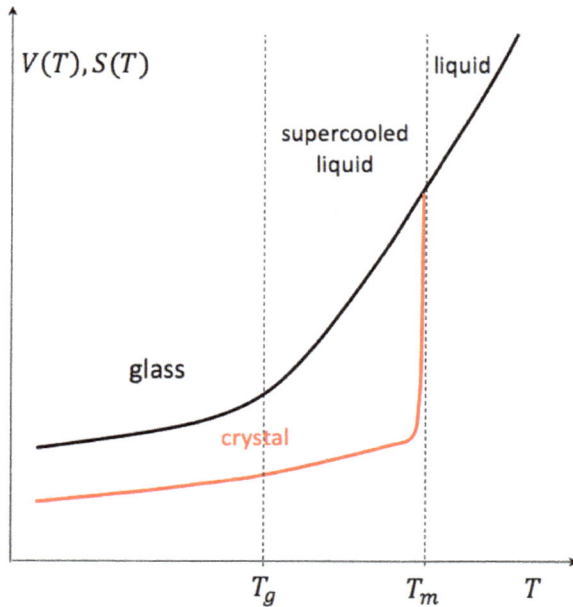

Fig. 3.32. The volume V or entropy S versus temperature for a material cooled from the liquid state. The two lines correspond to different speeds of cooling.

cooling, the molecules lose translational mobility and around the glass transition temperature T_g the system is frozen in a solid structure far from the regular arrangement typical of a crystal lattice. This solid and disordered structure, similar to that of a liquid, is distinctive of the glassy state typical of amorphous materials. In this case, the physical parameters have a smooth change versus temperature. This behavior is clearly shown in Fig. 3.32.

The reversible transition of a liquid into an amorphous material is called *glass transition*. It is a transition from a molten or rubberlike state to a solid and usually brittle state. It is still the subject of many investigations, since several aspects have not been clarified yet. Actually, the glass transition temperature T_g is not defined in a unique way, even because it changes depending on the cooling speed from the liquid state. Often it is defined by the fast change of viscosity as the liquid passes through T_g in the cooling process, as shown in Fig. 3.33, where the viscosity and specific heat of the supercooled liquid are plotted versus the temperature during cooling.

Glassy materials strongly differ from crystalline solids since they keep frozen in the solid state the same disordered structure of the supercooled

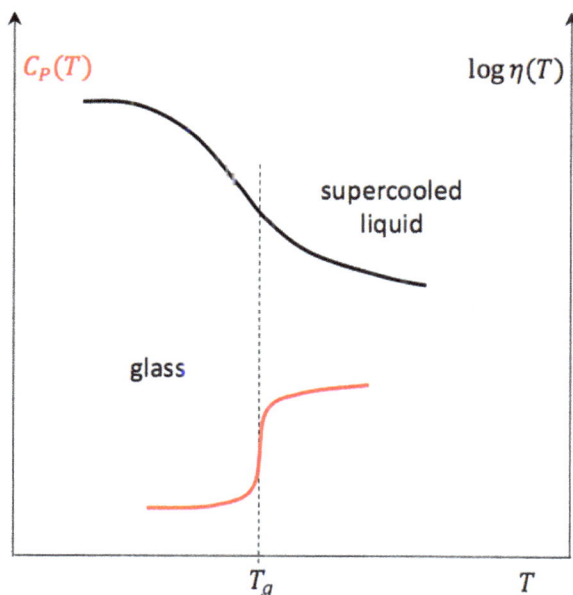

Fig. 3.33. Specific heat and viscosity versus temperature for a liquid passing through the *glass transition*.

liquid. Therefore, they resemble liquids as they have no long-range order, but they are similar to solids for what concerns mechanical properties, since no translation of molecules occurs (except on a long time scale) and only vibrations around fixed positions are allowed. On the other hand, the structural similarity to a liquid allows using the pair distribution function also in studying the structural properties of glassy materials.

3.11. Liquid crystals

Some materials present symmetries of the structure and physical properties intermediate between those of crystalline solids and liquids; for this reason, they are called *liquid crystals*. The liquid-crystalline state is actually a phase showing anisotropy of some physical properties such as most of crystals do, but at the same time it presents an aggregation state similar to liquids with a degree of fluidity dependent on the specific phase. Therefore, it is also called *mesophase* or *anisotropic liquid* to compare to the usual liquid state characterized by isotropy of the physical properties. The corresponding *mesomorphic materials* show this particular aggregation state by varying

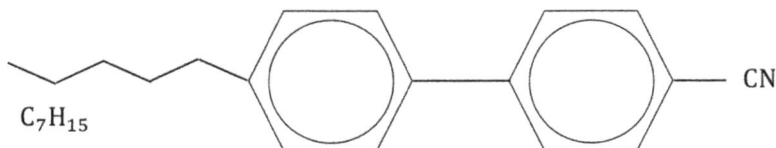

Fig. 3.34.   Chemical structure of 4-cyano-4′-pentylbiphenyl (5CB).

a thermodynamic parameter. At constant pressure and volume we call *thermotropics* the ones showing mesophases by changing temperature, while we call *lyotropics* the ones showing mesophases by changing concentrations of the components. In the following of this section, we discuss some general properties of thermotropics.

The basic structural difference between liquid crystals and isotropic liquids is the long-range orientational order present in the mesophase. Of course, in order to define a molecular orientation it is necessary to have a nonspherical shape of the molecule; in fact, typical molecules of thermotropic liquid crystals are elongated in one direction and can be represented by rigid rods whose length is in the range of 20–40 Å and width of 4–5 Å. One of the most common molecules giving rise to a mesophase is 4-cyano-4′-pentylbiphenyl (commonly known as 5CB), depicted in Fig. 3.34. It shows the typical feature of a mesogenic molecule: rodlike with a quasirigid core and two terminal groups at the sides of it.

The chemical structure determines the possible *polymorphism* of the compound, namely the possibility of getting different mesomorphic phases by changing temperature, with an increasing degree of order for lower and lower temperature.

The most common mesophases are summarized below.

In the *nematic phase*, molecules have long-range orientational order and no positional order. This means that molecules align their symmetry axis along a preferred direction, being randomly located in space. Besides the orientational order, the *smectic phases* have a layered structure with additional positional order possible in each layer; depending on the different positional order, several smectic phases can be identified (A, B, C, etc.). As a consequence, these phases have a higher degree of order (and lower symmetry) than the nematic phase, and therefore, when they are present in a compound showing the nematic phase, they appear at lower temperature. They are still fluid from a morphological point of view with an increasing viscosity as the order increases, approaching the phase transition temperature to the solid crystal state. In Fig. 3.35, the structures of an

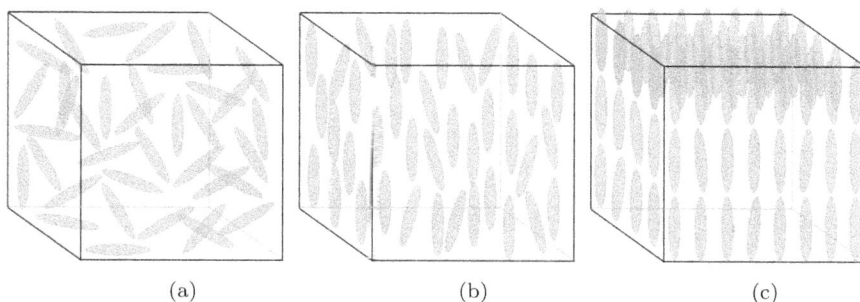

Fig. 3.35. In a finite volume is sketched the structure of (a) an isotropic liquid, (b) a nematic liquid crystal, and (c) a smectic A liquid crystal.

isotropic liquid, a nematic liquid crystal and a smectic A liquid crystal are schematically shown.

A very important property that liquid-crystalline phases may have is *chirality*.

Chirality

The concept of chirality is very important in chemistry, physics and biology, it being responsible for properties related to several fundamental processes. A clear definition is: "*The geometric property of a rigid object (or spatial arrangement of points or atoms) of being nonsuperposable on its mirror image.*"[19] This concept has been deeply investigated in chemistry for what concerns atom arrangement in molecular structure and became of paramount importance in biochemistry with the discovery of the helical structure of the DNA molecule. For what concerns supramolecular structures, chirality corresponds to the existence of a helical order with respect to a specific direction (the *helix axis*).

A nematic liquid crystal where the average orientation of molecules regularly rotates around an axis is called a *cholesteric liquid crystal*; in this case, the average molecular direction is perpendicular to the helical axis, being uniformly rotated around it moving along the axis direction. The distance corresponding to a 2π rotation defines the helical pitch. Chirality is usually a consequence of the molecular structure, which includes

---

[19]https://www.chemicool.com/definition/chirality.html

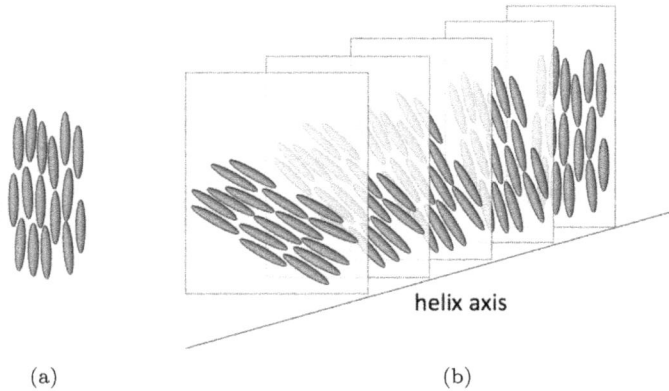

(a)   (b)

Fig. 3.36. The helical structure of a cholesteric liquid crystal (b) compared to the nematic phase (a).

a chiral moiety. It gives very important properties to smectics C, which become ferroelectrics owing to chirality. In smectics C the average molecular orientation is tilted with respect to the normal of each layer and chirality appears as a uniform rotation of the tilted orientation with respect to a helical axis normal to the smectic layers. In Fig. 3.36, the chiral structure of a cholesteric liquid crystal is depicted and compared to the corresponding nonchiral phase.

In any liquid crystal, a preferred molecular orientation is found in a volume large enough to include a large number of molecules, but small enough to allow defining a vector field $\mathbf{n}$, called the *director*, associated with each point in space. Since it represents the average molecular direction, the degree of order of the molecules can be given by a parameter taking into account the distribution $f(\theta)$ of the angle θ between the actual molecule's orientation and the director $\mathbf{n}$. This is done by defining a *scalar order parameter S*:

$$S = \frac{1}{2}\langle(3\cos^2\theta - 1)\rangle = \int_\Omega f(\theta)\left[\frac{1}{2}(3\cos^2\theta - 1)\right]d\Omega, \qquad (3.180)$$

where $d\Omega = \sin\theta\,d\theta\,d\phi$ is the infinitesimal solid angle, and θ and ϕ are the Euler angles defining the molecule's orientation with respect to $\mathbf{n}$.

In this way, $S = 0$ for the isotropic state and $S = 1$ when all the molecules are aligned along the same direction. Without going into details which can be found in texts devoted to liquid crystals, we notice that S does not depend on the local orientation of the director, but on the thermodynamic parameters, and first of all on the temperature. Defining

T_c as the phase transition temperature to the isotropic state, several experimental observations can be explained by the mean field theory,[20] where S is found to be a universal function of the parameter T/T_c, decreasing by increasing temperature down to the value $S(T_c) = 0.44$ for all the nematic–isotropic transitions and jumping to 0 in the isotropic state.

The director **n** plays a very important role in the description of the physical properties of a liquid crystal. For a real sample, its orientation is not uniform in the whole volume and it changes depending on boundary conditions and external excitations (electric, magnetic, optical fields or mechanical strains). Since very often the external factors affect the energy of a molecule for a tiny quantity as compared to the intermolecular potential, a remarkable variation of **n** occurs over distances much larger than the molecular scale. In this case, it is possible to consider the liquid crystal as a continuum elastic medium, taking into account in this way the long-range molecular correlation. This is the *continuum theory* approach, very effective in describing the collective properties of the medium through the director orientation in space **n(r)** determined by the Euler angles in the laboratory frame, as depicted in Fig. 3.37. Following this point of view, the elastic free energy density of the medium can be calculated as being dependent on the

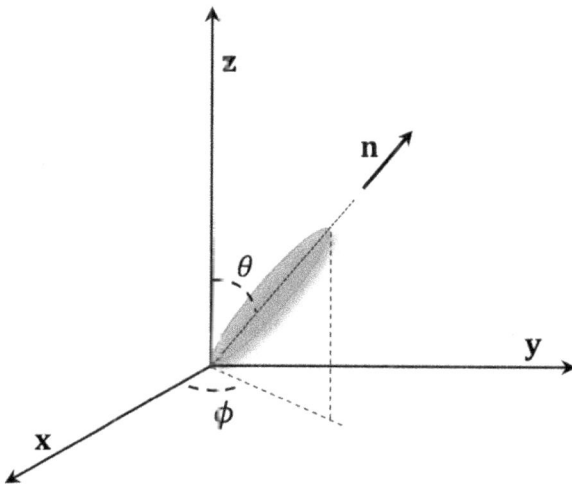

Fig. 3.37. The Euler angles determining the director orientation in space.

[20]W. Maier and A. Saupe, *Z. Naturforsch.* **A13**, 564 (1958); **A14**, 882 (1959); **A15**, 287 (1960).

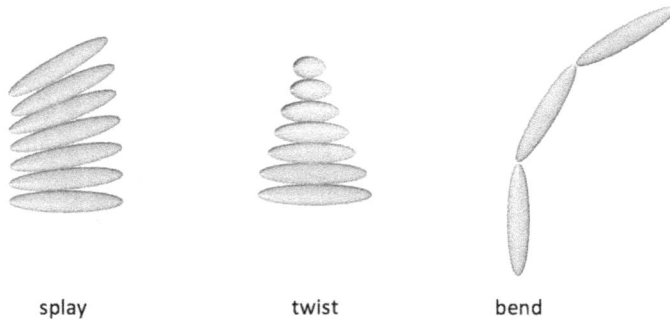

splay   twist   bend

Fig. 3.38. The possible deformations of **n** in a liquid crystal: (a) splay, (b) twist, (c) bend.

vector field **n** using three elastic constants, K_1, K_2, K_3, which correspond to the different possible deformations of **n**, namely splay, twist and bend, as shown in Fig. 3.38.

Accordingly, the bulk elastic energy density can be written as

$$F_K = \frac{K_1}{2}(\nabla \cdot \mathbf{n})^2 + \frac{K_2}{2}(\mathbf{n} \cdot \nabla \times \mathbf{n})^2 + \frac{K_3}{2}(\mathbf{n} \times \nabla \times \mathbf{n})^2. \qquad (3.181)$$

In the absence of external excitation, the stable director configuration is determined by the bulk elasticity expressed by Eq. (3.181) and by the boundary conditions given by the substrates affecting the liquid crystal orientation on the limiting surfaces.

It is very important that substrates can be properly treated in order to induce a uniform alignment on the surface over macroscopic dimensions — alignment that is transferred to the bulk owing to the long-range orientational correlation of the liquid crystal molecules. In such a way one gets a uniform orientation of macroscopic size, i.e. a *single crystal* of the considered liquid crystalline material. The most-used configurations for nematic liquid crystal samples are: (i) *planar* or *homogeneous* alignment — when the director is parallel to the boundary surfaces; (ii) *homeotropic* alignment — when the director is normal to the boundary surfaces; (iii) *twisted* alignment — when the director is parallel to the boundary surfaces, but they are rotated by $\pi/2$ with respect to each other. A sketch of planar and homeotropic cells is given in Fig. 3.39.

The interfacial interaction at the boundaries gives rise to a surface contribution to the free energy density that affects the equilibrium orientation of the director. Any other external excitation (electric, magnetic, optical fields or mechanical strains) produces an interaction energy that can

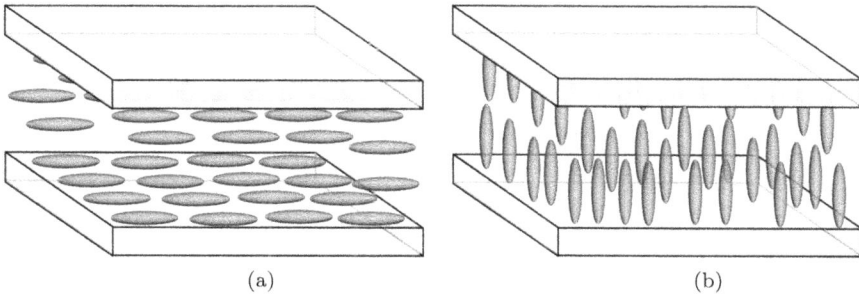

Fig. 3.39. Uniformly aligned samples of nematic liquid crystals: (a) planar, (b) homeotropic.

induce distortions from the initial orientation, and the collective response of the medium results in a modification of some physical property on a macroscopic scale. In particular, optical properties can be easily controlled by external fields and this makes liquid crystals very important for display fabrication.

In concluding this section, we note the basic difference between an isotropic liquid and a liquid crystal for what concerns the response to an external excitation. In isotropic liquids, the molecular orientation may change only on a microscopic scale: the orientation of only a few molecules around the ones excited can be affected owing to the short-range correlation; on the contrary, in liquid crystals any orientational modification at a specific location may affect the molecular orientation on a macroscopic length scale. Actually, this is the peculiarity of the collective behavior that enhances any local effect by involving a large number of molecules.

3.12. Polymers

Polymers together with liquid crystals are the main group of materials defined as *soft matter*. In fact, polymers appear as solid materials from a morphological point of view; however, their mechanical properties differ qualitatively from those of crystalline solids, owing to higher flexibility and possible rubber-like behavior, with higher mobility at the molecular level. Their peculiar chemical and physical properties allow realizing the plastic materials constituting the objects and devices in everyday use.

From a structural point of view, polymers are compounds made by many repeating units. Let us take the example of the most-used plastic material — polyethylene, obtained by polymerization of ethylene,

$$n(CH_2 = CH_2) \rightarrow$$

$$\ldots - CH_2 - CH_2 - CH_2 - CH_2 - CH_2 - CH_2 - \cdots \text{ or } [-CH_2 - CH_2 -]_n$$

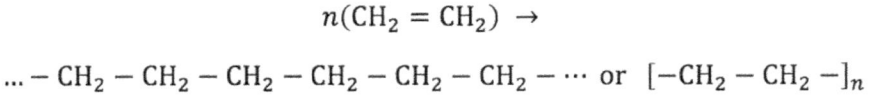

Fig. 3.40. Chemical structure of polyethylene: n molecules of ethylene ($CH_2 = CH_2$) polymerize into a long chain.

a hydrocarbon where four hydrogen atoms are bound to a pair of carbon atoms. In this case, ethylene is the *monomer* which forms the *polymer* when many units are joined by chemical bonds in a long chain. The *degree of polymerization* n is defined as the number of monomer units in the polymer chain. The chemical structure of polyethylene is shown in Fig. 3.40.

As a consequence, the molecular mass of the polymer M_p is the molecular mass of the monomer M_m multiplied by the number of monomers that play an active part in the polymerization,

$$M_p = nM_m, \tag{3.182}$$

where n may reach the value of 10,000 or more, so that the polymer molecular mass can be in the range of 10^4–10^6. However, in the polymer the molecules may differ from each other, being a mixture of species with different values of M_p, and therefore usually the average value of the molecular mass is considered. For any polymer, the *number average molecular mass* is defined as

$$M_n = \frac{\sum_{i=1}^{k} M_i N_i}{\sum_{i=1}^{k} N_i}, \tag{3.183}$$

where N_i is the number of molecules with molecular mass M_i, while k is the number of different species (in general they can be different in length of chain and/or monomer type).

The *weight average molecular mass* is defined as

$$M_w = \frac{\sum_{i=1}^{k} N_i M_i^2}{\sum_{i=1}^{k} N_i M_i}. \tag{3.184}$$

In this way, it is possible to define the polydispersity P to measure the degree of heterogeneity of the molecular size in the polymer:

$$P = \frac{M_w}{M_n}. \tag{3.185}$$

With this definition, $P \geq 1$, with $P = 1$ for monodisperse polymers that have all the molecules of the same size. Of course, polydispersity affects

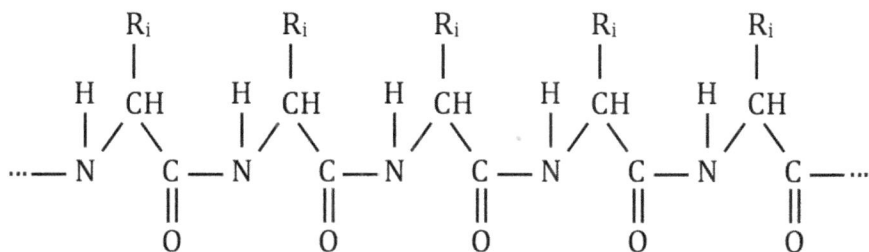

Fig. 3.41. Typical chemical structure of a protein: the nature of the amino acid residues R_i determines the specific protein.

the physical properties of the polymer, which are mainly determined by its molecular mass.

Actually, a polymer can appear also as a two- or three-dimensional network where the units repeat in different directions. Three-dimensional structures are characterized by chemical bonds between the chains, defining the *crosslinking* between the units. Such crosslinking increases the rigidity of the structure.

It is also possible that polymerization occurs between monomers of different chemical structure; in this case, one speaks of *copolymers*. Different kinds of copolymers exist depending on the sequence of the monomers in the structures: block, statistical, periodic copolymers, etc.

Among the large number of natural polymers, very important are the *biopolymers*, which are present in biological systems. Among them are proteins, DNA (deoxyribonucleic acid) and RNA (ribonucleic acid). In Fig. 3.41 is shown the chemical structure of a protein.

It is also worth noting that some polymers may present an orientational order of molecules similar to that of liquid crystals. In this case, one finds the typical liquid-crystalline properties in polymeric materials and speaks of *liquid-crystalline polymers*.

The polymerization process, i.e. the formation of chains of monomers and their propagation, can occur in many different ways, starting from the vapor, liquid or solid phase of the monomer. Chain and step polymerization are usually defined.

Chain polymerization consists of three steps (see Fig. 3.42): (i) *initiation* of the monomer molecule, i.e. the molecule becomes active to react with other monomers (activation can be induced by heat, light absorption, ionizing radiation or chemical reaction); (ii) *chain propagation*, i.e. the reaction between monomers starts and proceeds to link them in a long chain;

i) *monomer initiation*: $M \rightarrow M^*$

ii) *chain propagation*: $M^* + M \rightarrow M_1^*$

$$M_1^* + M \rightarrow M_2^*$$

$$\dots\dots\dots\dots$$

$$M_{n-1}^* + M \rightarrow M_n^*$$

iii) *chain termination*: $\left[\begin{array}{l} M_{n-1}^* + M^* \rightarrow M_n \\ M_{n-j}^* + M_j^* \rightarrow M_n \end{array}\right.$

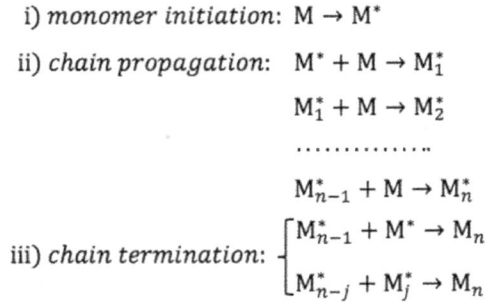

Fig. 3.42.   Steps of chain polymerization of a monomer M, leading to a chain with degree of polarization n.

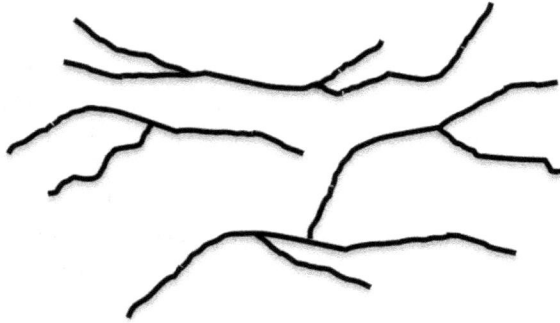

Fig. 3.43.   Chains in a branched polymer.

(iii) *chain termination*, the final process ending the chain propagation and leading to an inactive polymer molecule, via reaction between two active units (chain–chain or chain–monomer).

Step polymerization consists in the stepwise addition of bifunctional or multifunctional monomers to one another to form dimers, trimers, etc. that can react together to give rise to longer and longer chains. In this process, the interaction and link between different growing chains may occur; as a consequence, one can get branched polymers characterized by bifurcations of the polymer chain, as shown in Fig. 3.43.

Of course, there is a lot of chemistry involved in all these processes, and we leave the interested reader to find details in specialized books.

As an example of the polymerization process, we can mention the very common effect of hardening of an adhesive material. In this case, one starts from a liquid solution of monomers that polymerize in some cases owing to

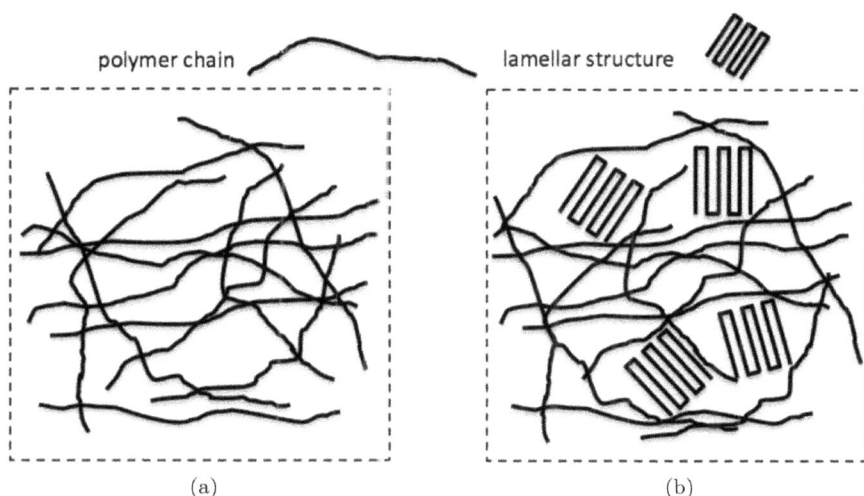

Fig. 3.44. Sketch of polymer structure: (a) amorphous polymer, (b) semicrystalline polymer.

mixture with another compound (chemically initiated polymerization); in other cases, owing to UV light (photoinitiated polymerization).

Actually, the final solid state of a polymeric material is found in two different forms: (i) a crystalline lamellar structure created by regular arrangement of the polymeric chains; (ii) an amorphous state where the polymeric chains are frozen in a disordered structure. Often, the solid structure is a mixture of crystalline and amorphous polymers where crystalline lamellae are dispersed into an amorphous matrix.

The appearance of the crystalline or amorphous state depends on the chemical structure of the material and on its thermal history.[21]

In Fig. 3.44, the structure of an amorphous polymer (a) and a semicrystalline polymer (b) are sketched.

Quite remarkable is the behavior of the amorphous polymer versus temperature. At low temperature the molecules are frozen like in glass, and therefore the polymer is in the *classy state*, being hard and rigid. By heating, the polymer becomes soft and flexible, i.e. rubber-like, and it is said to be in the *rubbery state*. The temperature characterizing the transition from the glassy state to the rubbery state is the glass transition temperature T_g.

---

[21] As usual, slow cooling gives rise to a crystal structure, and fast cooling to an amorphous state.

If the polymer also has crystalline regions, a further rise in temperature allows achieving a melting temperature, T_m. In this way, the volume of the polymer will show a change with temperature given by a superposition of the two curves as the ones (black and red) reported in Fig. 3.32 in discussing the glassy state.

The majority of the synthesized polymers are *thermoplastic*, meaning that once the polymer is formed it can be heated and re-formed many times. On the contrary, *thermosets* are polymers that after formation cannot be melted again. In this case, heating leads to degradation and permanent modification of the material properties.

3.13. Matter anisotropy

Anisotropy is strongly linked to the concept of "cause and effect", since it is related to the response of a material to a perturbation from its thermodynamic equilibrium and points out the dependence of this response on the direction of the excitation. It is usually considered from a macroscopic point of view, i.e. anisotropy is related to physical quantities taking into account the behavior of a large number of atoms or molecules.

The concept is easily understood if one considers an excitation described by a vector such as the electric field $\mathbf{E}$ which induces a polarization in a material described by the vector $\mathbf{P}$. We know that these two quantities are linked by the relation

$$\mathbf{P} = \underline{\chi}\mathbf{E}. \tag{3.186}$$

The well-known dielectric susceptibility $\underline{\chi}$ is just a scalar function of frequency in the case of an isotropic material, which must be represented by a matrix for an anisotropic material:

$$\underline{\chi} = \begin{pmatrix} \chi_{xx} & \chi_{xy} & \chi_{xz} \\ \chi_{yx} & \chi_{yy} & \chi_{yz} \\ \chi_{zx} & \chi_{zy} & \chi_{zz} \end{pmatrix}. \tag{3.187}$$

Then

$$\begin{aligned} P_x &= \chi_{xx}E_x + \chi_{xy}E_y + \chi_{xz}E_z, \\ P_y &= \chi_{yx}E_x + \chi_{yy}E_y + \chi_{yz}E_z, \\ P_z &= \chi_{zx}E_x + \chi_{zy}E_y + \chi_{zz}E_z. \end{aligned} \tag{3.188}$$

This means that each component of **P** depends on all the components of **E**, and hence the two vectors are not parallel.

By a proper axis transformation, the matrix (3.187) can be diagonalized. The new axes are called *principal axes* and represent symmetry axes of the system. In the new reference frame (XYZ), we have

$$\underline{\chi} = \begin{pmatrix} \chi_{XX} & 0 & 0 \\ 0 & \chi_{YY} & 0 \\ 0 & 0 & \chi_{ZZ} \end{pmatrix} \tag{3.189}$$

and

$$\begin{aligned} P_X &= \chi_{XX} E_X, \\ P_Y &= \chi_{YY} E_Y, \\ P_Z &= \chi_{ZZ} E_Z. \end{aligned} \tag{3.190}$$

From Eqs. (3.190) it is clear that $\chi_{XX} = \chi_{YY} = \chi_{ZZ}$ means isotropy. On the other hand, these expression clarify the meaning of anisotropy, since if $\chi_{XX} \neq \chi_{YY} \neq \chi_{ZZ}$ the same electric field $(E_X = E_Y = E_Z)$ gives rise to a different polarization $(P_X \neq P_Y \neq P_Z)$ of the material, depending on the direction of application.

In the above example, the anisotropy was discussed in relation to a vector quantity originated in the material by a vector excitation. The same behavior could be highlighted for other relations, such as

$$\begin{aligned} \mathbf{D} &= \underline{\varepsilon}\mathbf{E}, \\ \mathbf{B} &= \underline{\mu}\mathbf{H}, \\ \mathbf{J} &= \underline{\sigma}\mathbf{E}, \end{aligned} \tag{3.191}$$

and for any relation including material parameters. However, anisotropy is not limited to vector quantities. For instance, by using the last of the equations (3.191), one can easily show that a different current is induced in an anisotropic conductor by applying the same voltage along different directions. Let us write

$$\begin{aligned} J_X &= \sigma_{XX} E_X, \\ J_Y &= \sigma_{YY} E_Y, \\ J_Z &= \sigma_{ZZ} E_Z, \end{aligned} \tag{3.192}$$

with σ_{XX}, σ_{YY} and σ_{ZZ} being the electric conductivities along X, Y and Z. By considering a cubic conductor with sides parallel to these axes and length L, a voltage V applied to the cube along the different direction will produce the same electric field:

$$E_X = E_Y = E_Z = \frac{V}{L}. \tag{3.193}$$

As a consequence,

$$J_X = \sigma_{XX} \frac{V}{L},$$
$$J_Y = \sigma_{YY} \frac{V}{L}, \tag{3.194}$$
$$J_Z = \sigma_{ZZ} \frac{V}{L}.$$

We now use the definition of the electric current as $I = J \cdot A$, where $A = L^2$ is the cross section of the conductor. Therefore, we will measure three different values of the electric current along the three directions[22]:

$$I_1 = \sigma_{XX} V L,$$
$$I_2 = \sigma_{YY} V L, \tag{3.195}$$
$$I_3 = \sigma_{ZZ} V L.$$

In other words, the same cause produces a different effect. Moreover, in this case, it involves a scalar quantity as the current I.

This direction-dependent response of a material is obviously due to its structure; in general, anisotropy occurs as the long-range order increases. Gases, liquids and amorphous solid materials are highly disordered systems and have an isotropic behavior; in fact, even if liquids and amorphous materials may have a strong molecular correlation on a short-range scale, this correlation is lost on a long-range scale and therefore they are macroscopically isotropic. On the contrary, most of the crystalline solids and all liquid crystals are anisotropic, owing to the long-range molecular order in one or more directions.

Wave propagation in anisotropic media

Matter anisotropy strongly affects wave propagation. Owing to the relevance of this issue, we will give here some details starting from the tensorial

---

[22]We use the subindices $1, 2, 3$ to avoid misleading since the current is not a vector.

relationship between the vectors **D** and **E**:

$$D_x = \epsilon_{xx} E_x + \epsilon_{xy} E_y + \epsilon_{xz} E_z,$$
$$D_y = \epsilon_{yx} E_x + \epsilon_{yy} E_y + \epsilon_{yz} E_z, \qquad (3.196)$$
$$D_z = \epsilon_{zx} E_x + \epsilon_{zy} E_y + \epsilon_{zz} E_z.$$

In this case, the energy density given by Eq. (1.52) must be generalized and written as

$$u = \frac{1}{2} \mathbf{E} \cdot \mathbf{D} = \frac{1}{2} \sum_{kl} E_k \epsilon_{kl} E_l. \qquad (3.197)$$

By exploiting the Maxwell equations, it can be shown that $\epsilon_{kl} = \epsilon_{lk}$, i.e. the dielectric tensor is symmetric, which means that the corresponding matrix includes only six independent terms.[23] On the other hand, by a coordinate transformation to a frame based on the principal symmetry axes of the material, we have

$$\underline{\epsilon} = \begin{pmatrix} \epsilon_{XX} & 0 & 0 \\ 0 & \epsilon_{YY} & 0 \\ 0 & 0 & \epsilon_{ZZ} \end{pmatrix} \equiv \begin{pmatrix} \epsilon_1 & 0 & 0 \\ 0 & \epsilon_2 & 0 \\ 0 & 0 & \epsilon_3 \end{pmatrix}. \qquad (3.198)$$

Therefore,

$$u = \frac{1}{2}(\epsilon_1 E_X^2 + \epsilon_2 E_Y^2 + \epsilon_3 E_Z^2) \qquad (3.199)$$

or

$$2u = \frac{D_X^2}{\epsilon_1} + \frac{D_Y^2}{\epsilon_2} + \frac{D_Z^2}{\epsilon_3}. \qquad (3.200)$$

Given three orthogonal axis $\mathbf{D}_X, \mathbf{D}_Y, \mathbf{D}_Z$, Eq. (3.200) is an ellipsoid equation showing that the vertex of the **D** vector lies on an ellipsoid surface.

We can now consider a plane wave traveling in the anisotropic material, defining $\hat{\mathbf{k}}$ as the unit vector normal to the wave front, i.e. $\mathbf{k} = k\hat{\mathbf{k}}$. Then

$$\mathbf{E} = \mathbf{E}_0 e^{i(\mathbf{k}\cdot\mathbf{r}-\omega t)} = \mathbf{E}_0 e^{i(\frac{n\omega}{c}\hat{\mathbf{k}}\cdot\mathbf{r}-\omega t)} = \mathbf{E}_0 e^{i\omega(\frac{n}{c}\hat{\mathbf{k}}\cdot\mathbf{r}-t)}. \qquad (3.201)$$

---

[23]See for instance: M. Born and E. Wolf, *Principles of Optics*, 6th ed. (Pergamon, 1980).

Now we come back to Eqs. (1.3) and (1.4), keeping the condition $\mathbf{J} = 0$:

$$\nabla \times [\mathbf{E}_0 e^{i\omega(\frac{n}{c}\hat{\mathbf{k}}\cdot\mathbf{r}-t)}] = -\mu_0 \frac{\partial}{\partial t}[\mathbf{H}_0 e^{i\omega(\frac{n}{c}\hat{\mathbf{k}}\cdot\mathbf{r}-t)}], \qquad (3.202)$$

$$\nabla \times [\mathbf{H}_0 e^{i\omega(\frac{n}{c}\hat{\mathbf{k}}\cdot\mathbf{r}-t)}] = -\frac{\partial}{\partial t}[\mathbf{D}_0 e^{i\omega(\frac{n}{c}\hat{\mathbf{k}}\cdot\mathbf{r}-t)}]. \qquad (3.203)$$

Then

$$\mathbf{k} \times \mathbf{E} = \omega\mu_0\mathbf{H} \rightarrow \mathbf{H} = \frac{n}{c\mu_0}\hat{\mathbf{k}} \times \mathbf{E}, \qquad (3.204)$$

$$\mathbf{k} \times \mathbf{H} = -\omega\mathbf{D} \rightarrow \mathbf{D} = -\frac{n}{c}\hat{\mathbf{k}} \times \mathbf{H}. \qquad (3.205)$$

From Eq. (3.205) we see that $\mathbf{D}$ is perpendicular to both $\hat{\mathbf{k}}$ and $\mathbf{H}$, which is also perpendicular to $\hat{\mathbf{k}}$, as shown by Eq. (3.204). Therefore, $\mathbf{D}$ and $\mathbf{H}$ belong to the wave front. On the other hand, the propagation direction $\hat{\mathbf{k}}$ is different from the direction $\hat{\mathbf{s}}$ of energy propagation given by the Poynting vector $\mathbf{S} = \mathbf{E} \times \mathbf{H}$, since $\mathbf{E}$ and $\mathbf{D}$ are not parallel to each other. This is shown in Fig. 3.45.

Combining Eqs. (3.204) and (3.205), we get

$$\mathbf{D} = -\frac{n^2}{c^2\mu_0}\hat{\mathbf{k}} \times \hat{\mathbf{k}} \times \mathbf{E}. \qquad (3.206)$$

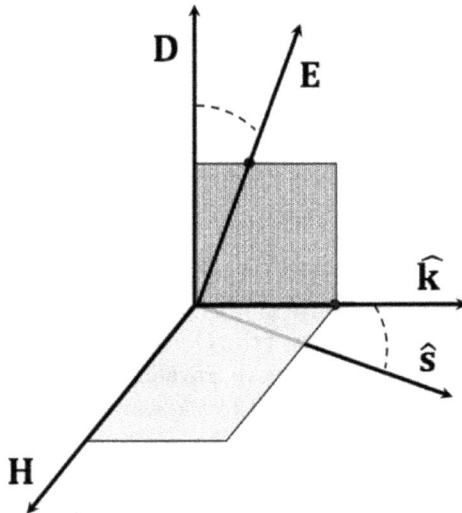

Fig. 3.45. Vector orientation for an electromagnetic wave propagating in an anisotropic medium.

We exploit the vector identity $\mathbf{a} \times (\mathbf{b} \times \mathbf{c}) = \mathbf{b}(\mathbf{a} \cdot \mathbf{c}) - \mathbf{c}(\mathbf{a} \cdot \mathbf{b})$ and write

$$\mathbf{D} = \frac{n^2}{c^2 \mu_0}[\mathbf{E} - \hat{\mathbf{k}}(\hat{\mathbf{k}} \cdot \mathbf{E})] = \frac{n^2}{c^2 \mu_0}\mathbf{E}_\perp, \tag{3.207}$$

where $\mathbf{E}_\perp$ is the field component perpendicular to $\hat{\mathbf{k}}$.[24]

For each vector component $D_i = \epsilon_i E_i = \epsilon_{ri}\epsilon_0 E_i$, we get

$$\epsilon_{ri}\epsilon_0 E_i = \frac{n^2}{c^2 \mu_0}[E_i - k_i(\hat{\mathbf{k}} \cdot \mathbf{E})] \rightarrow \epsilon_{ri} E_i = n^2[E_i - k_i(\hat{\mathbf{k}} \cdot \mathbf{E})], \tag{3.208}$$

where k_i are the components of the unit vector $\hat{\mathbf{k}}$. Then

$$E_i = \frac{n^2 k_i(\hat{\mathbf{k}} \cdot \mathbf{E})}{n^2 - \epsilon_{ri}}. \tag{3.209}$$

Finally, we multiply by k_i and sum over the three equations dividing by the common factor $(\hat{\mathbf{k}} \cdot \mathbf{E})$ to obtain

$$\frac{k_X^2}{n^2 - \epsilon_{r1}} + \frac{k_Y^2}{n^2 - \epsilon_{r2}} + \frac{k_Z^2}{n^2 - \epsilon_{r3}} = \frac{1}{n^2}. \tag{3.210}$$

This equation (a *Fresnel equation*) is quadratic in n, producing two solutions n_1 and n_2 (evidently, the solutions $-n_1$ and $-n_2$ are also allowed, but they correspond just to a reversal of the sign of the phase velocity, i.e. of the propagation direction).

With the given values of n_1 and n_2, we can solve the problem for the field component ratios $E_X : E_Y : E_Z$ and $D_X : D_Y : D_Z$. They are real quantities that correspond to linearly polarized waves.

In summary, we have found the fundamental properties of wave propagation in anisotropic media:

(a) Given an incident plane wave, two waves are originated in the medium traveling with different phase velocities corresponding to the different refractive indices n_1 and n_2;

(b) These two waves are linearly polarized; it can also be demonstrated that their polarizations are perpendicular to each other.[23]

By defining

$$X = \frac{D_X}{2u\sqrt{\epsilon_0}}, \quad Y = \frac{D_Y}{2u\sqrt{\epsilon_0}}, \quad Z = \frac{D_Z}{2u\sqrt{\epsilon_0}}, \tag{3.211}$$

---

[24]In fact, $\hat{\mathbf{k}}(\hat{\mathbf{k}} \cdot \mathbf{E}) = \mathbf{E}_\|$, the field component parallel to $\hat{\mathbf{k}}$, and obviously $[\mathbf{E} - \mathbf{E}_\|] = \mathbf{E}_\perp$.

we can write Eq. (3.200) as

$$\frac{X^2}{n_X^2} + \frac{Y^2}{n_Y^2} + \frac{Z^2}{n_Z^2} = 1. \tag{3.212}$$

This equation represents an ellipsoid with axes parallel to the **X**, **Y**, **Z** directions, where n_X, n_Y, n_Z are the lengths of semiaxes. It is called the *index ellipsoid* or *optical indicatrix*.

The index ellipsoid is a useful tool for finding the phase velocities and the refractive indices of the two waves traveling in the anisotropic medium. In fact, starting from a given propagation direction of the incident wave, we can: (i) determine the intersection between the planes through the origin normal to the propagation direction and the index ellipsoid; (ii) in this way, the semiaxes of the ellipse found by the intersection are the two refractive indices seen by the traveling waves, solutions to the Fresnel equations; (iii) these two axes are parallel to the oscillation direction of the **D** vectors which correspond to the two waves. A clear demonstration of these statements is given in the classic book by M. Born and E. Wolf.[23]

These basic results allow accounting for the peculiar optical properties of anisotropic materials and realizing basic optical devices such as polarizers, waveplates and phase compensators. Actually, the most important consequence of the anisotropic wave propagation is *birefringence*. In fact, since the Fresnel equation provides two solutions, namely n_1 and n_2, according to Snell's law we have, in general, two refracted rays, traveling in different directions corresponding to waves orthogonally polarized to each other. As a result, at the exit of the anisotropic media we may be able to have two light beams spatially separated and linearly polarized even if the incoming beam was not polarized. Separation does not occur when the incidence angle is zero; anyway, the two waves are delayed to each other owing to the different phase velocity. Therefore, by properly adjusting the material thickness with respect to the light wavelength, it is possible to control the polarization state of the outcoming wave due to the superposition of the two waves traveling in the medium, according to what was discussed in Section 1.3.

Many crystals have a rotational symmetry axis (*optic axis*); in this case, taking it to be coincident with the **Z** axis, we have $\epsilon_1 = \epsilon_2$ and consequently $n_X = n_Y$.

The ellipsoid equation reduces to

$$\frac{X^2 + Y^2}{n_o^2} + \frac{Z^2}{n_e^2} = 1, \tag{3.213}$$

where $n_o \equiv n_X = n_Y$ and $n_e \equiv n_Z$. These are called the ordinary (n_o) and the extraordinary (n_e) refractive index. In this case, one wave travels with a phase velocity given by $v_{po} = c/n_o$ and is called the *ordinary wave*, while the other (the *extraordinary wave*) travels with a phase velocity given by $v_{pe} = c/n(\theta)$ dependent on the direction of propagation of the incoming wave with respect to the optic axis. The refractive index $n(\theta)$ of the extraordinary wave can be determined[25] using Eq. (3.210) and taking into account the geometrical construction given in Fig. 3.46:

$$\frac{1}{n^2(\theta)} = \frac{\cos^2(\theta)}{n_o^2} + \frac{\sin^2(\theta)}{n_e^2}, \tag{3.214}$$

where $n(\theta = 0) = n_0$ and $n(\theta = \pi/2) = n_e$.

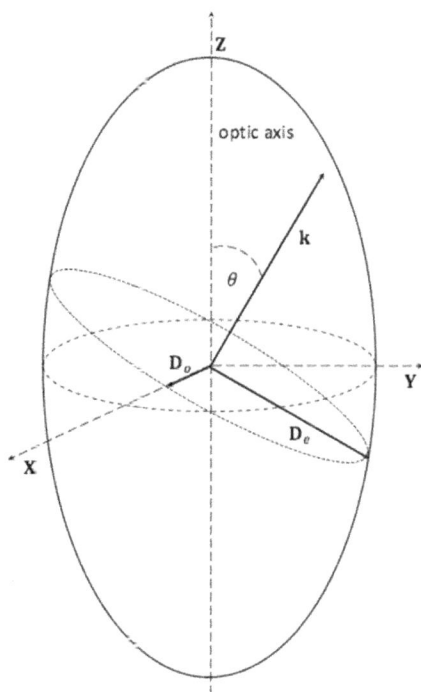

Fig. 3.46. The index ellipsoid and determination of the polarization of the ordinary ($\mathbf{D}_o$) and the extraordinary ($\mathbf{D}_e$) wave for a given wave vector $\mathbf{k}$ of the incident wave. The lengths of the axis of the ellipse coming from the intersection between the plane normal to $\mathbf{k}$ and the index ellipsoid are n_o and $n(\theta)$.

---

[25] Details of calculation can be found in: M. Born and E. Wolf, *Principles of Optics*, 6th ed. (Pergamon, 1980), pp. 670–680.

We speak of a uniaxial positive crystal when $n_o < n_e$ and a uniaxial negative crystal when $n_e < n_o$.

A few remarks can be made using Eq. (3.214) to highlight some special cases of interest:

(i) When the incoming wave has a propagation direction parallel to the optic axis $\mathbf{Z}$ ($\theta = 0$), the two waves traveling in the medium see the same refractive index n_o because the index ellipsoid has rotational symmetry around $\mathbf{Z}$. This occurs for any polarization of the incoming beam, and therefore the material behaves as an isotropic one.

(ii) When the incoming wave has a propagation direction perpendicular to the optic axis $\mathbf{Z}$ ($\theta = \pi/2$), a generic polarization produces two waves traveling with refractive indices n_o and n_e respectively, which are the semiaxes of the intersection ellipse.

(iii) When $\theta = \pi/2$ and the polarization direction is parallel to the optic axis, we have only one wave traveling in the medium as a pure extraordinary wave seeing the refractive index n_e. On the contrary, if the polarization is perpendicular to the optic axis, the single wave traveling in the medium is a pure ordinary wave seeing the refractive index n_o.

Chapter 4

Electrical Properties of Crystals

4.1. The free electron model

The electrical properties of matter include a wide range of phenomena that occur when an external electric field is applied in the space volume occupied by the considered material. Actually, electrical properties are mainly concerned with the induced charge movement which leads to current flow and/or to charge polarization. Additionally, the occurrence of electrical current or electrical polarization and their specific features depend on the material's structure, which also determines the different contributions of electrons and ions to these phenomena. In solid materials, electrical conduction is due to electrons while in fluids ions can play a major role. In this chapter, we focus on electrical conduction in crystalline solids, but many of the concepts, which we discuss here, can be extended to other materials.

We know that materials behave in a different way when submitted to an electric field. In some of them, the electrical stimulus gives rise to charge migration on a macroscopic length scale. These materials are called *conductors*.

Some basic properties of electrical conductors can be explained following the simple model of the *free electron*. This starts from considering the valence electrons of each atom participating in the metallic bonding, thus being delocalized in the solid volume where they are able to move on a macroscopic scale. In this volume, they have a constant potential energy (set $= 0$), and thus they possess only kinetic energy. In accordance with this model, Fig. 4.1 is drawn. It shows that in each lattice site of a metallic conductor is located a positive ion made up of the atomic nucleus (charge

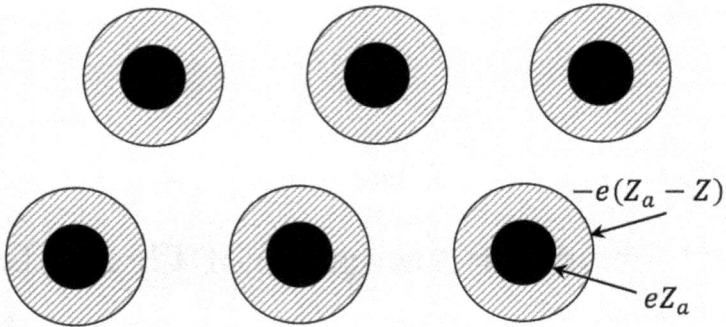

Fig. 4.1.   Scheme of the metallic conductor in the free electron model. The black circles represent the proton charge, surrounded by the tightly bound electrons (dashed circles). The gray background represents the cloud made by the conduction electrons.

eZ_a, with Z_a being the atomic number) and the tightly bound electrons $(Z_a - Z)$, while a negative charge $-eNZ$ is diffused over the crystal volume (Z is the number of valence electrons in each atom, and N the number of atoms in the lattice sites).

Then the number NZ is the number of the conduction electrons, i.e. the electrons participating in the electrical conduction process. It is quite interesting to note that most of the volume of the metallic conductor is occupied by the conduction electrons, leaving a small fraction of it to the positive ions. For instance, in sodium the ion Na^+ includes ten electrons occupying levels $1s$, $2s$, $2p$, with an average radius of the ion of $0.98\,\text{Å}$, while the distance from the nearest neighbor is $3.66\,\text{Å}$; then the positive ions fill only 15% of the total volume.

The basic assumptions of the free electron model are: (a) the interaction between electrons and ions is negligible; (b) the electron motion consists of segments changing direction suddenly owing to instantaneous collisions; (c) an average collision time τ is assumed, and as a consequence the collision probability per unit time is $1/\tau$; (d) collisions lead to equilibrium between electrons and the surrounding medium.

The density of conduction electrons can be easily evaluated. According to the law of Avogadro, we have $N_A = 6.022 \cdot 10^{23}$ atoms/mole and the density of moles/cm^3 is given by ρ/M_A, between the mass density ρ and the atomic mass M_A of the considered atom. Since each atom contributes with Z electrons, the density of conduction electrons is

$$n = ZN_A\frac{\rho}{M_A}. \tag{4.1}$$

Table 4.1. Density of conduction electrons for several metals [data from: N.W. Ashcroft and N.D. Mermin, *Solid State Physics* (Harcourt, 1976)].

Element	Z	$n(\text{cm}^{-3}) \cdot 10^{-22}$
Cu	1	8.47
Ag	1	5.86
Au	1	5.90
Be	2	24.7
Ca	2	4.61
Fe	2	17.0
Zn	2	13.2
Al	3	18.1
In	3	11.5
Sn	4	14.8
Pb	4	13.2
Bi	5	14.1
Sb	5	16.5

In Table 4.1, the free electron density is reported for a number of metals at room temperature and atmospheric pressure, resulting in the range of $10^{22} - 10^{23} \text{cm}^{-3}$.

This model was developed before the establishment of quantum mechanics. In the first version (Drude's model), the electron gas was considered as an ideal gas, and then its properties were described by the Maxwell–Boltzmann distribution function $f(E,T)$, giving the probability that at temperature T one conduction electron has energy between E and $E + dE$; using Eq. (3.141) we can write

$$f(E,T) = Be^{-\frac{E}{k_B T}}, \tag{4.2}$$

with B being a suitable factor that can be obtained by the development of statistical calculations.

Later on, the theory required proper correction consequent to the quantum properties of electrons, which make them obey Pauli's principle and, as a result, the Fermi–Dirac distribution function (3.144). For this reason, the free conduction electrons are denominated as a whole *Fermi gas*.

Therefore, each electron must fulfill the Schrödinger time-independent equation for a free particle:

$$\mathcal{H}\phi(\mathbf{r}) = -\frac{\hbar^2}{2m}\nabla^2\phi(\mathbf{r}) = E\phi(\mathbf{r}). \tag{4.3}$$

Since the bulk properties of a piece of metal, such as resistivity or specific heat, should be independent of the shape, we choose a cube with side length L ($V = L^3$).

In Chapter 2, this problem has been solved in one dimension (infinite potential well); we just need to extend the discussion to three dimensions. The solutions to Eq. (4.3) are plane waves (eigenfunctions of energy and momentum):

$$\phi_k(\mathbf{r}) = \frac{1}{V} e^{i\mathbf{k}\cdot\mathbf{r}} = \frac{1}{V} e^{i(k_x x + k_y y + k_z z)}, \tag{4.4}$$

where $1/V$ is the normalization factor.

The one-dimensional problem has been solved by choosing $\phi_k(\mathbf{r}) = 0$ at the boundaries of the potential well. This choice could also be done in this case, but the analytic procedure would be more complex. Since it can be shown that the bulk properties of a solid are not dependent on the chosen boundary conditions, it is more suitable to choose the ones suggested by Born and von Karmann, which require periodicity of the wave function at the boundaries:

$$\phi_k(x + L, y + L, z + L) = \phi_k(x, y, z). \tag{4.5}$$

This choice corresponds to a division of the whole volume of a solid in cubic subsections with the same physical properties (but not necessarily cubic, since one might choose different values of L along the three directions).

Applying the boundary condition (4.5) to the plane wave (4.4), we have

$$\exp i(k_x[x + L] + k_y[y + L] + k_z[z + L]) = \exp i(k_x x + k_y y + k_z z). \tag{4.6}$$

Then

$$k_x = q\frac{2\pi}{L}, \quad k_y = s\frac{2\pi}{L}, \quad k_z = t\frac{2\pi}{L}, \tag{4.7}$$

with q, s and t being integer numbers. In other words, the values of the electron wave vector are limited to the ones fulfilling the condition (4.7).

This quantization of the values of $\mathbf{k}$ leads to quantization of energy,

$$E = \frac{\hbar^2}{2m} k^2 = \frac{\hbar^2}{2m}(k_x^2 + k_y^2 + k_z^2), \tag{4.8}$$

and obviously of momentum,

$$\mathbf{p} = \hbar\mathbf{k} = \hbar(k_x, k_y, k_z). \tag{4.9}$$

The values of $\mathbf{k}$ can be displayed as an array of points in the space of momenta defined by the axes $\mathbf{k}_x, \mathbf{k}_y, \mathbf{k}_z$, showing the possible states of the free electron in metals. From Eqs. (4.7) we see that the distance between two neighbor points along each axis is $2\pi/L$; since L is a macroscopic length, this quantity is small and the number of possible states is extremely large. For this reason, it is more convenient to deal with $\mathbf{k}$ as a continuous quantity, taking into account that not all the values will be allowed for the electron. In fact, we have one value of $\mathbf{k}$ in each volume $(2\pi/L)^3$ in $\mathbf{k}$-space. Then we can calculate the number of allowed states $g(k)dk$, defining $g(k)$ as the density of states in the range of values between k and $k + dk$. In order to perform this calculation, we consider the volume between two spheres with the radii k and $k + dk$, respectively. This volume is simply obtained by differentiating the sphere volume:

$$d\left(\frac{4}{3}\pi k^3\right) = 4\pi k^2 dk.$$

Since the density of k values is $(L/2\pi)^3$, we have

$$g(k)dk = \left(\frac{L}{2\pi}\right)^3 4\pi k^2 dk = \left(\frac{Vk^2}{2\pi^2}\right)dk, \qquad (4.10)$$

where $V = L^3$ is the volume. Then, the density of states for each k is

$$g(k) = \left(\frac{Vk^2}{2\pi^2}\right). \qquad (4.11)$$

Since each state in $\mathbf{k}$ space can be filled with two electrons with opposite spins, the corresponding available states in energy are

$$g(E)dE = 2g(k)dk = 2\left(\frac{Vk^2}{2\pi^2}\right)dk, \qquad (4.12)$$

defining $g(E)$ as the density of state in energy.

The expression (4.12) allows writing

$$g(E) = 2g(k)\frac{dk}{dE}. \qquad (4.13)$$

On the other hand from Eq. (4.8) we have

$$k = \frac{\sqrt{2mE}}{\hbar}, \quad \frac{dk}{dE} = \frac{1}{2}\frac{\sqrt{2m}}{\hbar\sqrt{E}}, \qquad (4.14)$$

and finally

$$g(E) = 2\left(\frac{V}{2\pi^2}\right)\frac{2mE}{\hbar^2}\left[\frac{1}{2}\frac{\sqrt{2m}}{\hbar\sqrt{E}}\right] = \left(\frac{V}{2\pi^2\hbar^3}\right)(2m)^{3/2}\sqrt{E}. \qquad (4.15)$$

It is important to underline that the energy density of states has been
obtained using the function $E(k)$ given by Eq. (4.8), which is fulfilled
when the behavior of the electron can be approximated to that of a free
particle with zero energy at $k = 0$.

First of all, we should note that when the energy is written as

$$E(k) = E_n + \frac{\hbar^2 k^2}{2m}$$

(as is often used near the bottom of conduction bands in semiconductors), this corresponds to a shift of the energy zero, such that in
Eq. (4.15) the quantity $E - E_n$ must replace E.

However, it is more important to underline that the free electron
condition may not be fulfilled for electrons in energy bands (as we will
see later in this chapter), and therefore a different expression of the
density of states can be found. The calculation of the general expression
needed to obtain the density of states is beyond the aim of this book
and can be found in many basic texts on solid state physics. Anyway,
we recall the general expression, which can be written as

$$g(E) = \frac{1}{(2\pi)^3} \int_S^E \frac{dS}{|\nabla_k E|},$$

where S is a surface in the $\mathbf{k}$ space at constant E and the dispersion
relation $E(k)$ affects the result through the gradient $\nabla_k E$.

Given N free electrons at $T = 0\,\mathrm{K}$ (i.e. when the thermal energy
$k_B T = 0$), the electrons will fill the available energy states, increasing their
energy if all the states at lower energy are occupied, following in this way
Pauli's principle. As a consequence, the available energy states will be
filled from the lower energy to a certain value called the *Fermi energy*,
E_F. Therefore, at $T = 0\,\mathrm{K}$ the number of filled states N is equal to the sum
of all the available states from 0 to E_F:

$$N = \int_0^{E_F} g(E)dE = \frac{2}{3}\left(\frac{V}{2\pi^2\hbar^3}\right)(2m)^{3/2}E_F^{3/2} = \frac{2}{3}E_F g(E_F), \quad (4.16)$$

which can be used to get E_F:

$$E_F = \frac{\hbar^2}{2m}\left(\frac{3\pi^2 N}{V}\right)^{2/3}. \quad (4.17)$$

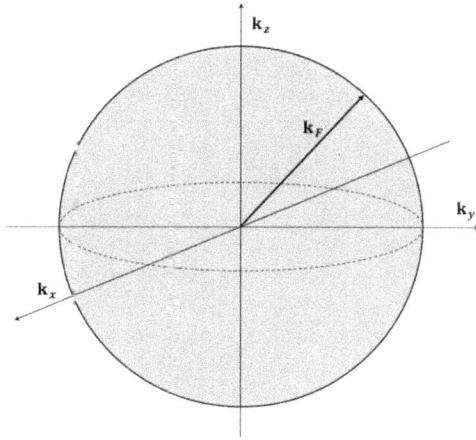

Fig. 4.2. The Fermi surface at $T = 0\,\mathrm{K}$ for free electrons.

Since for free electrons $E_F = \hbar^2 k_F^2 / 2m$, this means that all the **k** states from $k = 0$ to k_F are filled, with

$$k_F = \left(\frac{3\pi^2 N}{V} \right)^{1/3},$$ (4.18)

allowing one to get the maximum electron velocity at $T = 0\,\mathrm{K}$:

$$v_F = \frac{\hbar k_F}{m} = \frac{\hbar}{m} \left(\frac{3\pi^2 N}{V} \right)^{1/3}.$$ (4.19)

In other words, in the **k**-space, at $T = 0\,\mathrm{K}$, the filled states are the ones available in the sphere with the center at $\mathbf{k} = 0$ and radius k_F. The surface of the sphere is called the Fermi surface and is represented in Fig. 4.2.

It is important to be aware of the order of magnitude of these quantities. Typical values of E_F are a few eV, while k_F is of the order of the inverse of the lattice constants (the order of $10^{-8}\,\mathrm{cm}^{-1}$) and the corresponding v_F is of the order of $10^8\,\mathrm{cm/s}$ (it is worth underlining that for a "classic" ideal gas at $T = 0\,\mathrm{K}$ the kinetic energy and the velocity are zero!).

The *Fermi temperature* is usually defined from $E_F = k_B T_F$, i.e. $T_F = E_F / k_B$. It comes out that T_F is higher than $10^4\,\mathrm{K}$ — actually higher than the temperature of phase transition of metals between the liquid and the vapor state, the parameter showing the extremely high effective temperature of the electrons at the Fermi surface. Based on data on electron density reported in Table 4.1, it is possible to calculate the

Table 4.2. Parameters at the Fermi surface for some metals at room temperature, according to the free electron model [data from: C. Kittel, *Introduction to Solid State Physics*, 8th ed. (Wiley & Sons, 2005)].

Metal	electrons/cm^3	E_F(eV)	k_F(cm^{-1})	v_F(cm/s)	T_F(K)
Cu	$8.45 \cdot 10^{22}$	7.0	$1.36 \cdot 10^8$	$1.57 \cdot 10^8$	$8.1 \cdot 10^4$
Ag	$5.85 \cdot 10^{22}$	5.5	$1.20 \cdot 10^8$	$1.39 \cdot 10^8$	$6.4 \cdot 10^4$
Be	$24.2 \cdot 10^{22}$	14.1	$1.93 \cdot 10^8$	$2.23 \cdot 10^8$	$16.4 \cdot 10^4$
Zn	$13.1 \cdot 10^{22}$	9.4	$1.57 \cdot 10^8$	$1.82 \cdot 10^8$	$10.9 \cdot 10^4$
In	$11.5 \cdot 10^{22}$	8.6	$1.50 \cdot 10^8$	$1.74 \cdot 10^4$	$10.0 \cdot 10^4$
Pb	$13.2 \cdot 10^{22}$	9.4	$1.57 \cdot 10^8$	$1.82 \cdot 10^8$	$10.9 \cdot 10^4$
Sn	$14.5 \cdot 10^{22}$	10.0	$1.62 \cdot 10^8$	$1.88 \cdot 10^8$	$11.6 \cdot 10^4$

physical parameters at the Fermi surface. They are given in Table 4.2 for some metals.

Taking as room temperature the standard value of $T = 300\,\mathrm{K}$, we understand that $k_B T \ll k_B T_F$, in fact, $k_B T = 0.026\,\mathrm{eV} \ll E_F$. For this reason, we expect that only the electrons with energy close to E_F have the opportunity to be thermally excited to a higher energy state when the temperature increases. In fact, states with $E > E_F$ are the only ones available to be filled. On the contrary, electrons with energy $E \ll E_F$ cannot reach any available empty state when acquiring the additional thermal energy $k_B T \ll E_F$.

This process takes place in agreement with the Fermi–Dirac distribution followed by electrons, where we can replace μ with E_F:

$$f(E,T) = \frac{1}{e^{\frac{E - E_F}{k_B T}} + 1}. \tag{4.20}$$

As pointed out in Chapter 3, the Fermi–Dirac distribution changes by increasing T in an energy range of the order of a few $k_B T$, as shown in Fig. 4.3.

Accordingly, the concentration of conduction electrons (number/volume) at thermal equilibrium at a temperature T is obtained by multiplying the density of states $g(E)$ by the probability distribution $f(E,T)$,

$$n(E,T) = g(E)f(E,T), \tag{4.21}$$

and for the free electrons Eq. (4.21) become

$$n(E,T) = \left(\frac{V}{2\pi^2 \hbar^3}\right)(2m)^{3/2}\sqrt{E}\,\frac{1}{e^{\frac{E - E_F}{k_B T}} + 1}, \tag{4.22}$$

which is plotted in Fig. 4.4.

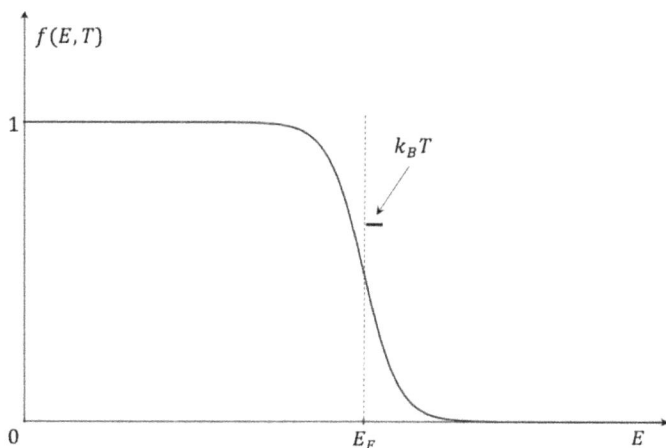

Fig. 4.3. The Fermi–Dirac probability distribution at temperature T. The range of energy where the modifications occur is shown.

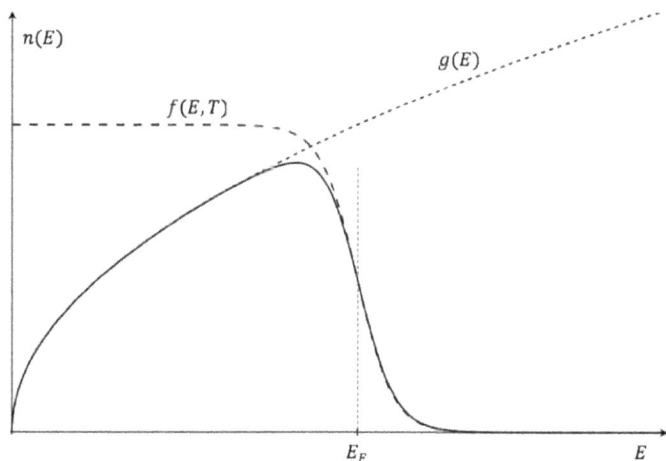

Fig. 4.4. The concentration of electrons at a given energy at $T > 0\,\mathrm{K}$.

This discussion has shown the strong difference between a classical ideal gas and the quantum electron gas. Only a small fraction of electrons of the order of $T/T_F \ll 1$ are able to gain thermal energy, and therefore we expect electrons to make a negligible contribution to the specific heat of a crystal. In fact, it can be demonstrated and measured that the main contribution to it comes from the thermal excitation of vibrations of the lattice, being the electron contribution relevant only at very low temperatures.

4.2. Electrical conduction in metals

Taking into account the properties of the free electron gas and the
quantum principle of correspondence allowing the use of the classical laws of
mechanics for the expectation values of the physical quantities, it is possible
to apply the classical equation of motion to the electron in metal to describe
the electrical conduction effect. Then the electron subject to an electric field
ε is driven by a force $-e\boldsymbol{\mathcal{E}}$, while impurities and lattice vibrations produce
a friction dependent on the velocity:

$$-e\boldsymbol{\mathcal{E}} - \gamma\mathbf{v} = m\frac{d\mathbf{v}}{dt}, \qquad (4.23)$$

where γ is the friction coefficient.

The electron velocity can be written as $\mathbf{v} = \mathbf{v}_f + \mathbf{v}_{\mathcal{E}}$, where $\mathbf{v}_f$ is the
free electron speed when no field is applied. It abruptly changes direction
after each collision against lattice ions, in such a way that free electrons are
running in all directions and the average value $\langle \mathbf{v}_f \rangle = 0$. On the contrary,
the applied electric field produces an additional *drift velocity* $\mathbf{v}_{\mathcal{E}}$, the same
for each electron that gives rise to a net electrons current. Its steady state
value is defined as $\mathbf{v}_D$. Therefore, considering the average electron velocity,
Eq. (4.23) can be rewritten as

$$-e\boldsymbol{\mathcal{E}} - \frac{m\mathbf{v}_{\mathcal{E}}}{\tau} = m\frac{d\mathbf{v}_{\mathcal{E}}}{dt}. \qquad (4.24)$$

Under steady state conditions, $d\mathbf{v}_{\mathcal{E}}/dt = 0$. Then

$$\mathbf{v}_D = -\frac{e\tau}{m}\boldsymbol{\mathcal{E}}, \qquad (4.25)$$

with τ being the average time interval between two collisions, when the
electric field is effective on the electron.

It is easy to show that the friction coefficient $\gamma = m/\tau$.

Since the average velocity has the same direction of the field, we can
project the motion equation along this direction, taking into account
that the electron velocity is opposite to the field direction:

$$-e\mathcal{E} + \gamma\mathrm{v} = m\frac{d\mathrm{v}}{dt}.$$

Under steady state conditions,

$$\frac{dv}{dt} = 0 - v \equiv v_D \rightarrow \gamma = \frac{e\mathcal{E}}{v_D}.$$

Then

$$-e\mathcal{E} + \frac{e\mathcal{E}}{v_D}v = m\frac{dv}{dt} \rightarrow -\frac{e\mathcal{E}}{v_D}(v_D - v) = m\frac{dv}{dt} \rightarrow -\frac{e\mathcal{E}}{mv_D}dt = \frac{dv}{v_D - v}.$$

Integration gives

$$-\frac{e\mathcal{E}}{mv_D}t + C = -\ln(v_D - v) \rightarrow v - v_D = C'e^{-\frac{e\mathcal{E}}{mv_D}t}.$$

Since at $t = 0$ and $v = 0$, we have $v(0) - v_D = C' \rightarrow -v_D = C' \rightarrow$
$v = v_D - v_D e^{-\frac{e\mathcal{E}}{mv_D}t} = v_D\left(1 - e^{-\frac{t}{\tau}}\right).$

This expression shows that v asymptotically approaches the steady state value v_D with a characteristic time

$$\tau = \frac{mv_D}{e\mathcal{E}}.$$

By comparing it to $\gamma = e\mathcal{E}/v_D$, we have

$$\gamma = \frac{m}{\tau}.$$

We note that $\mathbf{v}_D =$ when the field is absent. Using Eq. (4.25), we can evaluate v_D to be of the order of 10^{-2} m/s; then $v_D \ll v_f$. However, as already mentioned, the average over all the possible directions gives $\langle \mathbf{v}_f \rangle = 0$, while $\mathbf{v}_D$ is the same for all the electrons, thus producing a nonzero flow of charges in the opposite direction of the field (owing to the negative charge of the electron).

With n being the concentration of conduction electrons, the current density becomes

$$\mathbf{J} = -ne\,\mathbf{v}_D = \frac{ne^2\tau}{m}\mathcal{E}, \qquad (4.26)$$

which allows defining the electrical conductivity as

$$\sigma = \frac{ne^2\tau}{m} \qquad (4.27)$$

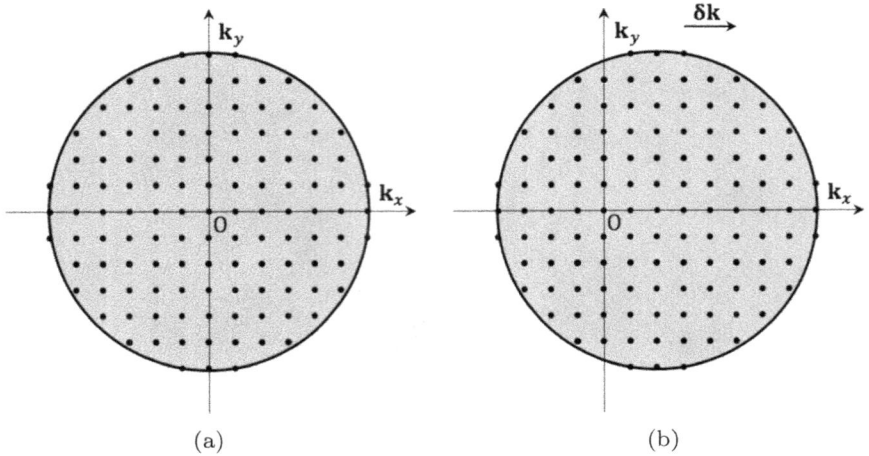

(a)   (b)

Fig. 4.5. A shift of the Fermi sphere projected on the $k_x k_y$ plane when an electric field is applied along the **x** axis. The black dots represent a visualization of the available quantized states of the electron wave vector.

or the electrical resistivity as

$$\varrho = \frac{m}{ne^2\tau}. \tag{4.28}$$

Before discussing this result, it is worth showing the effect of the electric field on the distribution of **k** vectors in the **k**-space. The additional drift velocity $\mathbf{v}_D$ increases the wave vector of each electron by the quantity

$$\delta\mathbf{k} = \frac{m}{\hbar}\mathbf{v}_D. \tag{4.29}$$

Thus, by taking the **x** axis in the direction of $\mathbf{v}_D$ and looking at the projection of the Fermi sphere in the **k**-space on the $k_y k_x$ plane, the result is a shift of the sphere as shown in Fig. 4.5.

Equation (4.26) shows the linear dependence of the current density on the electric field. It can be easily rewritten in the usual form of the Ohm law, taking into account the section A of the conductor normal to the electron flow to define the current $I = J \cdot A$ and the electric potential difference between two sections at distance d, $V = \mathcal{E}d$:

$$V = \frac{m}{ne^2\tau}\frac{d}{A}I \equiv RI, \tag{4.30}$$

with

$$R = \frac{m}{ne^2\tau}\frac{d}{A} = \varrho\frac{d}{A}, \tag{4.31}$$

the electrical resistance of the conductor with length d and cross section A, characterized by the material parameter ρ and by the geometrical factor d/A.

The electrical resistivity of most of metals at room temperature is affected by collisions of conduction electrons with lattice vibrations, while at low temperatures collisions with impurities and mechanical imperfections dominate.[1] The first component ρ_L is temperature-dependent, because the amplitude of vibrations increases with $k_B T$, while the second contribution ρ_i depends only on the morphology of the crystal:

$$\varrho = \varrho_L(T) + \varrho_i. \tag{4.32}$$

This behavior is known as the *Matthiessen rule* and becomes clear by looking at the example shown in Fig. 4.6, where the relative resistivity is plotted versus temperature for different samples of the same metal. At low value of T resistivity is nearly constant in all the samples, but is higher for the impure ones owing to the higher contribution of ϱ_i. By increasing T, all the samples show the same temperature behavior, since they have the

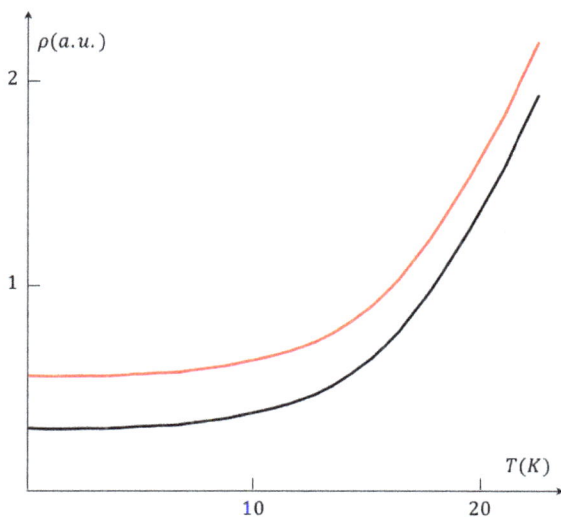

Fig. 4.6. Typical dependence of the resistivity of a metal on temperature. The black curve refers to a material with lower concentrations of impurities with respect to the one whose resistivity is given by the red line.

[1] See for instance. D.K.C. MacDonald and K. Mendelssohn. *Proc. R. Soc.* **A202**, 103 (1950).

Table 4.3.   Resistivity of some metals at two temperature values
[data from: N.W. Ashcroft and N.D. Mermin, *Solid State Physics*
(Harcourt, 1976)].

Metal	$\varrho(\Omega \cdot cm) \cdot 10^6 (T = 77$ K)	$\varrho(\Omega \cdot cm) \cdot 10^6 (T = 273$ K)
Na	0.8	4.2
Cu	0.2	1.56
Ag	0.3	1.51
Au	0.5	2.04
Mg	0.62	3.9
Fe	0.66	8.9
Zn	1.1	5.5
Al	0.3	2.45
In	1.8	8.0
Sn	2.1	10.6

same lattice structures then they are affected by the same function, $\varrho_L(T)$.
In Table 4.3, the values of resistivity of several metals for two different
temperatures are reported.

Superconductivity

In 1911, it was discovered that the resistivity of mercury at very
low temperature (4.15 K) drops approximately to zero. After this first
obervation, several metals and alloys were demonstrated to reach a state
showing a negligible electrical resistivity when approaching a critical
temperature T_c close to the absolute zero temperature. This was defined
as the *superconducting state*. The critical temperatures related to the
superconducting state of some pure metals are listed in Table IN4.1.

Starting from 1986, a new class of *superconductors* were discovered based on ceramic materials including copper oxides showing the
superconducting state at much higher temperature: from 40 K to above
130 K. These high T_c superconductors opened up a number of potential
applications, driving a strong effort in basic and technological research.

The physical origin of this effect has previously been explained by
the BCS theory (Bardeen, Cooper and Schrieffer), which has lately been
modified to account for high temperature superconductivity. It is based
on the formation of the *Cooper pair* of electrons interacting with the ion
vibrations positively, thus reducing drastically scattering and leading to
a cooperative drift of conduction electrons through the crystal.

Table IN4.1. Critical temperature of the super-conducting state of some metals [data from: B.T. Matthias, T.H. Geballe and V.B. Compton, *Rev. Mod. Phys.* **35**, 1 (1963)].

Element	$T_c(\text{K})$
Al	1.18
Ti	0.39
Mo	0.92
Zn	0.85
Nb	9.1
In	3.41
Hg	4.15

The typical behavior of resistivity ρ versus temperature near T_c is shown in Fig. IN4.1.

The superconducting state is strongly related to magnetic phenomena, since it can be quenched by an applied magnetic field over a critical value. Very remarkable is the occurrence in superconductors of the *Meissner effect*, which is the repulsion of the magnetic flux; this effect may lead to *magnetic levitation* when a magnetic material gets close to a material in the superconducting state.

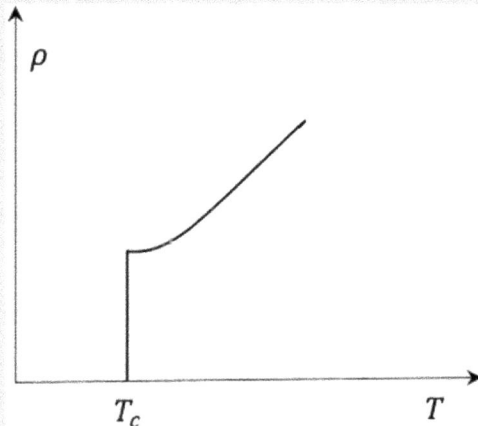

Fig. IN4.1. Typical dependence of resistivity versus temperature for a superconductor near the critical temperature.

Time-dependent electrical conductivity

In the case where the applied electric field is oscillating at an angular frequency ω, we write[2] $\boldsymbol{\mathcal{E}}(t) = \boldsymbol{\mathcal{E}}_0 e^{-i\omega t}$; by looking for a time-dependent solution to Eq. (4.24) as $\mathbf{v}_{\mathcal{E}}(t) = \mathbf{v}_{\mathcal{E}} e^{-i\omega t}$, we get

$$-e\boldsymbol{\mathcal{E}} - \frac{m\mathbf{v}_{\mathcal{E}}}{\tau} = -i\omega m \mathbf{v}_{\mathcal{E}}. \tag{4.33}$$

Then

$$\mathbf{v}_{\mathcal{E}} = -\frac{e\tau}{m(1 - i\omega\tau)}\boldsymbol{\mathcal{E}} \tag{4.34}$$

and the current density becomes

$$\mathbf{J} = -ne\mathbf{v}_{\mathcal{E}} = \frac{ne^2\tau}{m(1 - i\omega\tau)}\boldsymbol{\mathcal{E}}. \tag{4.35}$$

Defining $\sigma_0 = \sigma(\omega = 0) = ne^2\tau/m$, the quantity given in Eq. (4.27), we have

$$\mathbf{J} = \sigma_0 \frac{1}{1 - i\omega\tau}\boldsymbol{\mathcal{E}} \tag{4.36}$$

and

$$\sigma = \sigma_0 \frac{1}{1 - i\omega\tau} = \sigma_0 \frac{1 + i\omega\tau}{1 + (\omega\tau)^2} = \sigma_0 \left[\frac{1}{1 + (\omega\tau)^2} + i\frac{\omega\tau}{1 + (\omega\tau)^2}\right]. \tag{4.37}$$

In this way, we have a component of the current oscillating in phase with the exciting field (the real part) and a component oscillating with a phase difference of $\phi = \pi/2$ (the imaginary part):

$$\mathbf{J}' = \sigma_0 \frac{1}{1 + (\omega\tau)^2}\boldsymbol{\mathcal{E}}_0 e^{-i\omega t}, \tag{4.38}$$

$$\mathbf{J}'' = i\sigma_0 \frac{\omega\tau}{1 + (\omega\tau)^2}\boldsymbol{\mathcal{E}}_0 e^{-i\omega t} = \sigma_0 \frac{\omega\tau}{1 + (\omega\tau)^2}\boldsymbol{\mathcal{E}}_0 e^{-i\left(\omega t - \frac{\pi}{2}\right)}. \tag{4.39}$$

At high frequency, $\omega\tau \gg 1$, and Eq. (4.37) becomes

$$\sigma \approx \sigma_0 \left[\frac{1}{(\omega\tau)^2} + i\frac{1}{\omega\tau}\right]. \tag{4.40}$$

Then the imaginary part dominates and conductivity is independent on τ:

$$\sigma = \frac{ne^2}{m\omega^2\tau} + i\frac{ne^2}{\omega m} \approx i\frac{ne^2}{\omega m}. \tag{4.41}$$

---

[2] We recall that the actual physical quantity is the real part of the complex expression.

The Weidemann–Franz law

The free electron model succeeds in verifying the *Weidemann–Franz law*, which establishes the ratio between the thermal and the electrical conductivity of metals as linearly dependent on the temperature T.

The thermal conductivity of free particles can be written as

$$K = \frac{1}{3}c_V\, vl, \tag{4.42}$$

with c_V being the specific heat at constant volume, V the particle velocity and l the mean free path between two collisions. Since the only electrons to be thermally excited are the ones close to the Fermi surface, we can put $v \approx v_F$. Then, by calculating the specific heat,[3] we get

$$K = \frac{\pi^2 n k_B^2 T \tau}{3m}. \tag{4.43}$$

Therefore, the ratio

$$\frac{K}{\sigma_0} = \frac{\pi^2 k_B^2}{3e^2} T \tag{4.44}$$

is linearly dependent on T.

The Lorenz number is defined as

$$L = \frac{K}{\sigma_0 T} = \frac{\pi^2 k_B^2}{3e^2} = 2.45 \cdot 10^{-8}\ \mathrm{W \cdot \Omega/K^2}, \tag{4.45}$$

which is constant for any metal.

Table 4.4 shows the measured values of L for different metals. The result is quite satisfactory, with a maximum deviation below 10% from the value expected from Eq. (4.45).

4.3. The Hall effect

A very important effect occurs when a conductor carrying a current density **J** is subjected to a magnetic field whose induction is **B**. In this case, a transverse electric field appears (the Hall field), given by

$$\mathcal{E}_H = R_H \mathbf{B} \times \mathbf{J}, \tag{4.46}$$

where R_H is called the *Hall constant*.

A simple geometry accounting for Eq. (4.46) is shown in Fig. 4.7.

---

[3]The calculation can be found in any text on solid state physics.

Table 4.4.   Lorenz number L for different metals at room temperature [data
from: C. Kittel, *Introduction to Solid
State Physics*, 8th ed., (Wiley & Sons,
2005)].

Metal	$L(W \cdot \Omega/K^2) \cdot 10^8$ $(T = 273$ K$)$
Ag	2.31
Au	2.35
Cu	2.23
Ir	2.49
Pb	2.47
Sn	2.52
Zn	2.31

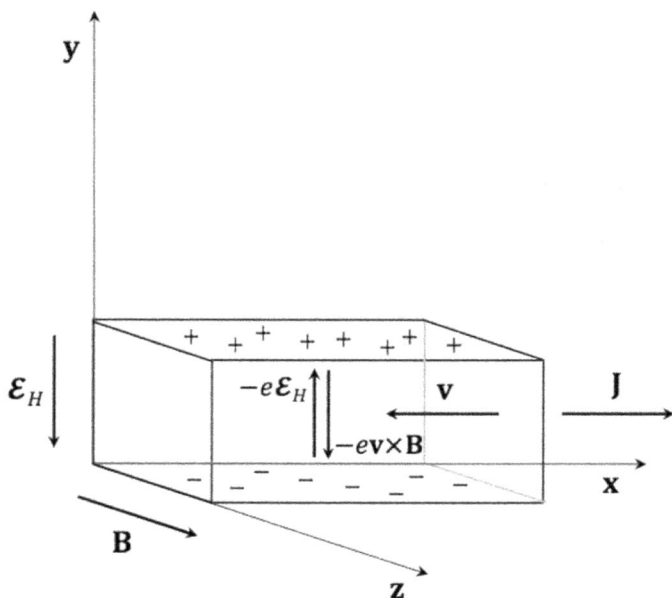

Fig. 4.7.   Generation of the Hall field $\mathcal{E}_H$ in a conductor; $\mathbf{v}$ is the velocity of electrons, and $\mathbf{B}$ the induction of the applied magnetic field.

The magnetic force on the electron $-e\mathbf{v} \times \mathbf{B}$ will increase the electron density on one side of the conductor, thus depleting the negative charge on the other side. In this way, an electric field $\mathcal{E}_H$ will rise pointed toward the high electron density side, acting with a force $-e\mathcal{E}_H$ on the electron opposite to the magnetic force. The field will increase until its force balances the magnetic one:

$$\mathcal{E}_H = \mathbf{v} \times \mathbf{B}. \tag{4.47}$$

In the simple geometry of Fig. 4.7, this means that

$$\mathcal{E}_y = v_x \, B = -\frac{J_x}{ne} B, \tag{4.48}$$

where $B \equiv B_z$.

By comparison with Eq. (4.46) we get

$$R_H = -\frac{1}{ne}. \tag{4.49}$$

If N is the concentration of atoms, with n being the concentration of conduction electrons, the ratio n/N should give the number of conduction electrons given by each atom. Therefore, we expect this value to be an integer such as 1, 2 or 3, depending on the atomic structure of the considered metal from Eq. (4.49) we can write

$$\frac{n}{N} = -\frac{1}{R_H N e}. \tag{4.50}$$

This quantity should be a positive number, since R_H is defined as negative from Eq. (4.49).

Measurements of n/N for different metals are reported in Table 4.5, showing that in some cases a noninteger number is found and even with an unexpected negative sign, which corresponds to positive conducting charges! This result cannot find any explanation in the theory of electrical

Table 4.5. The number of conduction electrons per atom calculated from Eq. (4.50) using experimental data obtained at high magnetic field and low temperature [data from: N.W Ashcroft and N.D. Mermin, *Solid State Physics* (Harcourt, 1976)].

Metal	n/N
Li	+0.8
K	+1.1
Cu	+1.3
Au	+1.5
Be	−0.2
In	−0.3
Al	−0.3

conduction presented in the previous section, and thus it requires a more advanced description of this phenomenon.

4.4. The work function of metals and related effects

Each conduction electron moves as a free particle in the bulk of the metal but, at the same time, it is bonded through the covalent metallic bonding to the lattice. This means that it is necessary to provide a certain amount of energy to extract it from the crystal. This energy is called the *work function ϕ* and can be regarded as the difference between the Fermi energy of electrons (i.e. the highest energy of conduction electrons) and the energy of the electron outside the conductor with $V = 0$ (zero kinetic energy). The energy scheme is sketched in Fig. 4.8.

In Table 4.6, the values of the work function for some polycrystalline samples of some metals are reported.[4]

The work function affects several phenomena in metals; a few of them are summarized below.

Photoelectric effect

We have already discussed this phenomenon, whose explanation by Einstein was a milestone, opening the way to a quantum-mechanical interpretation of the microscopic world.

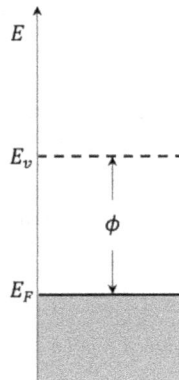

Fig. 4.8.   Scheme of energy levels pointing out the Fermi energy and work function for a conduction electron. E_v is the electron vacuum energy.

---

[4]In single crystals, the work function depends on the crystallographic direction.

Table 4.6. The measured value of the work function ϕ for some conductors. The data are from polycrystalline samples [data from: H.B. Michaelson, *J. Appl. Phys.* **48**, 4729 (1977)].

Metal	ϕ (eV)
Ag	4.7
Al	4.1
Cu	4.5
Fe	4.7
Ni	5.0
Zn	4.3
Na	2.3
K	2.2
Li	2.3

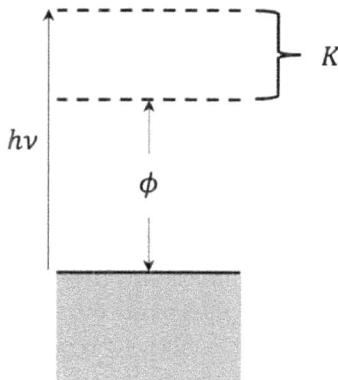

Fig. 4.9. Scheme of energy balance in the photoelectric effect (see text).

When one photon impinges on a metal surface, it will be able to extract one electron from the surface if its energy $h\nu$ is higher than the work function ϕ,

$$h\nu > \phi, \tag{4.51}$$

and therefore the kinetic energy of the outcoming electron will be

$$K = \frac{1}{2}mv^2 = h\nu - \phi, \tag{4.52}$$

as sketched in Fig. 4.9.

Thermoionic emission

When a metallic wire is heated, emission of electrons can be observed. It is easy to understand the occurrence of this effect by thinking of the Fermi–Dirac statistic followed by electrons. In fact, if K is the energy of the electron when it comes out of the metal, we can write the total electron energy in the statistical distribution as $E = K + \phi + E_F$ and

$$f(E,T) = \frac{1}{e^{\frac{E-E_F}{k_B T}} + 1} = \frac{1}{e^{\frac{K+\phi}{k_B T}} + 1} \approx e^{-\frac{K+\phi}{k_B T}} = e^{-\frac{\phi}{k_B T}} e^{-\frac{K}{k_B T}}, \quad (4.53)$$

since $K + \phi \gg k_B T$.

The quantity in Eq. (4.53) is the probability of finding one electron with energy K outside the metal at the temperature T and it shows how this probability increases with temperature and exponentially decreases for increasing values of the work function.

Volta effect

When two different metals are in contact, an electrical potential difference is found at their sides, as shown in Fig. 4.10.

The origin of the effect is the different work function values ϕ_A and ϕ_B of the two metals. The difference between ϕ_A and ϕ_B corresponds to the different values of the Fermi energies in the two metals. As will be explained in detail in Section 4.12, the equilibrium conditions at the interface lead to the Fermi energy of the two metals being at the same level; this necessarily induces a potential difference between the two metals, as shown in Fig. 4.11, where the energy levels are sketched before (a) and after (b) the contact.

The onset of this equilibrium condition can be easily understood by considering that, when in contact, electrons near the interface from the metal having higher Fermi energy spontaneously flow to the other side to reduce their energy. The process stops when the two Fermi energies are equal. In this way, negative charges decrease on one side and increase on the other side, with the result of establishing a potential difference between

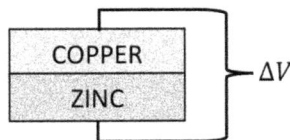

Fig. 4.10. Volta effect: contact between two metals and the measured potential difference.

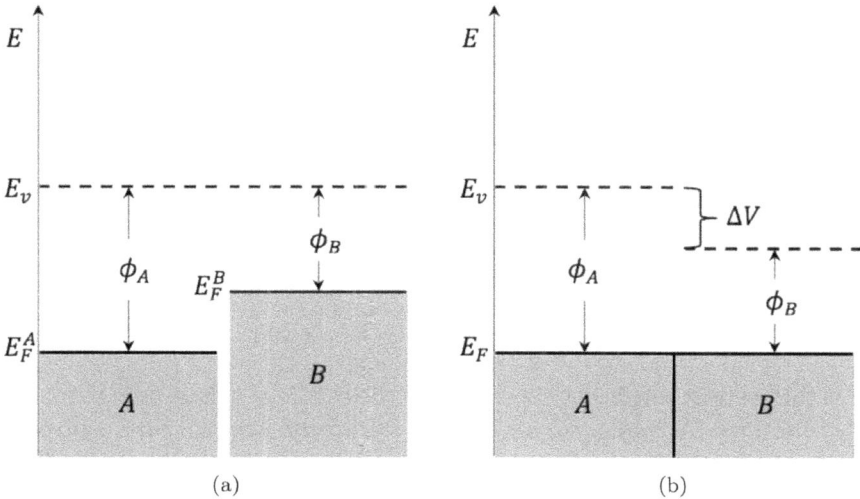

Fig. 4.11.   Energy levels before (a) and after (b) the contact between two metals.

the two metals, which prevents further migration of electrons through the interface.

This potential difference cannot be used to generate a current, since it is a consequence of a steady state condition. In order to give rise to electromotive force, the two metals must be separated by salt solutions necessary for producing ions, as realized by Volta (*voltaic pile*), demonstrating for the first time the concept of the modern batteries.

Seebeck and Peltier effects

The Volta effect allows explaining two important consequences widely exploited for precise temperature measurements and thermoelectric refrigeration. The *Seebeck or thermoelectric effect* is the observation of a potential difference on the side of two pieces of the same metal in contact with a different metal (as shown in Fig. 4.12), when the two interfaces are at a temperature difference $\Delta T = T_2 - T_1$. It is due to exceeding thermal excitation to energy above the Fermi energy in the hot interface with consequent migration of these extra electrons toward the cold interface, thus producing unbalancing of the Volta effect of the two interfaces with an overall potential difference measured on the sides of the external conductor proportional to the temperature difference of the interfaces:

$$\Delta V = S\Delta T. \tag{4.54}$$

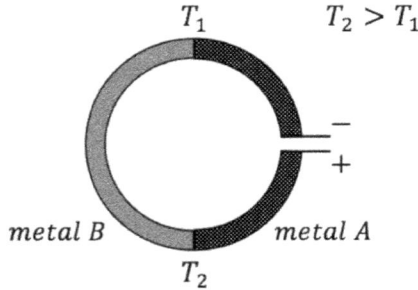

Fig. 4.12.   Scheme of the Seebeck effect.

S is the Seebeck coefficient dependent on the couple of metals used to realize the interface. A common system used for precise temperature control is made of copper (Cu) and the alloy constantan (Cu–Ni) with a pretty high value of $S = 43\,\mu\text{V/K}$ and a wide working range (from $-180°\text{C}$ to $+400°\text{C}$).

The Peltier effect is the reverse: a direct electric current through the two interfaces makes one colder and the other hotter.

4.5. Electrons in a periodic potential: the Bloch theorem

In order to improve the description of the properties of electrons in crystal, we have to consider that even the electrons able to travel along macroscopic distances, i.e. not bound to a specific atom, are not really free particles, but they feel the interaction forces originated by the crystal lattices where positive ions are left. Owing to the ordered lattice structure, the corresponding potential energy originated by it can be considered periodic with the same periodicity of the lattice. In other words, the valence electrons move in a space region where this condition holds:

$$\mathcal{V}(\mathbf{r}) = \mathcal{V}(\mathbf{R} + \mathbf{r}), \tag{4.55}$$

with $\mathbf{R}$ being a lattice vector.

In this case, it can be demonstrated that electron wave functions (solutions to the Schrödinger equation with periodic potential) can be written as

$$\phi_{n\mathbf{k}}(\mathbf{r}) = e^{i\mathbf{k}\cdot\mathbf{r}} u_{n\mathbf{k}}(\mathbf{r}), \tag{4.56}$$

where the functions $u_{n\mathbf{k}}(\mathbf{r})$ have the same lattice periodicity:

$$u_{n\mathbf{k}}(\mathbf{r}) = u_{n\mathbf{k}}(\mathbf{R} + \mathbf{r}). \tag{4.57}$$

From Eq. (4.56) we can write

$$\phi_{n\mathbf{k}}(\mathbf{R} + \mathbf{r}) = e^{i\mathbf{k}\cdot(\mathbf{R}+\mathbf{r})}u_{n\mathbf{k}}(\mathbf{R} + \mathbf{r}) = e^{i\mathbf{k}\cdot\mathbf{R}}e^{i\mathbf{k}\cdot\mathbf{r}}u_{n\mathbf{k}}(\mathbf{r})$$

$$= e^{i\mathbf{k}\cdot\mathbf{R}}\phi_{n\mathbf{k}}(\mathbf{r}). \tag{4.58}$$

That is to say, for any lattice vector we can write

$$\phi_{n\mathbf{k}}(\mathbf{R} + \mathbf{r}) = e^{i\mathbf{k}\cdot\mathbf{R}}\phi_{n\mathbf{k}}(\mathbf{r}). \tag{4.59}$$

The expressions (4.56)–(4.57) and (4.59) are the two equivalent formulations of the *Bloch theorem*. The functions fulfilling this property are called *Bloch functions*.

The demonstration of the Bloch theorem is easily derived. Let us define a *translation operator* T_R for any lattice vector $\mathbf{R}$ such that

$$T_R f(\mathbf{r}) = f(\mathbf{r} + \mathbf{R}).$$

Since the Hamiltonian has the lattice periodicity, we can write

$$T_R\mathcal{H}(\mathbf{r})\phi(\mathbf{r}) = \mathcal{H}(\mathbf{r} + \mathbf{R})\phi(\mathbf{r} + \mathbf{R}) = \mathcal{H}(\mathbf{r})\phi(\mathbf{r} + \mathbf{R}) = \mathcal{H}(\mathbf{r})T_R\phi(\mathbf{r}).$$

Therefore,

$$T_R\mathcal{H}(\mathbf{r}) = \mathcal{H}(\mathbf{r})T_R \quad \text{or} \quad [T_R, \mathcal{H}(\mathbf{r})] = 0.$$

As a consequence, T_R and $\mathcal{H}(\mathbf{r})$ have the same eigenfunctions ϕ:

$$\mathcal{H}(\mathbf{r})\phi(\mathbf{r}) = E\phi(\mathbf{r}),$$

$$T_R\phi(\mathbf{r}) = c(\mathbf{R})\phi(\mathbf{r}),$$

with $c(\mathbf{R})$ being the eigenvalue of the translation operator T_R. On the other hand, by applying two translations we also get

$$T_R T_{R'}\phi(\mathbf{r}) = T_{R'}T_R\phi(\mathbf{r}) = \phi(\mathbf{r} + \mathbf{R} + \mathbf{R'}),$$

and then

$$T_R T_{R'} = T_{R'}T_R = T_{R+R'}.$$

Therefore,

$$T_{R'}T_R\phi(\mathbf{r}) = c(\mathbf{R})T_{R'}\phi(\mathbf{r}) = c(\mathbf{R})c(\mathbf{R'})\phi(\mathbf{r})$$

$$= T_{R+R'}\phi(\mathbf{r}) = c(\mathbf{R} + \mathbf{R'})\phi(\mathbf{r}),$$

i.e.

$$c(\mathbf{R})c(\mathbf{R}') = c(\mathbf{R} + \mathbf{R}').$$

This expression will apply to any couple of lattice vectors, and thus also to the primitive vectors $\mathbf{a}_i$. Then, by using the Euler notation for the complex number, $c(\mathbf{a}_i) = e^{2\pi i x_i}$.

Since

$$\mathbf{R} = n_1\mathbf{a}_1 + n_2\mathbf{a}_2 + n_3\mathbf{a}_3,$$

this means that $\mathbf{R}$ can be obtained starting from any lattice site with n_1 translations of the primitive vector $\mathbf{a}_1$, followed by n_2 translations of the primitive vector $\mathbf{a}_2$ and n_3 translations of the primitive vector $\mathbf{a}_3$:

$$T_{n_3\mathbf{a}_3}T_{n_2\mathbf{a}_2}T_{n_1\mathbf{a}_1}\phi(\mathbf{r}) = c(\mathbf{a}_1)^{n_1}c(\mathbf{a}_2)^{n_2}c(\mathbf{a}_3)^{n_3}\phi(\mathbf{r}) = c(\mathbf{R})\phi(\mathbf{r}).$$

Then

$$c(\mathbf{a}_1)^{n_1}c(\mathbf{a}_2)^{n_2}c(\mathbf{a}_3)^{n_3} = c(\mathbf{R}) = e^{2\pi i n_1 x_1}e^{2\pi i n_2 x_2}e^{2\pi i n_3 x_3} = e^{i\mathbf{k}\cdot\mathbf{R}},$$

with

$$\mathbf{k} = x_1\mathbf{b}_1 + x_2\mathbf{b}_2 + x_3\mathbf{b}_3,$$

where $\mathbf{b}_i$ are the primitive vectors of the reciprocal lattice that fulfill the condition

$\mathbf{b}_i \cdot \mathbf{a}_j = 2\pi\delta_{ij}$, with δ_{ij} being the usual Kronecker symbol.

In fact,

$$e^{i\mathbf{k}\cdot\mathbf{R}} = e^{i(x_1\mathbf{b}_1 + x_2\mathbf{b}_2 + x_3\mathbf{b}_3)\cdot(n_1\mathbf{a}_1 + n_2\mathbf{a}_2 + n_3\mathbf{a}_3)}$$

$$= e^{2\pi i n_1 x_1}e^{2\pi i n_2 x_2}e^{2\pi i n_3 x_3}.$$

Going back to the eigenvalue equation $T_R\phi(\mathbf{r}) = c(\mathbf{R})\phi(\mathbf{r})$, we have shown that $c(\mathbf{R}) = e^{i\mathbf{k}\cdot\mathbf{R}}$. Then

$$T_R\phi(\mathbf{r}) = \phi(\mathbf{r} + \mathbf{R}) = c(\mathbf{R})\phi(\mathbf{r}) = e^{i\mathbf{k}\cdot\mathbf{R}}\phi(\mathbf{r}),$$

as stated in Eq. (4.59).

The expression (4.59) allows one to restrict the analysis of the electron behavior to a range of k between 0 and the value corresponding to its maximum inside the primitive cell of the reciprocal lattice. It can be easily shown in one dimension, where

$$\phi_{nk}(x + na) = e^{ikna}\phi_{nk}(x), \tag{4.60}$$

with a being the lattice primitive vector and n an integer number. By considering a new wave vector defined by

$$k' = k + n'\frac{2\pi}{a}, \tag{4.61}$$

with n' being an integer number, Eq. (4.60) becomes

$$\phi_{nk'}(x + na) = e^{ik'na}\phi_{nk'}(x) = e^{i(k+n'\frac{2\pi}{a})na}\phi_{nk'}(x) = e^{ikna}\phi_{nk'}(x), \tag{4.62}$$

which is equivalent to Eq. (4.60). This means that any different electronic state can be described within a k vector range of $2\pi/a$. The same concept can be applied to three dimensions. In a more general way, we can state that if $\mathbf{G}$ is a reciprocal lattice vector, any wave vector $\mathbf{k}'$ can be written as

$$\mathbf{k}' = \mathbf{k} + \mathbf{G}. \tag{4.63}$$

We note that $e^{i\mathbf{k}\cdot\mathbf{R}} = e^{i\mathbf{k}'\cdot\mathbf{R}}$, because $e^{i\mathbf{k}\cdot\mathbf{G}} = 1$.

The index n in Eq. (4.56) means that for any $\mathbf{k}$ value the solution to the Schrödinger equation provides many different wave functions with many different values of energy $E_n(\mathbf{k})$ associated with them, and therefore it indicates a specific set of energy values, forming a *band*, as will be clear later in this chapter.

Finally, another important property can be demonstrated, which appears as a generalization of the already given group velocity for an electronic wave:

$$\mathbf{v}_n(\mathbf{k}) = \frac{1}{\hbar}\nabla_{\mathbf{k}}E_n(\mathbf{k}). \tag{4.64}$$

In fact, in one dimension it simply states that

$$v_n(k) = \frac{1}{\hbar}\frac{\partial}{\partial k}E_n(k). \tag{4.65}$$

When $E_n(k) = \hbar\omega_n$, this means that

$$v_n(k) = \frac{\partial\omega_n}{\partial k}, \tag{4.66}$$

corresponding to Eq. (2.32).

Several properties of electrons in crystals can be inferred from the general form of the electron wave functions in a periodic potential, i.e. from the result that electron wave functions are Bloch functions.

4.6. Energy bands: the nearly free electron model

A possible approach to studying the motion of electrons in a periodic potential landscape is the one considering such potential $\mathcal{V}$ as a small perturbation of the total energy of the electron, i.e. $\mathcal{V} \ll \mathcal{H}_0$, with $\mathcal{H}_0$ being the free electron Hamiltonian. Moreover, this perturbation can be considered effective only very near the lattice sites, owing to the electrostatic interaction of the positive ions. Actually, the perturbation might not be small, and therefore this approach can only be qualitatively correct; however, it is convenient to figure out the energy band structure in crystalline solids.

In order to simplify the discussion, we will consider a one-dimensional case, taking into account that the real, three-dimensional situation has a higher complexity but is similar from a qualitative point of view. A possible potential energy fulfilling the condition (4.55) is

$$\mathcal{V}(x) = -\sum_{1}^{\infty} V_n \cos\left(\frac{2\pi n x}{a}\right), \tag{4.67}$$

where V_n are positive coefficients, each of them corresponding to a different integer n, and a is the lattice constant defining the smallest distance between two lattice sites. The expression (4.67) fulfills the requirement of a potential with minima on the lattice sites ($x = na$) where the electrostatic interaction between ions and electrons is stronger.

The perturbation theory allows calculating the new eigenvalues of the electron energy as modified by the perturbation. According to Eq. (2.213), at the first order of perturbation we have

$$W_k^{(1)} = E_k + \mathcal{H}'_{kk}, \tag{4.68}$$

where [see Eq. (2.207)]

$$\mathcal{H}'_{kk} = \frac{\int_0^a \phi_k^* \mathcal{H}' \phi_k \, dx}{\int_0^a \phi_k^* \phi_k \, dx}, \tag{4.69}$$

written in the general case, where ϕ_k are not normalized. The integrals in Eq. (4.69) are limited to the interval $(0, a)$ owing to the periodicity of the potential energy. In our case, $\mathcal{H}' \equiv \mathcal{V}$, and therefore the correction in the energy value is

$$\Delta E = W_k - E_k = \mathcal{H}'_{kk} = \frac{\int_0^a \phi_k^* \mathcal{V} \phi_k dx}{\int_0^a \phi_k^* \phi_k dx}. \qquad (4.70)$$

The wave functions to be considered in Eq. (4.70) are those of the free electron. Since in the unperturbed case it has only kinetic energy,

$$E = \frac{\hbar^2 k^2}{2m},$$

this means that different values of k give rise to different wave functions and two different wave functions with opposite values k and $-k$ are associated with the same value of energy. On the other hand, the general solution to the Schrödinger equation for the free particle is written in the form

$$\phi_k = A e^{ikx} + B e^{-ikx}, \qquad (4.71)$$

which is the superposition of two plane waves traveling in opposite directions.

The only two functions fulfilling all these requirements are the ones obtained from Eq. (4.71) using

$$A = B \rightarrow \phi_1 = 2A \cos(kx), \qquad (4.72)$$

$$A = -B \rightarrow \phi_2 = 2A \sin(kx),$$

where $2A \equiv c$, a constant to be determined by the normalization condition. It can be easily verified that they fulfill the ortogonality condition. Therefore, we can use these two functions to evaluate ΔE by Eq. (4.70), substituting in one case $\phi_1 = c \cos(kx)$ and in the other case $\phi_2 = c \sin(kx)$.

Let us first consider

$$\phi_1 = c \cos(kx).$$

Then

$$\Delta E_1 = \frac{\int_0^a \phi_k^* \mathcal{V} \phi_k dV}{\int_0^a \phi_k^* \phi_k dV} = \frac{\int_0^a \cos(kx) \mathcal{V} \cos(kx) dx}{\int_0^a \cos^2(kx) dx}$$

$$= -\frac{\int_0^a \cos^2(kx) \sum_1^\infty V_n \cos\left(\frac{2\pi nx}{a}\right) dx}{\int_0^a \cos^2(kx) dx}$$

---

and by the substitution $\cos(2kx) = 2\cos^2(kx) - 1$ we get

$$\Delta E_1 = -\frac{\frac{1}{2}\int_0^a [1 + \cos(2kx)] \sum_1^\infty V_n \cos(\frac{2\pi nx}{a}) dx}{\frac{1}{2}\int_0^a [1 + \cos(2kx)] dx}$$

$$= -\frac{\int_0^a \sum_1^\infty V_n \cos\left(\frac{2\pi nx}{a}\right) dx + \int_0^a \cos(2kx) \sum_1^\infty V_n \cos(\frac{2\pi nx}{a}) dx}{\int_0^a dx + \int_0^a \cos(2kx) dx}.$$

We take into account that for any variable α the integral $\int_0^{2\pi} \cos\alpha \, d\alpha = 0$, and therefore the above expression reduces to

$$\Delta E_1 = -\frac{\int_0^a \cos(2kx) \sum_1^\infty V_n \cos(\frac{2\pi nx}{a}) dx}{\int_0^a dx}.$$

Considering that the integral including summation over n is always zero except for

$$k = \pm \frac{n\pi}{a}$$

and setting

$$\xi = \frac{2\pi nx}{a},$$

we have

$$\Delta E_1 = -\frac{\frac{a}{2\pi n} V_n \int_0^{2\pi n} \cos^2(\xi) d\xi}{a} = -\frac{V_n}{2\pi n} \left[\frac{\xi + \sin\xi\cos\xi}{2}\right]_0^{2\pi n}$$

$$= -\frac{V_n}{2\pi n}[\pi n] = -\frac{V_n}{2}.$$

The same calculation performed with

$$\phi_2 = c\sin(kx)$$

gives

$$\Delta E_2 = +\frac{V_n}{2}.$$

Therefore, the perturbation changes the energy of the states described by the unperturbed wave functions in a different way by increasing one by a quantity $V_n/2$ and decreasing the other one by the same quantity when $k = \pm n\pi/a$. Thus, at these values of k the electron energy has the values

$$E_n = \frac{\hbar^2 k^2}{2m} \pm \frac{V_n}{2}.$$ (4.73)

In other words, at these values of k there is a discontinuous jump in the dispersion curve $E(k)$ from the value $E_n = \frac{\hbar^2 k^2}{2m} - \frac{V_n}{2}$ to the value $E_n = \frac{\hbar^2 k^2}{2m} + \frac{V_n}{2}$ and this happens for each value of the integer n.

A lower energy for the wave function $\cos(kx)$ with respect to $\sin(kx)$ could be expected. In fact, on the lattice sites, $x = na$, and therefore at $k = \pm n\pi/a$ we have

$$\cos\left(\frac{n\pi}{a} na\right) = \pm 1, \quad \sin\left(\frac{n\pi}{a} na\right) = 0.$$ (4.74)

This means that in the first case we have the maximum probability of finding the electron in the lattice site location where the electrostatic interaction with the ion is expected to be maximum. On the contrary, in the other case the probability of finding the electron in this location is zero, i.e. the interaction with the positive ion is weaker and the total electron energy is higher.

The modification of the free electron dispersion curve $E(k)$ due to the periodic perturbation is shown in Fig. 4.13: the energy of the free electron given by the dashed parabolic curve is modified around the values $k = \pm n\pi/a$, opening in this way n gaps in the allowed energy values.

Then the effect of the periodic potential landscape is the creation of *bands* of allowed energy values for the electron and corresponding *band gaps* with forbidden energy values. The k axis is divided into different regions called *Brillouin zones*, corresponding to the different energy bands, i.e. from $-\pi/a$ to π/a we have the first Brillouin zone, from $-\pi/a$ to $-2\pi/a$ and from π/a to $2\pi/a$ we have the second Brillouin zone, etc.

It is important to note that electron wave functions at $k = \pm n\pi/a$ represent standing waves, while for any other k value they are traveling waves within the crystal. In fact, as shown in Eq. (4.74), the two wave functions $\cos(kx)$ and $\sin(kx)$ have always the same value at the lattice sites (respectively ± 1 or 0). As a consequence, we expect the velocity to be zero in these locations. Then

$$v_n(k = \pm n\pi/a) = \frac{1}{\hbar} \frac{\partial}{\partial k} E_n(k = \pm n\pi/a) = 0,$$ (4.75)

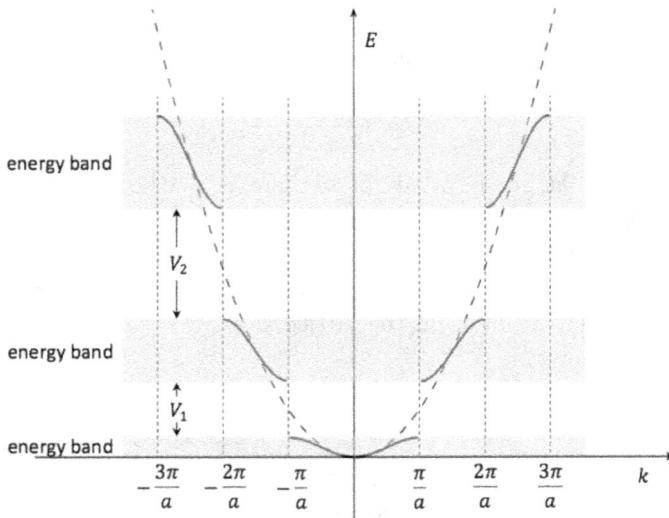

Fig. 4.13.   The dispersion curve $E(k)$ in the nearly free electron model in one dimension. The dashed curve is the one related to the free electron.

which means a horizontal geometrical tangent for the function $E_n(k)$ in these locations.

In order to better understand why the electron is not able to travel with wave vector $k = \pm\, n\pi/a$, we can consider the reciprocal lattice vectors. In one dimension, they are defined as

$$G = n\frac{2\pi}{a}. \qquad (4.76)$$

On the other hand, the Bragg condition for diffraction is given by [see Eqs. (3.163) and (3.164)]

$$\Delta\mathbf{k} = \mathbf{k}' - \mathbf{k} = \mathbf{G} \quad \text{or} \quad \mathbf{k}' = \mathbf{k} + \mathbf{G}. \qquad (4.77)$$

Since $k = |\mathbf{k}| = |\mathbf{k}'|$, we can write $|\mathbf{k} + \mathbf{G}|^2 = |\mathbf{k}'|^2 = k^2$ and therefore

$$k^2 + 2kG + G^2 = k^2, \qquad (4.78)$$

i.e.

$$k_n = \pm\frac{G}{2} = \pm\, n\frac{\pi}{a}, \qquad (4.79)$$

corresponding to the limits of the Brillouin zones. This means that at these k values the electron wave is totally diffracted by the lattice and cannot propagate through it — as a consequence we have the energy gap!

A very simple remark allows pointing out the different behavior of materials with respect to applied electric fields just based on the energy band structure, thus explaining why a material can be a good electrical conductor or electrical insulator. In one dimension, with N being the number of lattice sites, the length of the crystal will be

$$L = Na. \tag{4.80}$$

On the other hand, the minimum distance between two allowed values of k in the k space is $2\pi/L$, and therefore the total number of allowed k states in the range $2\pi/a$ is

$$n_k = \frac{2\pi/a}{2\pi/L} = \frac{L}{a} = \frac{Na}{a} = N. \tag{4.81}$$

Since for each value of k we have two states corresponding to the possible values of the electron spin, the total number of available energy states in the linear lattice of length L is $2N$. These energy states will be completely filled or partially filled, depending on the total number of valence electrons in the crystal. For instance, if each atoms has only one valence electron, we have N electrons, and therefore only half of the energy band will be filled. On the contrary, in the case of two valence electrons per atom, the total number of available electrons is $2N$, thus filling completely the band. In the first case of a half-filled band, the electrons on the Fermi surface can find available free states next to their original states when an applied electric field provides the additional wave vector $\delta\mathbf{k}$ given by Eq. (4.29) and the whole Fermi sphere can be shifted in the direction of the electric field, as already discussed. In other words, the half-filled band corresponds to a good electrical conductor. On the contrary, when the band is completely filled the electrons on the Fermi surface cannot change their state since the closest states are forbidden, falling in the energy gap; as a consequence, no electron can gain the additional wave vector and the material is an insulator. This very empirical example explains the basic concept related to the nature of conductors and insulators. In the first case the Fermi energy (the highest energy of a filled state) falls in the middle of an energy band, while in the second case the Fermi energy corresponds to the top of an energy band.

Of course, when the real, three-dimensional case is considered, this easy correspondence with the valence of each atom in the lattice site cannot be carried on; nevertheless, the relation between the location of the Fermi energy in the band and the conductivity of the material holds, thus explaining the origin of the corresponding behavior of the material.

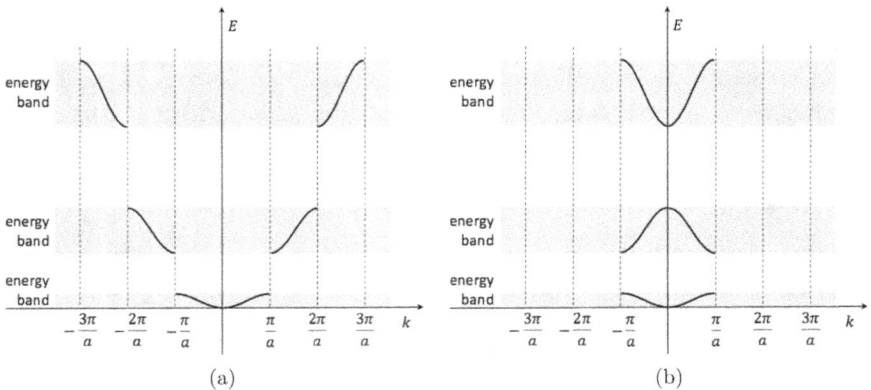

Fig. 4.14. (a) Actual energy band dispersion curve $E(k)$; (b) the same function in the reduced zone scheme.

Finally, we can exploit the property of the electron Bloch function shown by Eqs. (4.60) and (4.62) to represent in a more compact way the band structure of a crystal. Those relations say that for each electron state it is equivalent to consider the actual wave vector k or the wave vector $k + G = k \pm n\pi/a$ (in one dimension), and therefore all the different physical situations for the electron in the crystal can described by k values in a range from $-\pi/a$ to π/a, in the *reduced zone scheme*, as shown in Fig. 4.14. Under this scheme, the whole dispersion curve $E(k)$ is plotted using the values of the wave vector k in the *first Brillouin zone*. This is accomplished by a translation $\pm 2n\pi/a$ applied to each wave vector k that originally has a value outside this zone.

4.7. Energy bands: the Kronig–Penney model

A traditional way to justify the occurrence of energy bands in crystalline solids is given by the Kronig–Penney model, which exploits the property of the electron wave function as a Bloch function.

The periodic potential energy is chosen as a square well periodic potential, such as the one reported in Fig. 4.15. Since the interaction of electrons with the positive ions located in the lattice sites is limiting the freedom of their motion along the crystal, this model considers the electron traveling through potential barriers appearing with the lattice periodicity.

In this way, the electron wave function must obey two different Schrödinger equations in region I (length a), where it has only kinetic

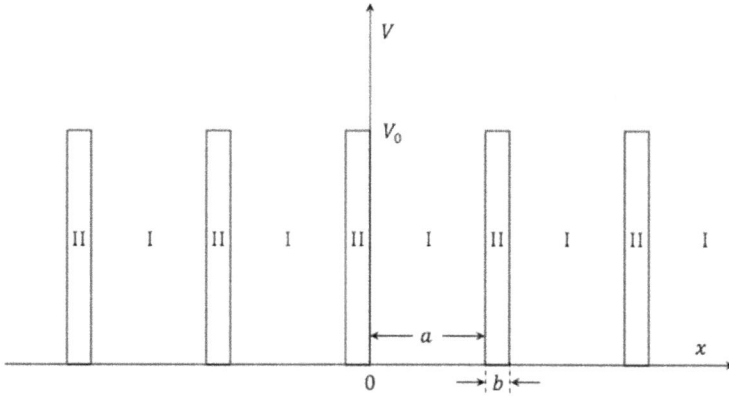

Fig. 4.15. Periodic distribution of potential barriers in one dimension: in region I, $V = 0$; in region II, $V = V_0$.

energy, and in region II (length b), where a potential barrier V_0 is present:

$$\frac{d^2\phi_k}{dx^2} + \frac{2m}{\hbar^2} E_k \phi_k = 0 \quad \text{(region I)}, \tag{4.82}$$

$$\frac{d^2\phi_k}{dx^2} + \frac{2m}{\hbar^2}(E_k - V_0)\phi_k = 0 \quad \text{(region II)}, \tag{4.83}$$

which belong to the general form

$$\frac{d^2\phi_k}{dx^2} + c\phi_k = 0, \tag{4.84}$$

where for the two regions we define

$$c = \alpha^2 = \frac{2m}{\hbar^2} E_k \quad \text{(region I)}, \tag{4.85}$$

$$c = -\beta^2 = \frac{2m}{\hbar^2}(E_k - V_0) \quad \text{(region II)}. \tag{4.86}$$

On the other hand, we know that ϕ_k must be Bloch functions fulfilling the condition

$$\phi_k(x) = e^{ikx} u_k(x), \quad \text{with} \quad u_k(x + a + b) = u_k(x). \tag{4.87}$$

Therefore, we get

$$\frac{d\phi_k}{dx} = ike^{ikx} u_k + e^{ikx}\frac{du_k}{dx}, \tag{4.88}$$

$$\frac{d^2\phi_k}{dx^2} = e^{ikx}\left[\frac{d^2u_k}{dx^2} + 2ik\frac{du_k}{dx} - k^2 u_k\right]. \tag{4.89}$$

That is to say, in the two regions we have

$$\frac{d^2 u_k}{dx^2} + 2ik\frac{du_k}{dx} - (k^2 - \alpha^2)u_k = 0 \quad \text{(region I)}, \tag{4.90}$$

$$\frac{d^2 u_k}{dx^2} + 2ik\frac{du_k}{dx} - (k^2 + \beta^2)u_k = 0 \quad \text{(region II)}. \tag{4.91}$$

These equations correspond to the well-known equation of a damped oscillator,

$$\frac{d^2 f}{dx^2} + c_1\frac{df}{dx} + c_2 f = 0, \tag{4.92}$$

whose solution is

$$f = e^{-\frac{c_1}{2}x}[Ae^{i\gamma x} + Be^{-i\gamma x}], \quad \text{with} \quad \gamma = \sqrt{c_2 - \frac{c_1^2}{4}}. \tag{4.93}$$

In our case, $c_1 = 2ik$ and

$$c_2 = -(k^2 - \alpha^2), \quad \text{then} \quad \gamma = \sqrt{\alpha^2 - k^2 + \frac{4k^2}{4}} = \alpha \quad \text{(region I)}; \tag{4.94}$$

$$c_2 = -(k^2 + \beta^2), \quad \text{then} \quad \gamma = \sqrt{-\beta^2 - k^2 + \frac{4k^2}{4}} = i\beta \quad \text{(region II)}. \tag{4.95}$$

As a consequence, the solutions are

$$u_k = e^{-ikx}(Ae^{i\alpha x} + Be^{-i\alpha x}) \quad \text{(region I)}, \tag{4.96}$$

$$u_k = e^{-ikx}(Ce^{-\beta x} + De^{\beta x}) \quad \text{(region II)}, \tag{4.97}$$

where the constants A, B, C, D have to be determined according to the boundary conditions, which require continuity of the solutions at the interfaces of the two regions.

The four equations necessary for getting the unknown parameters A, B, C, D are obtained by setting the following boundary conditions:

(1) $u_k(0)_\mathrm{I} = u_k(0)_\mathrm{II}$,

(2) $\dfrac{du_k}{dx}(0)_\mathrm{I} = \dfrac{du_k}{dx}(0)_\mathrm{II}$,

(3) $u_k(a)_\mathrm{I} = u_k(-b)_\mathrm{II}$,

(4) $\dfrac{du_k}{dx}(a)_\mathrm{I} = \dfrac{du_k}{dx}(-b)_\mathrm{II}$.

It is straightforward to obtain from Eqs. (1)–(4) the following:

$$A + B = C + D,$$
$$A(i\alpha - ik) + B(-i\alpha - ik) = C(-\beta - ik) + D(\beta - ik),$$
$$Ae^{(i\alpha - ik)a} + Be^{(-i\alpha - ik)a} = Ce^{(ik+\beta)b} + De^{(ik-\beta)b},$$
$$A(i\alpha - ik)e^{(i\alpha - ik)a} + (-i\alpha - ik)Be^{(-i\alpha - ik)a}$$
$$= (-\beta - ik)Ce^{(ik-\beta)b} + (\beta - ik)De^{(ik-\beta)b}.$$

Solutions exist if the determinant of the coefficients of A, B, C, D is equal to zero.

Taking into account the definitions

$$\sinh \xi = \frac{1}{2}(e^{\xi} - e^{-\xi}) = -i\sin(i\xi), \quad \cosh \xi = \frac{1}{2}(e^{\xi} + e^{-\xi}) = \cos(i\xi),$$

after a lengthy calculation the condition on the determinant gives

$$\frac{\beta^2 - \alpha^2}{2\alpha\beta} \sinh(\beta b) \sin(\alpha a) + \cosh(\beta b) \cos(\alpha a) = \cos k(a + b).$$

Then, the usual approximation is to consider the potential barriers extremely narrow with the potential expressed by a periodic sequence of Dirac delta functions (*Dirac's comb*). The approximation means that the attractive force of the ion on the electron is considered effective only very close to the lattice site. This assumption can be taken in a more intuitive way by considering $V_0 \gg E$ and b very small, leading however to a finite value of $V_0 b$.

Under these assumptions in Eq. (4.86) E can be neglected:

$$\beta^2 = \frac{2m}{\hbar^2} V_0, \quad \beta b = \sqrt{\frac{2m}{\hbar^2}(V_0 b)} \sqrt{b}.$$

Then $\beta b \to 0$ as $b \to 0$ and accordingly $\cosh(\beta b) \to 1$ and $\sinh(\beta b) \to \beta b$. Moreover, $\beta^2 - \alpha^2 \to \beta^2$ and $a + b \to a$. In this way, the above equation obtained by equating the determinant of the coefficient to zero becomes

$$\frac{\beta^2}{2\alpha\beta}\beta b \sin(\alpha a) + \cos(\alpha a) = \cos k(a)$$

Finally, taking into account the definition of α and β given in Eqs. (4.85) and (4.86), in the approximation $V_0 \gg E$ and $b \to 0$ (but with $V_0 b$ a finite number) we find the condition to be fulfilled for the existence of the solutions (4.96) and (4.97):

$$\frac{mV_0}{\alpha \hbar^2} b \sin(\alpha a) + \cos(\alpha a) = \cos(ka), \tag{4.98}$$

which can be written as

$$\left[\frac{mV_0 ab}{\hbar^2}\right] \frac{\sin(\alpha a)}{\alpha a} + \cos(\alpha a) = \cos(ka). \tag{4.99}$$

Since $-1 \le \cos(ka) \le 1$, this means that

$$-1 \le \left[\frac{mV_0 ab}{\hbar^2}\right] \frac{\sin(\alpha a)}{\alpha a} + \cos(\alpha a) \le 1. \tag{4.100}$$

This condition can be numerically analyzed by plotting the function

$$Y(\alpha) = \left[\frac{mV_0 ab}{\hbar^2}\right] \frac{\sin(\alpha a)}{\alpha a} + \cos(\alpha a) = \chi \frac{\sin(\alpha a)}{\alpha a} + \cos(\alpha a), \tag{4.101}$$

defining

$$\chi = \frac{mV_0 ab}{\hbar^2}. \tag{4.102}$$

The function $Y(\alpha)$ is shown in Fig. 4.16, for $\chi = 5$.

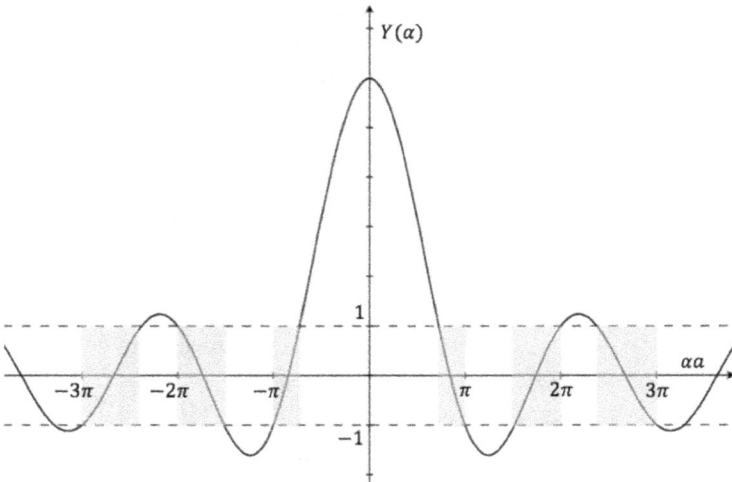

Fig. 4.16. The function $Y(\alpha)$ calculated for $\chi = 5$. The horizontal dashed lines limit the range of the function fulfilling (4.100) and the corresponding allowed values of the variable αa (gray regions).

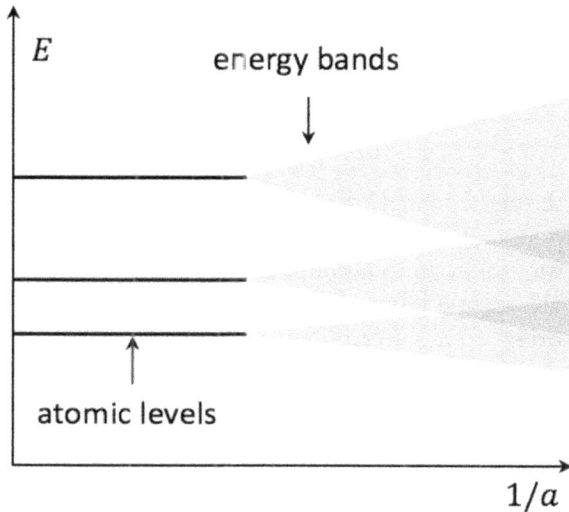

Fig. 4.17. Sketch of the transition from atomic levels to energy bands by decreasing the interatomic distance a.

The figure shows that the condition (4.100) is fulfilled only for certain ranges of values of α. Since $\alpha \propto \sqrt{E}$, this means that solutions to the problem exist only for definite ranges of values of the energy separated by gaps of forbidden energy values. The size of these bands depends on the parameter χ. Higher χ means steeper maxima and minima of $Y(\alpha)$ and then narrower energy bands; or, the contrary, lower χ means wider energy bands. Since χ is proportional to the interatomic distance a, this means that by decreasing the value of the interatomic distance we have the transition from sharp energy levels to bands as sketched in Fig. 4.17.

It is also worth noting that in the absence of the potential barrier Eq. (4.85) gives, as expected, $\alpha = k$, which is the well-known dispersion relation of the free particle $k = \sqrt{2mE}/\hbar$.

In the other extreme case of very large $V_0 b$ ($\chi \to \infty$), the condition (4.100) is fulfilled if $\sin(\alpha a)/\alpha a \to 0$, i.e. $\alpha a = n\pi$, which is the same solution we got for the infinite well potential, leading to the same allowed energy states, $E_n = (\pi^2 \hbar^2/2ma^2)n^2$.

4.8. Effective mass and the concept of a hole

In the previous section, we have pointed out the difference between the dispersion relation $E(k)$ of the electron moving in a periodic potential and that of a free particle. According to it, we should expect the velocity of the electron moving in a crystal structure to be different from $\hbar k/m$. In fact, it must be calculated from Eq. (4.64) as

$$\mathbf{v}(\mathbf{k}) = \frac{1}{\hbar}\nabla_{\mathbf{k}}E(\mathbf{k}). \tag{4.103}$$

Nevertheless, the electron motion can be described by the classical equation of Newton if we use an *effective mass* that depends on the dispersion curve $E(\mathbf{k})$, in this way taking into account the interaction of the electron wave function with the crystal lattice. This point is easily shown, by considering the power absorbed by an electron when submitted to a force $\mathbf{F}$:

$$\frac{dE}{dt} = \mathbf{F}\cdot\mathbf{v}, \tag{4.104}$$

where the right-hand side term is the usual power absorbed in the time dt when a work $\mathbf{F}\cdot\mathbf{ds}$ is performed by the force. By using Eq. (4.103), we have

$$\frac{dE}{dt} \equiv \nabla_{\mathbf{k}}E\cdot\frac{d\mathbf{k}}{dt} = \hbar\mathbf{v}\cdot\frac{d\mathbf{k}}{dt}, \tag{4.105}$$

and therefore from Eq. (4.104) we have

$$\mathbf{F} = \hbar\frac{d\mathbf{k}}{dt}. \tag{4.106}$$

On the other hand, we can write the electron acceleration as

$$\mathbf{a} = \frac{d\mathbf{v}}{dt} \equiv \nabla_{\mathbf{k}}\mathbf{v}\cdot\frac{d\mathbf{k}}{dt} = \frac{1}{\hbar}\nabla_{\mathbf{k}}\nabla_{\mathbf{k}}E(\mathbf{k})\cdot\frac{1}{\hbar}\mathbf{F}, \tag{4.107}$$

i.e.

$$\mathbf{F} = m_e\mathbf{a}, \tag{4.108}$$

where the *effective mass* is defined as

$$m_e = \frac{\hbar^2}{\nabla_{\mathbf{k}}\nabla_{\mathbf{k}}E(\mathbf{k})}. \tag{4.109}$$

This means an anisotropic effective mass whose actual value depends on the direction of the applied force:

$$(m_e)_{ij} = \frac{\hbar^2}{\partial^2 E/\partial k_i \partial k_j}, \quad \text{with} \quad i,j = x,y,z. \tag{4.110}$$

In one dimension,

$$m_e = \frac{\hbar^2}{\partial^2 E/\partial k^2}. \tag{4.111}$$

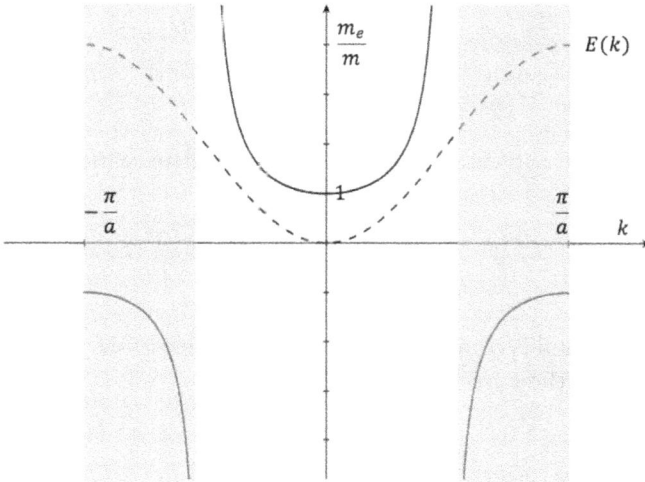

Fig. 4.18. The energy dispersion curve $E(k)$ (dashed line) and the corresponding normalized effective mass m_e/m (full line).

Thus, the effective mass is a function of k and depends on the slope of $E(k)$, and therefore it can have values very different from the mass of the electron, assuming negative values around the maxima of $E(k)$. The most interesting behavior occurs near the edge of the band, i.e. near the boundary of the Brillouin zone. Figure 4.18 shows an energy band and the corresponding values of the effective mass m_e normalized to the electron mass m. The ratio $m_e/m \approx 1$ in the center of the band, while it is divergent when approaching the edge and becoming negative.

This fact suggests the description of the behavior of an electron near the maximum of an energy band as that of a particle with opposite charge and opposite mass. In order to go further in this approach, it is necessary to discuss the conduction behavior of many electrons in partially filled bands.

The current density due to electrons with the wave vector in the range between k and $k + dk$ is given by

$$d\mathbf{J} = -en_k\mathbf{v}(\mathbf{k})d^3k, \tag{4.112}$$

with n_k being the density of electrons with the wave vector in the considered range. Since in $\mathbf{k}$ space we have one allowed value of k in the volume $V/8\pi^3$ and for each value of $\mathbf{k}$ we may have two electrons with opposite spins, we get

$$n_k = \frac{1}{4\pi^3}. \tag{4.113}$$

Then

$$dJ = -e\mathbf{v}(\mathbf{k})\frac{d^3k}{4\pi^3}, \tag{4.114}$$

and therefore the current density due to all the filled states (f.s.) in a band is given by

$$\mathbf{J} = -e\int_{\text{f.s.}} \mathbf{v}(\mathbf{k})\frac{d^3k}{4\pi^3}. \tag{4.115}$$

Of course, if a band is completely filled (f.b.), the electrons cannot be driven by the field and the current density is zero:

$$\mathbf{J} = -e\int_{\text{f.b.}} \mathbf{v}(\mathbf{k})\frac{d^3k}{4\pi^3} = 0. \tag{4.116}$$

We can write Eq. (4.116) as the sum of the contributions of two subbands b_1 and b_2:

$$0 = -e\int_{\text{f.b.}} \mathbf{v}(\mathbf{k})\frac{d^3k}{4\pi^3} = -e\int_{b_1} \mathbf{v}(\mathbf{k})\frac{d^3k}{4\pi^3} - e\int_{b_2} \mathbf{v}(\mathbf{k})\frac{d^3k}{4\pi^3}, \tag{4.117}$$

i.e.

$$-e\int_{b_1} \mathbf{v}(\mathbf{k})\frac{d^3k}{4\pi^3} = e\int_{b_2} \mathbf{v}(\mathbf{k})\frac{d^3k}{4\pi^3}. \tag{4.118}$$

Therefore, if b_1 represents the filled part of the band, b_2 will be the empty one and Eq. (4.118) tells us that the current density of a partially filled band [the left-hand side of Eq. (4.118)] can be written as the current density due to positive charges (with charge e) filling the empty states of the band. In other words, instead of considering the real current density due to electrons filling the band, we can consider the current of *holes* in the empty part of the band if we associate with these virtual positive charges in each state $\mathbf{k}$ the same velocity that the electron would have in this state: $\mathbf{v}_{\text{hole}}(\mathbf{k}) = \mathbf{v}_{\text{electron}}(\mathbf{k}) \equiv \mathbf{v}(\mathbf{k})$.

It is now possible to complete the definition of the hole taking into account the result obtained for the effective mass of the electron on the top of an energy band where $m_e < 0$. We define

$$m_h = -m_e, \tag{4.119}$$

and as a consequence

$$\mathbf{k}_h = \frac{m_h \mathbf{v}_h}{\hbar} = -\frac{m_e \mathbf{v}_e}{\hbar} = -\mathbf{k}_e. \qquad (4.120)$$

This result is quite easy to understand by thinking of a symmetric energy band. If it is completely filled, the total wave vector is

$$\sum_{\text{band}} (\mathbf{k}_e)_i = 0, \qquad (4.121)$$

since for each wave vector $(\mathbf{k}_e)_j$ the opposite $-(\mathbf{k}_e)_j$ exists. If we imagine that one electron of the wave vector $\mathbf{k}_e^*$ is missing, the total wave vector of the band will be

$$\sum_{\text{band}} (\mathbf{k}_e)_i - \mathbf{k}_e^* = -\mathbf{k}_e^* = \mathbf{k}_h^*, \qquad (4.122)$$

which is the wave vector of the hole associated with the empty state!

In summary, the behavior of an "almost" filled band (usually called the *valence band*) will be more efficiently described by considering the holes as charge carriers, whose properties are summarized in Table 4.7.

One important remark to be made concerns the energy. Since the effective mass of the hole is opposite to that of electrons this means by Eq. (4.109) that the energy has an opposite slope. That is to say,

$$E_h = \frac{\hbar^2 k_h^2}{2m_h} = -\frac{\hbar^2 k_e^2}{2m_e} = -E_e, \qquad (4.123)$$

since $k_h^2 = k_e^2$ and $m_h = -m_e$.

This must be taken into account when we are dealing with holes, even if we usually represent the dispersion curve $E(k)$ of electrons also when holes play an important role. For instance, Fig. 4.19 shows the top of a valence energy band for electrons and the corresponding energy band for holes, which is symmetric to the former with respect to the horizontal axis.

Table 4.7. The link between the hole properties and the corresponding vacant electron.

	Vacant electron	Hole
Charge	$-e$	e
Wave vector	$\mathbf{k}_e$	$-\mathbf{k}_e$
Velocity	$\mathbf{v}_e$	$\mathbf{v}_e$
Effective mass	m_e	$-m_e$

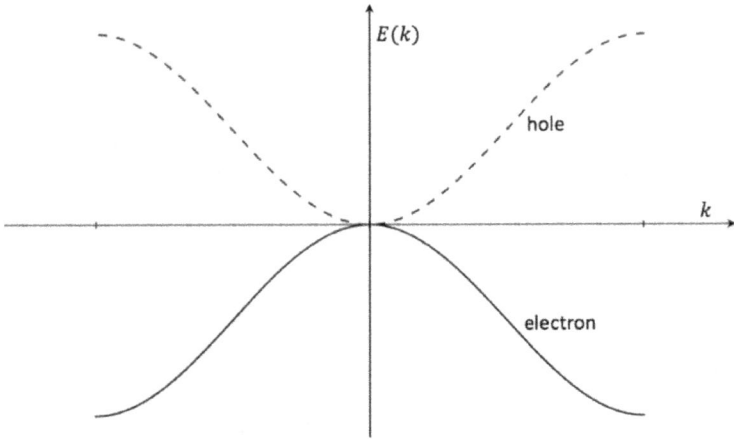

Fig. 4.19. The top of an energy band for electrons (full line) and the corresponding band for holes (dashed line).

4.9. Electronic properties of semiconductors

Semiconductors are materials showing a room temperature resistivity between 10^{-2} and 10^9 $\Omega \cdot$ cm, larger than in good conductors ($\sim 10^{-6}$ $\Omega \cdot$cm) and much lower than in insulators (in the range of $10^{14} - 10^{22} \Omega \cdot$ cm). The typical values of the band gap at 0 K between the last band completely filled (*valence band*) and the next one empty (*conduction band*) is around 1 eV, with 1.17 for silicon and 0.74 for germanium. Setting to zero the energy on the top of the valence band, this means that the bottom of the conduction band is at energy $E_G = 1.17$ for silicon and $E_G = 0.74$ for germanium. In Table 4.8, the values of E_G are reported for some semiconductors at room temperature. This parameter is very important, first of all because the conductivity of a pure semiconductor is controlled by the ratio $E_G/k_B T$. Since at room temperature $k_B T \approx 0.026$ eV, $E_G \gg k_B T$ and the Fermi distribution becomes

$$f(E,T) = \frac{1}{e^{\frac{E-\mu}{k_B T}} + 1} \approx e^{\frac{\mu - E}{k_B T}}, \qquad (4.124)$$

and in the conduction band $E \gg \mu$; then Eq. (4.124) corresponds to the Boltzmann distribution:

$$f(E,T) \approx e^{-\frac{E}{k_B T}}, \qquad (4.125)$$

Table 4.8. Measured energy gap E_G of some
semiconductors measured at 300 K [data from: H.T.
Grain, *Introduction to Semiconductors Physics*
(World Scientific, 1999)].

Crystal	E_G (eV)
Si	1.12
Ge	0.66
InSb	0.18
InAs	0.35
InP	1.34
GaAs	1.42
GaP	2.27
GaN	3.44
GaSb	0.75

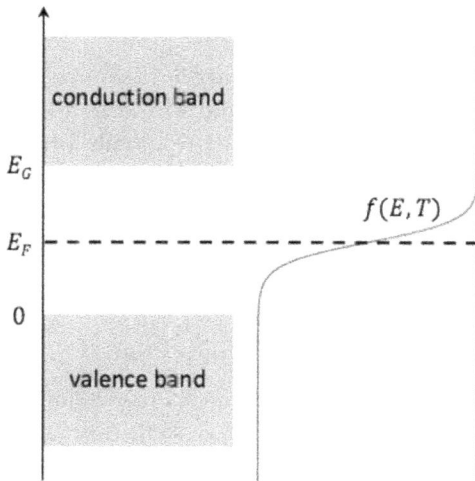

Fig. 4.20. Sketch of the valence and conduction bands in a pure semiconductor and the corresponding Fermi–Dirac distribution. In the figure, $E_F = \mu$ is conventionally used.

where $E \geq E_G \gg k_B T$, and therefore we expect a very small number of electrons to be excited in the conduction band, being able to contribute to electrical current.

This is shown in Fig. 4.20, where a sketch of the valence and conduction band in a semiconductor is given together with a plot of the Fermi distribution, where we anticipate a result that we will find in the next section, namely the location of the chemical potential of the distribution just in the middle between the two bands at energy $E_G/2$, for a pure semiconductor at $0\,\mathrm{K}$. As a consequence, we get an *intrinsic conductivity* strongly dependent on temperature, as we will show later by calculating the density of charge carriers in this case.

On the other hand, impurities can strongly affect conductivity in semiconductors by inducing changes in the allowed energy states of electrons that may favor their excitation in the conduction band. As a matter of fact, the controlled inclusion of impurities in a semiconductor crystal is the well-known technique used to drive the conductivity of these materials. This is easily understood by considering what happens, for instance, when an atom with valence 5 — i.e. five electrons in the outer shell, like phosphorus (P), arsenic (As), antimony (Sb) — is embedded in a lattice site in a crystal made of Si or Ge atoms whose valence is 4.

In this case, four of the five electrons will participate in the covalent bonds responsible for the crystal structure, while the fifth electron will remain weakly bonded to the impurity atom, under the electrostatic attraction of the additional positive charge of this nucleus. In this way, the unpaired electron and the impurity atom will form a hydrogen-like system with one negative charge bonded by the Coulomb force to a positive one. The main difference with the hydrogen atom is the presence in this case of the other electrons of the crystal that screen the interaction. This is taken into account by the dielectric susceptibility to be included in the potential energy:

$$\mathcal{V} = -\frac{1}{4\pi\epsilon}\frac{1}{r}, \qquad (4.126)$$

where $\epsilon \gg \epsilon_0$, the value in vacuum. Then we can estimate the energy levels of these unpaired electrons by using the result obtained for the hydrogen atom with the substitution $\epsilon_0 \to \epsilon$ (SI units),

$$E_n = -\frac{1}{2(4\pi\epsilon)^2}\frac{m_e e^4}{\hbar^2}\frac{1}{n^2}, \qquad (4.127)$$

with the corresponding radii of the electron orbits:

$$r_n = \frac{(4\pi\epsilon)\hbar^2}{m_e e^2}n^2. \qquad (4.128)$$

It is useful to evaluate the quantity E_1 and r_1 for the fundamental state of the bonded electron. Actually, the result depends on the chosen value for the effective mass, which is generally strongly anisotropic. For this reason, we choose the example of indium arsenide (InAs), where this anisotropy is negligible ($m_e/m = 0.024$ at $300\,\mathrm{K}$). Additionally, we choose the static value of the relative dielectric permittivity $\epsilon = 14.55$. Then for InAs we have

$$E_1 \approx -0.0015\,\mathrm{eV}, \quad r_1 \approx 320\,\text{Å},$$

and thus we have an electron cloud in a range much wider than the lattice constant (of the order of a few angstroms) and a bonding energy lower than the thermal energy at room temperature, $k_B T \approx 0.026\,\mathrm{eV}$. For other semiconductors with higher effective mass, this bonding energy can be just slightly higher than $k_B T$ at room temperature. This example shows that the electrons bonded on the impurity site can be easily ionized, i.e. excited to the conduction band. Since usually $E_1 < k_B T$, a large portion of them are naturally excited to the conduction band, increasing the actual number of charge carriers contributing to electrical current.

The impurity atoms providing these additional electrons are called *donors*. Thus, by controlling the concentration of donors in the doping process, it is possible to control the material conductivity.

A similar discussion can be carried out if dopants are atoms with valence 3 — the so-called *acceptors* [like gallium (Ga), indium (In), boron (B), aluminum (Al)]. In this case, one electron is missing to create the covalent bond. It must be taken from the filled valence band, leaving a positive hole able to be linked to the acceptor atom now appearing as a negative ion with respect to the host atoms of the crystal. Hydrogen-like energy levels are created close to the top of the valence band, whose value can be estimated as for the donor levels. A sketch of the band structure for a semiconductor including the first donor and acceptor levels is shown in Fig. 4.21.

By setting $E = 0$ at the top of the valence band, E_A is the binding energy of the acceptor level, E_G the energy gap, and $E_G - E_D$ the energy of the donor level, if E_D is its binding energy. As in the case of the donor, $E_A < k_B T$ and electrons from the top of the valence band are easily excited to the acceptor level, thus providing additional holes as positive charge carriers for electrical conduction.

In summary, doping a semiconductor with donors increases the concentration of electrons in the conduction band, while doping a semiconductor with acceptors increases the concentration of holes in the valence band. Both processes increase the concentration of charge carriers, contributing to the current in the presence of an electric field. If the donor concentration overcomes that of the acceptor, we speak of an *n-type* semiconductor; in

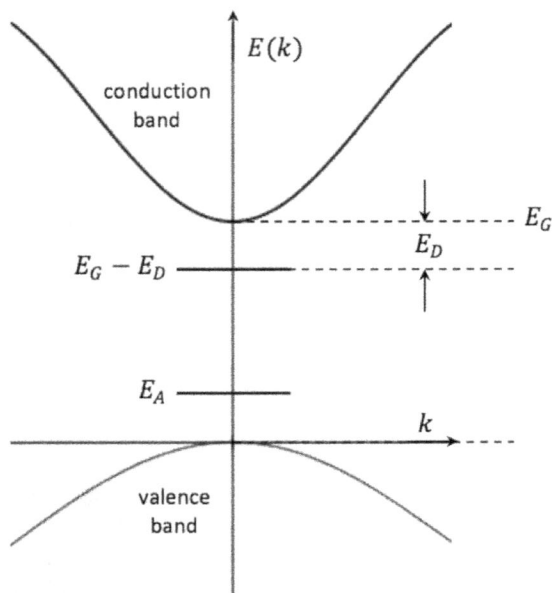

Fig. 4.21.   The valence and conduction bands in a semiconductor and donor and acceptor energy levels.

the opposite case, we speak of a *p-type* semiconductor. Indeed, electrons are called *majority carriers* in an *n-type* semiconductor and *minority carriers* in a *p-type* semiconductor. The opposite applies to holes.

4.10.   The mass action law for semiconductors

An important issue is the knowledge of the charge carrier concentrations able to contribute to electrical conductivity in semiconductors, taking into account that electrons in the conduction band and holes in the valence band must both be evaluated. In order to carry out the calculation, we consider the simplified band structure shown in Fig. 4.21. The number of charge carriers in each band will be given by the product between the energy density of states and the Fermi distribution, expressing the probability of such states to be filled. Then such a product must be integrated over all the possible energies in the band. In Fig. 4.22, the densities of states versus E are plotted for our case, indicating also the possible number of states corresponding to the donors (concentration N_D) and acceptors (concentration N_A).

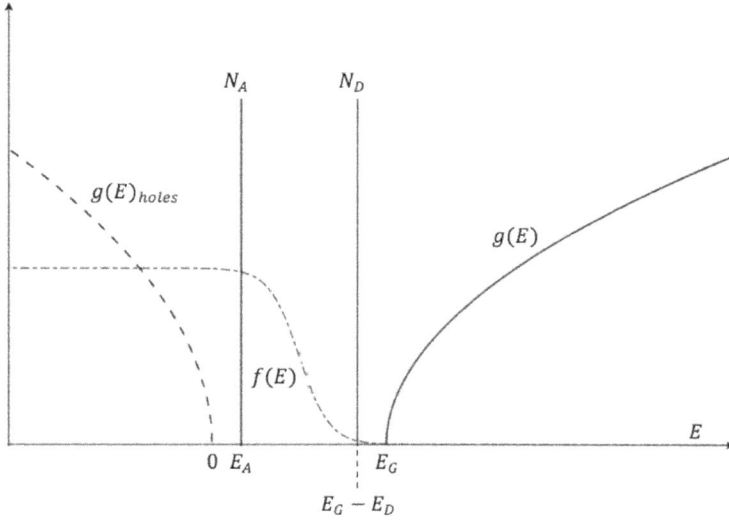

Fig. 4.22. Densities of states versus electron energy in the valence and conduction bands of a semiconductor. The Fermi distribution is also shown.

Accordingly, the electron density is given by

$$n = \frac{1}{V} \int_{E_G}^{\infty} f(E,T)g(E)dE, \qquad (4.129)$$

where integration has been extended to ∞ and not to the limit of the band since $f(E,T)$ is negligibly small for $E \gg E_G$, and therefore the result of integration is not affected by this approximation.

With $E - \mu \gg k_B T$, for $E > E_G$, the distribution function becomes

$$\dot{f}(E,T) = e^{\frac{\mu - E}{k_B T}} \qquad (4.130)$$

and the density of states in the conduction band is given by Eq. (4.15) translating the origin at $E = E_G$:

$$g(E) = \left(\frac{V}{2\pi^2\hbar^3}\right)(2m_e)^{3/2}\sqrt{E - E_G}. \qquad (4.131)$$

Then

$$n = \frac{1}{V} \int_{E_G}^{\infty} \left(\frac{V}{2\pi^2\hbar^3}\right)(2m_e)^{3/2}\sqrt{E - E_G}\, e^{\frac{\mu - E}{k_B T}}\, dE, \qquad (4.132)$$

to be rewritten as

$$n = \frac{(2m_e)^{3/2}}{2\pi^2\hbar^3} e^{\frac{\mu - E_G}{k_B T}} \int_0^{\infty} \sqrt{E - E_G}\, e^{-\frac{E - E_G}{k_B T}}\, d(E - E_G). \qquad (4.133)$$

In this way, integration is performed using the variable

$$\zeta = \frac{E - E_G}{k_B T}, \tag{4.134}$$

i.e.

$$\int_0^\infty \sqrt{E - E_G} e^{-\frac{E - E_G}{k_B T}} d(E - E_G) = (k_B T)^{3/2} \int_0^\infty \sqrt{\zeta} e^{-\zeta} d\zeta$$

$$= (k_B T)^{3/2} \left(\frac{\sqrt{\pi}}{2} \right). \tag{4.135}$$

Therefore, we get

$$n = N_C e^{\frac{\mu - E_G}{k_B T}}, \tag{4.136}$$

where

$$N_C = 2 \left[\frac{m_e k_B T}{2 \pi \hbar^2} \right]^{3/2}. \tag{4.137}$$

In order to perform the same calculation to evaluate the concentration of holes in the valence band, we must take into account that the probability of occupancy for a hole corresponds to the probability of vacancy for an electron, i.e. we must use $1 - f(E, T)$ in place of $f(E, T)$:

$$1 - f(E, T) = 1 - \frac{1}{e^{\frac{E - \mu}{k_B T}} + 1} = \frac{1}{e^{\frac{\mu - E}{k_B T}} + 1} \approx e^{\frac{E - \mu}{k_B T}} \tag{4.138}$$

when $\mu - E \gg k_B T$. Then the hole density becomes

$$p = \frac{1}{V} \int_{-\infty}^0 [1 - f(E, T)] g(E) dE = \frac{1}{V} \int_{-\infty}^0 e^{\frac{E - \mu}{k_B T}} g(E) dE. \tag{4.139}$$

Since the energy of the holes is opposite to that of the electron and according to Fig. 4.22 we have $E = 0$ on the top of the valence band, we get

$$g(E) = \left(\frac{V}{2 \pi^2 \hbar^3} \right) (2 m_h)^{3/2} \sqrt{-E}. \tag{4.140}$$

Then

$$p = \frac{(2 m_h)^{3/2}}{2 \pi^2 \hbar^3} \int_{-\infty}^0 \sqrt{-E} e^{\frac{E - \mu}{k_B T}} d(-E)$$

$$= \frac{(2 m_h)^{3/2}}{2 \pi^2 \hbar^3} e^{-\frac{\mu}{k_B T}} \int_0^\infty \sqrt{E_h} e^{-\frac{E_h}{k_B T}} dE_h, \tag{4.141}$$

where the energy of the hole $E_h = -E$.

We have the same integral as in Eq. (4.135):

$$p = N_V e^{-\frac{\mu}{k_B T}}, \tag{4.142}$$

where

$$N_V = 2 \left[\frac{m_h k_B T}{2\pi \hbar^2} \right]^{3/2}. \tag{4.143}$$

The product of Eqs. (4.136) and (4.142) gives

$$np = N_C N_V e^{-\frac{E_G}{k_B T}} = 4[m_h m_e]^{3/2} \left[\frac{k_B T}{2\pi \hbar^2} \right]^3 e^{-\frac{E_G}{k_B T}}, \tag{4.144}$$

which is known as the *law of mass action for semiconductors*.

It is very important, since it connects the concentrations of electrons and holes participating in the conductivity of the semiconductor. It states that this product is a characteristic of the material as np depends on the energy gap E_G and on the effective masses of the electron (m_e) in the conduction band and of the hole (m_h) in the valence band. Obviously, this product depends on the temperature. However, the most important feature of Eq. (4.144) is the absence of μ, which means that the product np is not affected by doping, which affects the location of the chemical potential in the energy gap.

Intrinsic semiconductors

An intrinsic (i) semiconductor is one where impurities play a negligible role, and therefore in this case we should have $n_i = p_i$. In fact, at $T = 0\,\mathrm{K}$, $n_i = p_i = 0$ (conduction band empty and valence band filled), and by increasing the temperature some electrons are excited from the valence band to the conduction band, leaving vacancies in the valence band. Therefore, from Eq. (4.144), we get

$$n_i p_i = n_i^2 = p_i^2 = N_C N_V e^{-\frac{E_G}{k_B T}}, \tag{4.145}$$

i.e.

$$n_i = p_i = \sqrt{N_C N_V} e^{-\frac{E_G}{2k_B T}}. \tag{4.146}$$

This expression explicitly shows the basic parameter affecting the concentration of charge carriers, determining the expected conductivity of an intrinsic semiconductor, namely the value of the energy gap E_G as compared to the thermal energy $k_B T$. As already outlined at room temperature,

$E_G \gg 2k_BT$, and therefore Eq. (4.146) explains why we should expect a very low conductivity for intrinsic semiconductors. This is clear by looking at the properties of germanium and silicon. In the first case, $E_G = 0.66\,\text{eV}$ at room temperature; then a concentration of conduction electrons $n_i = 1.7 \cdot 10^{13}\,\text{cm}^{-3}$ can be calculated from Eq. (4.146) using average values for the effective masses. Under the same conditions, the energy gap of silicon is $E_G = 1.12\,\text{eV}$, leading to a concentration $n_i = 6.9 \cdot 10^9\,\text{cm}^{-3}$ — lower by more than three orders of magnitude with respect to that of germanium!

Looking at Eqs. (4.136) and (4.142), we can also remark that N_C and N_V are the asymptotic values of the electrons' and holes' concentrations obtained for $T \to \infty$; this means that the carriers' concentration will be lower than either N_C or N_V and they also represent an upper limit of any reasonable doping. Typical values of N_C and N_V are in the range $10^{18} - 10^{19}\,\text{cm}^{-3}$.

By using both of the Equations (4.136) and (4.146), we get

$$N_C e^{\frac{\mu - E_G}{k_B T}} = \sqrt{N_C N_V}\, e^{-\frac{E_G}{2k_B T}}, \qquad (4.147)$$

and substituting the expressions (4.137) and (4.143),

$$e^{\frac{\mu}{k_B T}} = \left(\frac{m_h}{m_e}\right)^{3/4} e^{\frac{E_G}{2k_B T}}. \qquad (4.148)$$

Then

$$\mu = \frac{1}{2}E_G + \frac{3}{4}k_B T \ln\left(\frac{m_h}{m_e}\right), \qquad (4.149)$$

showing that when $m_h = m_e$ or at $T = 0\,\text{K}$, $\mu = \frac{1}{2}E_G$ is falling just in the middle of the forbidden band (this result has been anticipated in Fig. 4.20). The term $\ln(\frac{m_h}{m_e})$ is often close to 1, and therefore by increasing the temperature μ moves from the value $\frac{1}{2}E_G$ for quantities of the order $k_B T$.

Extrinsic semiconductors

The concentration of charge carriers of intrinsic semiconductors is too low to give rise to significant electrical conduction; however, it can be enhanced and easily controlled by doping the material with impurities. We describe as an *extrinsic semiconductor* the one where conductivity is due to impurities that strongly affect the position of the chemical potential of the Fermi distribution, thus increasing the number of charge carriers.

In fact, the neutrality condition of the material gives

$$n + N_A^- = p + N_D^+, \tag{4.150}$$

where N_A^- and N_D^+ are the concentrations of ionized acceptors and donors, respectively. Since according to Fig. 4.21 the donor energy level is at $E_G - E_D$, the concentration of donor electrons in this state is

$$N_D^0 = N_D f(E_G - E_D, T), \tag{4.151}$$

with N_D being the concentration of donor atoms. Therefore, the ionized donor electrons are

$$N_D^+ = N_D - N_D^0 = N_D - N_D f(E_G - E_D, T) = N_D \left\{ 1 - \frac{1}{e^{\frac{E_G - E_D - \mu}{k_B T}} + 1} \right\}. \tag{4.152}$$

On the other hand, the electron excited to the acceptor level at E_A over the valence band can be evaluated starting from the concentration N_A of acceptor atoms:

$$N_A^- = N_A f(E_A, T) = N_A \left\{ \frac{1}{e^{\frac{E_A - \mu}{k_B T}} + 1} \right\}. \tag{4.153}$$

Using Eqs. (4.152) and (4.153) together with Eqs. (4.136) and (4.142) in the neutrality condition (4.150), we can obtain μ for the extrinsic semiconductors.

Usually, doping is made in order to have either $n \gg p$ or $p \gg n$, to have n-type materials in the first case and p-type materials in the second.

Considering $E_A, E_D \ll k_B T$, we can consider all the donor and acceptor atoms to be ionized. Then for n-type materials we have

$$n = p + N_D^+ - N_A^- \approx N_D - N_A, \tag{4.154}$$

where we have neglected the concentration p of minority carriers. From Eq. (4.136),

$$N_D - N_A = N_C e^{\frac{\mu_n - E_G}{k_B T}}, \tag{4.155}$$

and therefore

$$\mu_n = E_G - k_B T \ln \left(\frac{N_C}{N_D - N_A} \right) \tag{4.156}$$

In the same way, we get for a p-type material

$$p \approx N_A - N_D \tag{4.157}$$

and

$$\mu_p = k_B T \ln\left(\frac{N_V}{N_A - N_D}\right). \tag{4.158}$$

We can also consider $N_A \ll N_D$ in n-type materials and $N_D \ll N_A$ in p-type materials, thus rewriting Eqs. (4.156) and (4.158) as

$$\mu_n = E_G - k_B T \ln\left(\frac{N_C}{N_D}\right), \tag{4.159}$$

$$\mu_p = k_B T \ln\left(\frac{N_V}{N_A}\right). \tag{4.160}$$

Taking into account that typical doping concentrations are in the range of $10^{16} - 10^{18}\,\text{cm}^{-3}$, the former expressions tell us that μ_n is below E_G for an amount of a few $k_B T$, and μ_n is above the top of the valence band ($E = 0$) for a similar quantity. On the other, and at $T = 0\,\text{K}$, $\mu_n = E_G$ and $\mu_p = 0$.

Anyway, we should note the clear dependence of the chemical potential on the dopant concentration.

4.11. Electrical conductivity and the Hall effect in semiconductors

In semiconductors, electrons in the conduction band and holes in the valence band both contribute to the electrical conductivity. Then we must write the current density under an electric field $\boldsymbol{\varepsilon}$ as

$$\mathbf{J} = -ne\mathbf{v}_e + pe\mathbf{v}_h = \left(\frac{ne^2\tau_e}{m_e} + \frac{ne^2\tau_h}{m_h}\right)\boldsymbol{\varepsilon} = \sigma\boldsymbol{\varepsilon}, \tag{4.161}$$

where we have used the result obtained in Eq. (4.25).

We can define the electron and hole *mobility* as

$$\mu_e = \frac{e\tau_e}{m_e}, \tag{4.162}$$

and

$$\mu_h = \frac{e\tau_h}{m_h}. \tag{4.163}$$

Then we write

$$\mathbf{J} = (ne\mu_e + pe\mu_h)\boldsymbol{\varepsilon} \tag{4.164}$$

and

$$\sigma = \frac{1}{\rho} = (ne\mu_e + pe\mu_h). \tag{4.165}$$

Often, $\mu_e \approx \mu_h$. In this case, the conductivity is basically determined by the charge carriers' concentrations n and p. In doped semiconductors, these values are determined by the concentration of impurities; by increasing either n or p, according to the law of mass action, the product np keeps a constant value but the sum $n + p$ increases. In the case of $\mu_e \approx \mu_h \equiv \mu$, we have

$$\sigma = (ne\mu_e + pe\mu_h) = (n+p)e\mu. \tag{4.166}$$

Obviously, for n-type semiconductors $(n \gg p)$ the conductivity is dominated by electrons, while for p-type semiconductors $(p \gg n)$ it is dominated by holes. Besides the carriers' concentration, mobility is the material parameter that determines the conductivity and by the definition (4.162) or (4.163) it is lower and lower as the carrier path freedom decreases (i.e. when τ_e and τ_h become shorter). As a consequence, mobility decreases with temperature, which produces more scattering events; for the same reason, it will be higher for high-quality crystals including a low number of impurities or defects. In Table 4.9, the electron and hole mobility for a number of semiconductors are reported.

Using the values for the concentrations of intrinsic germanium and silicon at room temperature reported in the previous section and data on the mobility in Table 4.9, from Eq. (4.165) we can calculate the room temperature intrinsic conductivities to be $\rho_{\text{Ge}} = 64.5\,\Omega \cdot \text{cm}$ for germanium and $\rho_{\text{Si}} = 5.0 \cdot 10^5\,\Omega \cdot \text{cm}$ for silicon.

Table 4.9. Electron and hole mobility of some semiconductors measured at 300 K [data from: H.T. Grain, *Introduction to Semiconductors Physics* (World Scientific, 1999)].

Crystal	μ_e (m^2/V·s)	μ_h (m^2/V·s)
Si	0.145	0.037
Ge	0.390	0.180
InSb	7.000	0.085
InAs	2.500	0.030
InP	0.500	0.015
GaAs	0.840	0.045
GaP	0.016	0.014
GaSb	0.375	0.068

In semiconductors, the possibility of having both positive and negative charge carriers requires a more general treatment for the Hall effect. For this reason, we need to consider the motion equation (4.23) both for holes (h) and for electrons (e) including also the action of the applied magnetic field:

$$\pm e\mathcal{E} \pm e\mathbf{v}_{h,e} \times \mathbf{B} - m_{h,e}\frac{\mathbf{v}_{h,e}}{\tau_{h,e}} = m_{h,e}\frac{d\mathbf{v}_{h,e}}{dt}, \tag{4.167}$$

where $+$ is for holes and $-$ for electrons.

Steady state conditions lead to $d\mathbf{v}_{h,e}/dt = 0$. Using the simple geometry of Fig. 4.7 (with $\mathbf{J} = J_x\hat{\mathbf{x}}$ and $\mathbf{B} = B_z\hat{\mathbf{z}}$), we have

$$v_{ex} = -\frac{e\tau_e}{m_e}\mathcal{E}_x - \frac{e\tau_e}{m_e}v_{ey}B_z = -\mu_e\mathcal{E}_x - \mu_e v_{ey}B_z, \tag{4.168}$$

$$v_{hx} = \mu_h\mathcal{E}_x + \mu_h v_{hy}B_z, \tag{4.169}$$

and

$$v_{ey} = -\mu_e\mathcal{E}_y + \mu_e v_{ex}B_z, \tag{4.170}$$

$$v_{hy} = \mu_h\mathcal{E}_y - \mu_h v_{hx}B_z. \tag{4.171}$$

We note here that $\mathcal{E}_x$ represents the externally applied field inducing the longitudinal current ($v_{h,e}$), and $\mathcal{E}_y$ the field arising owing to the displacement of the charge carriers induced by the magnetic field. Using Eqs. (4.168)–(4.171), we can write the density current:

$$\begin{aligned} J_x &= ev_{hx}p - ev_{ex}n = e(\mu_h\mathcal{E}_x + \mu_h v_{hy}B_z)p + e(\mu_e\mathcal{E}_x + \mu_e v_{ey}B_z)n \\ &= e(p\mu_h + n\mu_e)\mathcal{E}_x + e(p\mu_h^2 - n\mu_e^2)B_z\mathcal{E}_y + e(-\mu_h^2 v_{hx}B_z^2 + \mu_e^2 v_{ex}B_z^2) \\ &\approx e(p\mu_h + n\mu_e)\mathcal{E}_x + e(p\mu_h^2 - n\mu_e^2)B_z\mathcal{E}_y, \end{aligned} \tag{4.172}$$

where the final expression is obtained by neglecting the terms depending on B_z^2.

While at equilibrium $J_y = 0$ with the onset of the Hall field:

$$J_y = ev_{hy}p - ev_{ey}n = e(\mu_h\mathcal{E}_y + \mu_h v_{hx}B_z)p + e(\mu_e\mathcal{E}_y + \mu_e v_{ex}B_z) = 0. \tag{4.173}$$

Performing the same calculation as above, we have

$$J_y = e(p\mu_h + n\mu_e)\mathcal{E}_y - e(p\mu_h^2 - n\mu_e^2)B_z\mathcal{E}_x = 0. \tag{4.174}$$

Then the Hall field is given by

$$\mathcal{E}_y = \frac{(p\mu_h^2 - n\mu_e^2)B_z\mathcal{E}_x}{(p\mu_h + n\mu_e)}. \tag{4.175}$$

Keeping the definition of the Hall constant,

$$R_H = \frac{\mathcal{E}_H}{\mathbf{B} \times \mathbf{J}} \quad \text{in our geometry becomes} \quad R_H = \frac{\mathcal{E}_y}{J_x B_z}. \tag{4.176}$$

By neglecting the magnetic field effects along the **x** direction, we get the approximate expression

$$R_H = \frac{p\mu_h^2 - n\mu_e^2}{e(p\mu_h + n\mu_e)^2}. \tag{4.177}$$

We first note that Eq. (4.177) has the usual form in the case of an n-type semiconductor ($n \gg p$),

$$R_H = \frac{-n\mu_e^2}{e(n\mu_e)^2} = -\frac{1}{ne}, \tag{4.178}$$

and a similar one for a p-type semiconductor ($p \gg n$):

$$R_H = \frac{p\mu_h^2}{e(p\mu_h)^2} = +\frac{1}{pe}. \tag{4.179}$$

In this way, we can explain the observation of positive values for the Hall constant in some materials: the meaning is that conduction is dominated by holes in the valence band rather than by electrons in the conduction band.

Additionally, Eq. (4.177) for intrinsic semiconductors ($n = p$) becomes

$$R_H = \frac{\mu_h^2 - \mu_e^2}{en_i(\mu_h + \mu_e)^2}, \tag{4.180}$$

being $=0$ when $\mu_h = \mu_e$.

4.12. The p–n junction

The p–n junction is the interfacial region between two semiconductors with different types of majority carriers, i.e. an n-type semiconductor and a p-type semiconductor. Since its properties have a basic importance in the development of transistors and in all the developments of microelectronics, we will devote this section to a simple description of the phenomenology leading to its peculiar electrical behavior. The analysis of the different configurations which have been developed to get specific functions in electrical circuits can be found in textbooks on semiconductor devices.

When a surface contact between an n-type semiconductor and a p-type semiconductor is realized, a migration of electrons near the interface from the n-type semiconductor is expected to fill vacancies present in the p-type semiconductor. Before we discuss in detail this process, a simple remark can be made from a thermodynamic point of view.

We observe that at equilibrium the migration of electrons changes their concentration in the n-type semiconductor by a quantity $\delta n = -\delta p$ (since in the n-type semiconductor there is a decrease in electron concentration n and an increase in hole concentration p), and the same relation holds in the p-type semiconductor where a reduction in the hole concentration corresponds to an increase in the electron concentration. In this case, the best thermodynamic potential to be considered for constant volume transformations is the *free energy*, F.

The free energy of the system is the sum of the free energy of the two subsystems made by the n-type and by the p-type semiconductor, in such a way that the equilibrium condition is given by

$$\delta F = \delta F_n + \delta F_p = 0. \tag{4.181}$$

At constant temperature and volume, the change of the free energy is given only by the change of the concentration c_i of each component times the related chemical potential μ_i:

$$\delta F = \sum_i \mu_i \delta c_i. \tag{4.182}$$

In our case, we can write:

$$\delta F = \mu_n' \delta n + \mu_p' \delta p. \tag{4.183}$$

Here μ_n' and μ_p' are the chemical potentials (i.e. the Fermi levels) of the n-type and the p-type semiconductor respectively, measured on the same energy scale (such as fixing a common zero level).

A thermodynamic infinitesimal transformation of a multicomponent system produces a change of the free energy F (see Section 3.7):

$$\delta F = \delta U - T\delta S - S\delta T + \sum_i \mu_i \delta c_i,$$

where U is the internal energy, T the absolute temperature, S the entropy, and c_i the concentration of the components with their respective chemical potentials μ_i.

The equilibrium condition $\delta F = 0$ gives

$$\delta F = \delta U - T\delta S - S\delta T + \sum_i \mu_i \delta c_i$$

$$= T\delta S - p\delta V - T\delta S - S\delta T + \sum_i \mu_i \delta c_i = 0. \quad \text{Then}$$

$$\delta F = -p\delta V - S\delta T + \sum_i \mu_i \delta c_i = 0$$

at constant temperature and volume $\delta V = \delta T = 0$, i.e.

$$\delta F = \sum_i \mu_i \delta c_i = 0.$$

Therefore, the equilibrium condition (4.181) becomes

$$\mu'_n \delta n = -\mu'_p \delta p. \tag{4.184}$$

As discussed above, the migration of electrons means a change δn on one side opposite to that on the other side (a decrease in electrons on the n side and an increase on the p side):

$$\delta n = -\delta p. \tag{4.185}$$

Then, from Eq. (4.184), we get the equilibrium condition

$$\mu'_n = \mu'_p. \tag{4.186}$$

Since the Fermi level of n-type semiconductors is close to the bottom of the conduction band while for p-type semiconductors is close to the top of the valence band, fulfillment of the condition (4.186) requires a shift of the bands on one side of the junction with respect to the other side. In the case of the same material with different types of doping, we can sketch the energy band through the junction as depicted in Fig. 4.23. The zero energy value is now fixed on the top of the valence band on the n side. This means that the electron energy bands on the p side are shifted upward with respect to the bands on the n side. The difference corresponds to the energy step $e\Delta\phi$ necessary for fulfilling the condition (4.186), in order to have the same value for the chemical potentials of the two sides.

junction

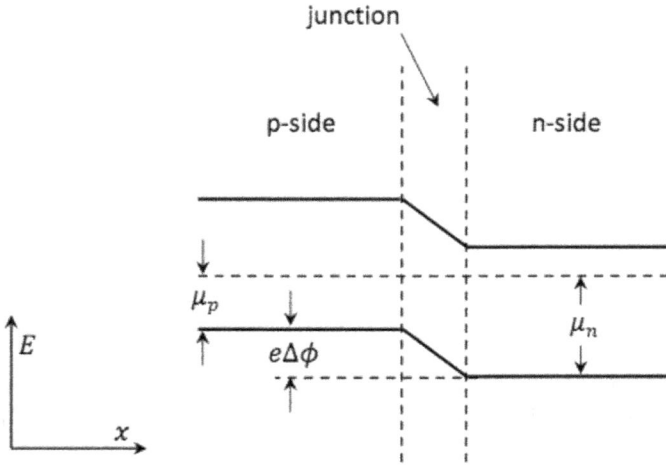

Fig. 4.23.   Sketch of energy bands and Fermi levels through a $p-n$ junction.

Taking into account the expressions (4.159) and (4.160) for μ_n and μ_p, the equilibrium condition leads to

$$\mu'_p = \mu_p + e\Delta\phi, \quad \mu'_n = \mu_n \tag{4.187}$$

and

$$\mu_n = \mu_p + e\Delta\phi \rightarrow E_G - k_B T \ln\left(\frac{N_C}{N_D}\right) = k_B T \ln\left(\frac{N_V}{N_A}\right) + e\Delta\phi. \tag{4.188}$$

Then

$$e\Delta\phi = E_G + k_B T \ln\left(\frac{N_D N_A}{N_C N_V}\right) = E_G + k_B T[\ln(N_D N_A) - \ln(N_C N_V)]. \tag{4.189}$$

Since Eq. (4.146) gives

$$N_C N_V = n_i^2 e^{\frac{E_G}{k_B T}}, \tag{4.190}$$

where n_i is the intrinsic concentration of electrons,

$$e\Delta\phi = k_B T\left[\ln\left(\frac{N_D N_A}{n_i^2}\right)\right]. \tag{4.191}$$

Looking at Fig. 4.24, it is easy to understand the phenomenology leading to the onset of this energy difference between the two sides of the junction.

The electrons from a layer $0 \leq x \leq w_n$ on the n side migrate to the p side, filling an equal number of holes in a layer $-w_p \leq x \leq 0$; in this

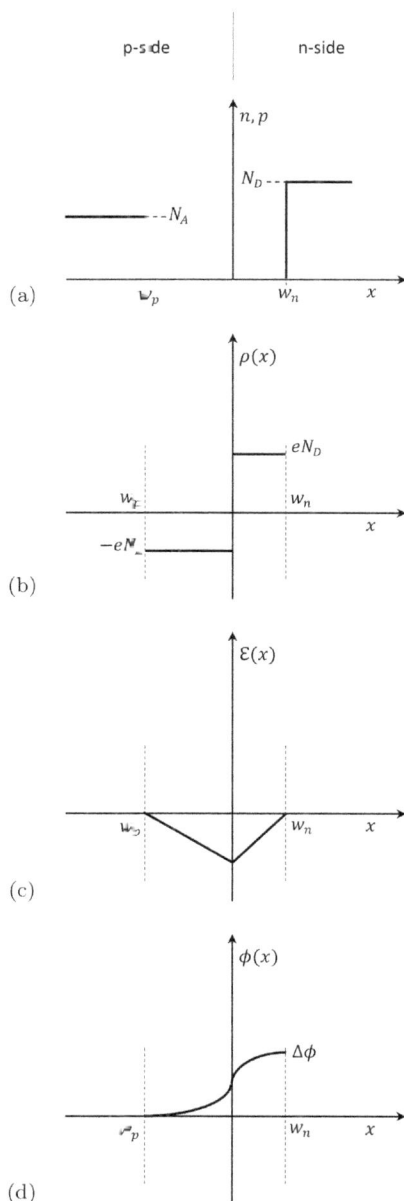

Fig. 4.24. Sketches of the following quantities through the junction: (a) charge carriers' concentrations, (b) localized charge density, (c) electric field, (d) electrostatic potential.

way, in a layer $w_p + w_n$ there are no free charge carriers. In the other part of the semiconductor, the density of charge carriers is considered equal respectively to the donor density N_D (n side) and the acceptor density N_A (p side) in the approximation that all the donor and acceptor atoms are ionized.

This is shown in Fig. 4.24(a) and can be summarized as follows:

$$
\begin{aligned}
n = p = 0 \quad & \text{for } -w_p \leq x \leq w_n; \\
n = N_D, \ p = 0 \quad & \text{for } x > w_n; \\
n = 0, \ p = N_A \quad & \text{for } x < -w_p.
\end{aligned}
\tag{4.192}
$$

Owing to the electron–hole recombination, the neutrality conditions are no longer fulfilled in the junction layer since the ionized donors and acceptors are no longer balanced by the charge carriers. For this reason, as shown in Fig. 4.24(b), the local charge density is

$$
\begin{aligned}
\rho(x) = eN_D \quad & \text{for } 0 \leq x \leq w_n; \\
\rho(x) = -eN_A \quad & \text{for } -w_p \leq x \leq 0; \\
\rho(x) = 0 \quad & \text{for } x < -w_p, \quad x > w_n.
\end{aligned}
\tag{4.193}
$$

Under these boundary conditions, we can solve the Poisson equation for the electrostatic potential ϕ:

$$
\frac{d^2\phi}{dx^2} = -\frac{\rho(x)}{\epsilon}.
\tag{4.194}
$$

The first integration provides the electric field

$$
\mathcal{E} = -\frac{d\phi}{dx},
\tag{4.195}
$$

i.e.

$$
\begin{aligned}
\mathcal{E}(x) = -\int_{-w_p}^{x} \frac{d^2\phi}{dx^2} dx = -\frac{eN_A}{\epsilon}(x + w_p) \quad & \text{for } -w_p \leq x \leq 0; \\
\mathcal{E}(x) = -\int_{x}^{w_n} \frac{d^2\phi}{dx^2} dx = \frac{eN_D}{\epsilon}(x - w_n) \quad & \text{for } 0 \leq x \leq w_n,
\end{aligned}
\tag{4.196}
$$

and $\mathcal{E}(x) = 0$ elsewhere.

Continuity at $x = 0$ provides

$$
-\frac{eN_A w_p}{\epsilon} = -\frac{eN_D w_n}{\epsilon}.
\tag{4.197}
$$

Then

$$
\frac{w_p}{w_n} = \frac{N_D}{N_A}.
\tag{4.198}
$$

The behavior of the electric field is shown in Fig. 4.24(c). It is negative (i.e. it is directed from the n side to the p side) and has the maximum at $x = 0$. The onset of this field prevents a further migration of electrons, since it acts with a force $-e\mathcal{E}$ on electrons in the positive x direction.

A further integration allows getting the potential. Setting equal to 0 the potential on the p side and $\Delta\phi$ the maximum value occurring at $x = w_n$, integration of Eq. (4.195) gives

$$\phi(x) = -\int_{-w_p}^{x} \mathcal{E}(x)dx = \frac{eN_A}{2\epsilon}(x + w_p)^2 \quad \text{for } -w_p \leq x \leq 0;$$

(4.199)

$$\phi(x) = \Delta\phi - \int_{x}^{w_n} \mathcal{E}(x)dx$$

$$= \Delta\phi - \frac{eN_D}{2\epsilon}(x - w_n)^2 \quad \text{for } 0 \leq x \leq w_n.$$

Continuity at $x = 0$ provides

$$\frac{eN_A}{2\epsilon}w_p^2 = \Delta\phi - \frac{eN_D}{2\epsilon}w_n^2$$

(4.200)

and

$$\Delta\phi = \frac{e}{2\epsilon}(N_A w_p^2 + N_D w_n^2).$$

(4.201)

Using Eq. (4.198), we have

$$\Delta\phi = \frac{e}{2\epsilon}\left(\frac{N_D^2}{N_A}w_n^2 + N_D w_n^2\right).$$

(4.202)

Then

$$w_n = \left[\frac{2\epsilon N_A \Delta\phi}{eN_D(N_A + N_D)}\right]^{1/2},$$

(4.203)

$$w_p = \left[\frac{2\epsilon N_D \Delta\phi}{eN_A(N_A + N_D)}\right]^{1/2}.$$

(4.204)

These results allow the calculation of the junction thickness,

$$\delta = w_p + w_n,$$

(4.205)

which obviously depends on the dopant concentration and usually ranges from 0.1 to 1.0 μm.

This description has been based on the ability of electrons to migrate through the junction to recombine with holes. This is actually possible if the diffusion length L_n of electrons is significantly larger than the junction thickness. In general, when for any reason the charge carrier concentration is modified with respect to the steady state value, electron–hole recombination leads to recovery of the steady state condition. This happens owing to diffusion of the charge carriers and follows a simple exponential law. For electrons, we can write (in one dimension)

$$n = n_0 e^{-\frac{x}{L_n}}, \tag{4.206}$$

with n_0 being the steady state value. The diffusion length L_n has been measured to be in the order of a few millimeters in semiconductors at room temperature, and therefore the required condition,

$$L_n \gg \delta, \tag{4.207}$$

is easily fulfilled.

4.13. Diode, transistor and quantum size effects

At constant temperature, when the equilibrium is reached a statistical flow of charge carriers crosses the junction in both directions, leading to electron–hole recombination. From the n side to the p side, the electron current density is simply due to the statistical probability of having electrons over the potential barrier $\Delta\phi$. For the same reason, on the p side a number of electrons will be excited from the valence to the conduction band and will flow to the n side, where the energy is lower.

The first effect contributes to a recombination current density J_{nr} and the second to a thermal generation current density J_{ng}. Of course, at thermal equilibrium we have

$$J_{nr} + J_{ng} = 0. \tag{4.208}$$

By applying an external voltage, we can increase or decrease the height of the potential barrier, thus affecting the recombination current density J_{nr}. On the contrary, the applied voltage will not affect the value of J_{ng} since the flow of electron from the p side to the n side occurs spontaneously and is due to the lower electron energy value of the conduction band on the n side.

First of all, we note the high resistivity of the junction due to the lack of free charges, and therefore we can assume the external voltage applied to the junction.

By applying a positive voltage V on the n side (*negative polarization*), we increase the potential barrier to $\Delta\phi + V$, thus decreasing the statistical recombination current by a Boltzmann factor $\exp(-eV/k_BT)$:

$$J_{nr}(V) = J_{nr}(0)e^{-\frac{eV}{k_BT}}. \tag{4.209}$$

The opposite occurs if the voltage is applied on the p side (*positive polarization*), since in this case the barrier height decreases to $\Delta\phi - V$. Then

$$J_{nr}(V) = J_{nr}(0)e^{\frac{eV}{k_BT}}. \tag{4.210}$$

On the other hand, the thermal generation current is not affected by the voltage V:

$$J_{ng}(V) = J_{ng}(0). \tag{4.211}$$

The overall contribution of electrons to the current density becomes

$$J_n(V) = J_{nr}(0)e^{\frac{eV}{k_BT}} - J_{ng}(0) = J_{ng}(0)\left(e^{\frac{eV}{k_BT}} - 1\right). \tag{4.212}$$

Since the same arguments are also valid for holes, we can write

$$J_h(V) = J_{hr}(0)e^{\frac{eV}{k_BT}} - J_{hg}(0) = J_{hg}(0)\left(e^{\frac{eV}{k_BT}} - 1\right). \tag{4.213}$$

We can sum over the contributions of electrons and holes and divide by the junction cross section A to get the current I through the junction

$$I(V) = I_T\left(e^{\frac{eV}{k_BT}} - 1\right), \tag{4.214}$$

where the constant I_T is given by

$$I_T = \frac{J_{ng}(0) + J_{hg}(0)}{A}. \tag{4.215}$$

The expression (4.214) is usually reported as the $I(V)$ characteristic of the $p-n$ junction showing the well-known rectifying properties: for increasing values of the applied voltage ($V > 0$) we have an exponential growth of the current, while for increasing negative values ($V < 0$) the current asymptotically approaches the small quantity $-I_T$, as can be seen in Fig. 4.25.

The properties of the $p-n$ junction are exploited in the *bipolar junction transistor* (BJT), whose invention has been the base of the development of microelectronics, allowing the amazing progress in miniaturization of computers and other electronic devices.

A BJT can be made by a p-type semiconductor (*base*) placed in between two n-type semiconductors (*emitter* and *collector*), and therefore its conduction behavior is affected by two $p-n$ junctions: *base-emitter*

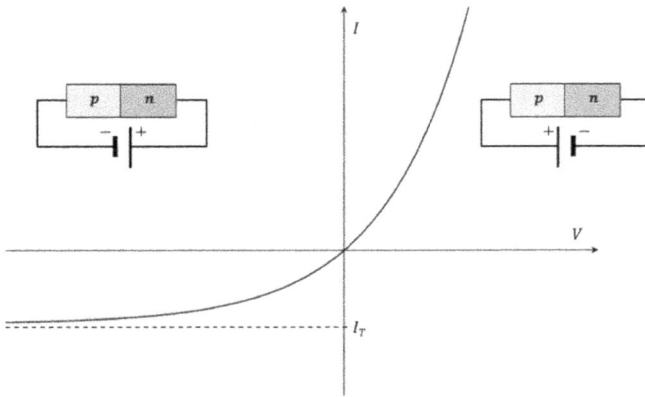

Fig. 4.25.   The characteristic $I(V)$ of the $p-n$ junction.

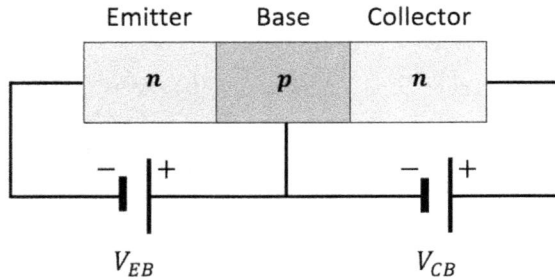

Fig. 4.26.   Sketch of a BJT in the direct polarization mode: positive polarization of the B–E junction and negative polarization of the B–C junction. V_{EB} — base–emitter polarization voltage; V_{CB} — base–collector polarization voltage.

(B–E) and *base–collector* (B–C). (A symmetric $p-n-p$ structure is also possible.)

The system is usually realized using a collector with high doping with respect to the emitter ($N_{DC} \gg N_{DE}$) and thickness of the base W much lower than the minority carriers'[5] diffusion length ($W \ll L_n$ for an $n-p-n$ BJT). With this configuration, the application of direct polarization to the B–E junction and inverse polarization to the B–C junction as depicted in Fig. 4.26 modifies the potential barriers, as shown in Fig. 4.27.

In this configuration, most of the electrons flowing from the emitter to the base are able to reach the collector since only a few of them recombine

---

[5]Electrons are minority carriers in the base (p-type semiconductor).

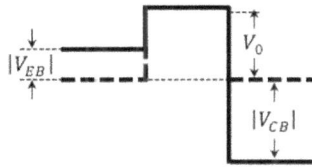

Fig. 4.27. Sketch of the bottom of conduction bands through the B–E and B–C junctions of a BJT. V_0: potential barrier before the transistor polarization.

with holes in the base owing to the larger diffusion length, thus arriving next to the B–C junction which they spontaneously cross as the energy of the conduction band in the collector is lower.

In this way, the collector current I_C is nearly equal to the emitter current I_E, with the significant difference being that I_E is flowing in a low-resistance circuit because the junction B–E has direct polarization while I_C is flowing in a high-resistance circuit because the junction B–C has indirect polarization Then a small modulation of the direct polarization voltage V_{BE} gives rise to a big change in I_E since the $V - I$ characteristic of the B–E junction corresponds to the right-hand side of Fig. 4.25. As a consequence, we have a big modulation of I_C. If this current crosses a high resistance, a large voltage modulation is measured. This simple phenomenology explains the amplifying properties of the BJT. The interested reader can find analytical details on the different configurations and applications of a BJT in technical textbooks on transistor electronics.

Quantum wells, wires, dots

The development of techniques such as molecular beam epitaxy (MBE) and metal–organic chemical vapor deposition (MOCVD) allowed the realization of customized structures with nanometric precision through a layer-by-layer deposition of chosen materials. In this way, it is possible to realize junctions of many different types, with controlled doping, nanometric thin interfaces and high purity, and to observe quantum phenomena owing to the structure sizes being comparable to the de Broglie electron wavelength.

In order to understand the origin of the *quantum size* effect, we consider a structure made by two junctions between a thin layer of a low-gap semiconductor (e.g. GaAs) and two high-gap semiconductors (e.g. AlGaAs). Depending on how many dimensions fulfill the condition to be of the order of the electron wavelength, the system realizes a *quantum well* (1D), a *quantum wire* (2D) or a *quantum dot* (3D).

When discussing the quantum size effect in one dimension, we can refer to Fig. IN4.2, which shows the top of the valence bands and the bottom of the conduction bands through the junctions. Looking at the upper part (conduction bands), the energy states of electrons in the quantum well are strongly modified with respect to the macroscopic situation, since they correspond to the discrete states of a finite potential well similar to the ones depicted in Chapter 2 for the case of the infinite potential barrier. The difference is the nonzero tails of the electron wave function through the barrier walls producing a nonzero probability of tunneling through it. In this way, defining L as the thickness of the quantum well, the discretization of the quasicontinuum density of state of a macroscopic conduction band due to the quantum size effect is easily understood. In fact, we have already noted that in a finite length L the k vector can only have discrete values:

$$k = p\frac{2\pi}{L}, \quad p = 1, 2, 3, \ldots$$

This means that the density of states in k decreases with size — a decrease by a factor of 10^6 going from $1\,\text{cm}$ to $10\,\text{nm}$! Since the energy is approximately proportional to k^2, the same effect is expected in the density of energy states. Using the infinite barrier approximation, the energy levels of the electron in the well can be written as

$$E_n = \frac{\hbar^2}{2m_e}\left[\frac{n\pi}{L}\right]^2, \quad n = 1, 2, 3, \ldots$$

This peculiar energy state structure strongly affects the transport properties, which can be additionally controlled by a proper sequence of semiconducting materials.

Multiple quantum wells are realized by a sequence of low-gap and high-gap semiconductors such as the one shown in Fig. IN4.3(a). In the case where the structure is achieved by a periodic arrangement of narrow high band gap materials, we speak of a *superlattice*, as depicted in Fig. IN4.3(b). In this case, the very thin potential barriers (1–10 nm) allow effective carrier tunneling and transport through the barriers.

In this way, different electronic functionalities can be obtained through a proper design of the structure, driving the electron flow from quasicontinuous conduction bands to discrete levels of the quantum wells by controlling heights of barrier and tunneling effects with proper voltage application. Several semiconductor devices can be realized based on these structures.

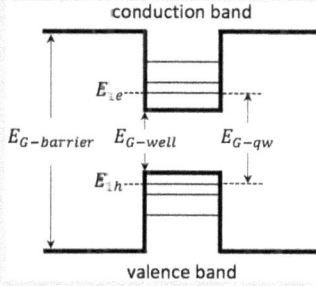

Fig. IN4.2. Energy scheme of a one-dimensional quantum well. E_{1e} is the lowest electron energy in the well; E_{1h} is the top energy of the hole in the well.

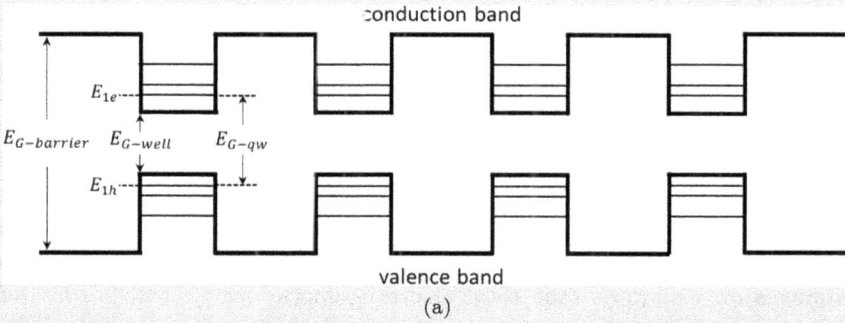

(a)

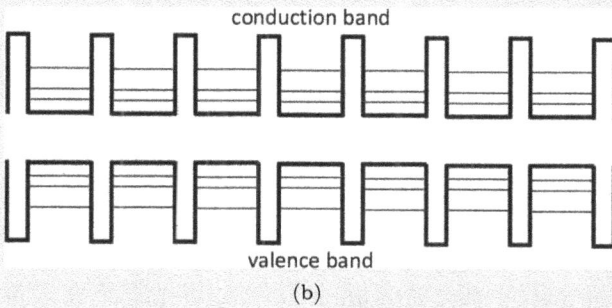

(b)

Fig. IN4.3. (a) Multiple quantum wells; (b) superlattice.

Chapter 5

Optical Properties

5.1. Lorentz–Drude theory

The discussion about the optical properties of matter can start in many different ways, depending whether the focus is on the macroscopic behavior of matter when it is "touched" by light or on the microscopic phenomena occurring in this case. Actually, an attempt will be made to clarify to the reader how phenomena on the microscale account for the macroscopic behavior. In terms of energy consideration, we can say that when light impinges on a material the total electromagnetic energy will be divided into three parts: the one reflected back in the half-space where the light is coming from; the one absorbed, which in turn can be transformed into another type of energy or can come back to the electromagnetic field after some delay; the one transmitted or scattered out of the material. The everyday experience tells us that this balance is strongly dependent on the nature of the material, on the geometry, and on the light characteristics (wavelength, propagation direction, polarization state). In the previous chapters, the optical properties of matter have been mentioned and have been summarized in the real part of the *refractive index* for what concerns propagation and in the *absorption coefficient* for what concerns energy dissipation in the medium. They are macroscopic parameters accounting for the most relevant optical properties of the material and are easy to measure. On the other hand, their value is the result of an average over a large number of microscopic contributions coming from the interaction of electrons with photons, and therefore in order to understand how this interaction affects their values it is necessary to work out the link between microscopic and macroscopic properties, as will be done in the next sections.

In addition note that light–matter interaction is also strongly affected by the aggregation state of the material particles. In fact, when one is considering atoms or molecules in a disordered state such as in vapors or isotropic liquids, the overall results of the interaction can be directly worked out from the one due to the single atom or molecule through a simple statistical summation, and therefore in this case the optical behavior of the material will be strictly related to that of the single atom or molecule. In contrast, when one is considering a crystalline solid or a material where long-range correlation is effective among atoms or molecules, the result of light–matter interaction will be affected by the material structure and cannot be discussed just in terms of atomic or molecular properties.

We have described in Chapter 1 the behavior of an e.m. wave when crossing an interface, taking into account the refractive indices of the two media in order to work out the reflection and transmission coefficients for the two polarizations s and p (perpendicular and parallel to the incidence plane, respectively). In doing that, we have temporarily not considered the frequency dependence of the material properties, which plays a fundamental role in the light–matter interaction. This leads to the well-known phenomenon of *dispersion* affecting light propagation and accounting for the frequency dependence of light absorption that leads to colors of the objects.

The classical picture due to Lorentz is very effective in describing this effect; it is able to get the frequency dependence of the material permittivity in dielectric materials, leading to final results formally close to the more rigorous ones obtained by a quantum-mechanical calculation.

In this model, each electron of a dielectric medium is bonded to the atomic nucleus by a springlike force; then its equilibrium position is kept by a harmonic restoring force. From this point of view, the classical motion equation in one dimension for the electron interacting with an electric field E is written as

$$-kx - eE(t) - \beta\frac{dx}{dt} = m\frac{d^2x}{dt^2}, \tag{5.1}$$

where on the left-hand side we have the elastic restoring force $(-kx)$, the electrostatic force $(-eE)$ and the viscous dissipative force $(-\beta\,dx/dt)$, which takes into account the interaction with the surrounding medium, where β is a friction coefficient. By considering a plane wave $E(t) = E_0 e^{-i\omega t}$ and the in-phase electron oscillation $x(t) = x_0 e^{-i\omega t}$, one gets

$$-\omega^2 x - i\gamma\omega x + \omega_0^2 x = -\frac{e}{m} E(t). \tag{5.2}$$

Then

$$x(t) = \left(-\frac{e}{m}\right) \frac{E(t)}{\omega_0^2 - \omega^2 - i\gamma\omega}, \tag{5.3}$$

where $\gamma = \beta/m$ and $\omega_0 = \sqrt{k/m}$ is the frequency resonance corresponding to the maximum amplitude of the oscillation.

This displacement gives rise to a dipole moment $\mu = -ex$, and therefore, with N being the volume density of these electrons, the consequent dipole moment per unit volume of the material is

$$P = -Nex, \tag{5.4}$$

which defines the polarization P. On the other hand, the polarization can be expressed by $P = \epsilon_0 \chi E$ and $\epsilon_r = 1 + \chi$. Then, from Eq. (5.3), we have

$$\chi(\omega) = \frac{Ne^2}{\epsilon_0 m} \frac{1}{\omega_0^2 - \omega^2 - i\gamma\omega} \tag{5.5}$$

or

$$\epsilon_r(\omega) = 1 + \frac{Ne^2}{\epsilon_0 m} \frac{1}{\omega_0^2 - \omega^2 - i\gamma\omega}. \tag{5.6}$$

Therefore, the dielectric permittivity is written as a complex quantity, $\epsilon_r(\omega) = \epsilon_r' + i\epsilon_r''$, with

$$\epsilon_r'(\omega) = 1 + \frac{Ne^2}{\epsilon_0 m} \frac{\omega_0^2 - \omega^2}{(\omega_0^2 - \omega^2)^2 + \gamma^2\omega^2}, \tag{5.7}$$

$$\epsilon_r''(\omega) = \frac{Ne^2}{\epsilon_0 m} \frac{\gamma\omega}{(\omega_0^2 - \omega^2)^2 + \gamma^2\omega^2}. \tag{5.8}$$

If we consider all the possible transitions at frequency ω_j, we have

$$\epsilon_r'(\omega) = 1 + \frac{Ne^2}{\epsilon_0 m} \sum_j \frac{\omega_j^2 - \omega^2}{(\omega_j^2 - \omega^2)^2 + \gamma^2\omega^2}, \tag{5.9}$$

$$\epsilon_r''(\omega) = \frac{Ne^2}{\epsilon_0 m} \sum_j \frac{\gamma\omega}{(\omega_j^2 - \omega^2)^2 + \gamma^2\omega^2}. \tag{5.10}$$

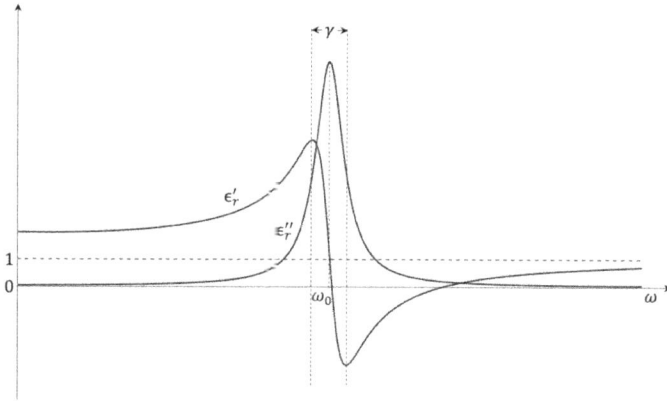

Fig. 5.1. Frequency dependence of the real (a) and imaginary (b) parts of the dielectric permittivity.

In Fig. 5.1, the real and imaginary parts of $\epsilon_r(\omega)$ are plotted as given by Eqs. (5.7) and (5.8). The real part $\epsilon'_r(\omega)$ increases with increasing frequency except in a narrow region around ω_0, corresponding to the *anomalous dispersion*. In this region, the imaginary part $\epsilon''_r(\omega)$ is maximum.

By looking at Eq. (5.8), it is easy to find the width of the curve to be linked to the coefficient γ, which takes into account the energy losses. In fact, near resonance $\omega \approx \omega_0$, and therefore[1]

$$\epsilon''_r(\omega) \approx \frac{Ne^2}{\epsilon_0 m} \frac{\gamma \omega_0}{4\omega_0^2(\omega_0 - \omega)^2 + \gamma^2 \omega^2}$$

$$= \frac{Ne^2}{\epsilon_0 m} \frac{\gamma}{4\omega_0(\omega_0 - \omega)^2 + \gamma^2 \omega_0}. \tag{5.11}$$

We have the maximum at $\omega = \omega_0$,

$$\epsilon''_r(\omega_0) = \frac{Ne^2}{\epsilon_0 m} \frac{1}{\gamma \omega_0}, \tag{5.12}$$

while the half width at half maximum corresponds to $\Delta \omega = \omega_0 - \omega$ when the denominator of Eq. (5.11) is twice its minimum value, i.e.

$$4(\omega_0 - \omega)^2 = 4(\Delta \omega)^2 = \gamma^2 \tag{5.13}$$

---

[1] Under this approximation, $(\omega_0^2 - \omega^2)^2 = (\omega_0^2 - \omega^2)(\omega_0^2 - \omega^2) = [(\omega_0 + \omega)(\omega_0 - \omega)]^2 \approx 4\omega_0^2(\omega_0 - \omega)^2$.

and

$$\Delta\omega = \pm\frac{\gamma}{2}. \tag{5.14}$$

In other words, γ is the full width at half maximum of $\epsilon_r''(\omega)$.

It can be shown that for the corresponding values of the frequency,

$$\omega_- = \omega_0 - \frac{\gamma}{2}, \quad \omega_+ = \omega_0 + \frac{\gamma}{2}, \tag{5.15}$$

the real part of the permittivity $\epsilon_r'(\omega)$ has its maximum and minimum, respectively; then γ is also the width of the anomalous dispersion region.

From the real and imaginary parts of the dielectric permittivity, it is possible to retrieve the real and imaginary parts of the refractive index from the definition (1.20):

$$\tilde{n}^2 = (n' + in'')^2 = n'^2 - n''^2 + i2n'n'' = \epsilon_r'(\omega) + i\epsilon_r''(\omega); \quad \text{then}$$

$$\epsilon_r'(\omega) = n'^2 - n''^2, \quad \epsilon_r''(\omega) = 2n'n''. \tag{5.16}$$

The frequency dependence of the real and imaginary parts of the refractive index that will be obtained from Eqs. (5.16) roughly follows that presented in Fig. 5.1 for the dielectric permittivity. In this way, the frequency dependence of the macroscopic optical properties such as reflectivity and absorbance versus frequency can be obtained.

The Kramers–Kronig relations

The Kramers–Kronig relations connect the real and imaginary parts of the response function of a linear system. These relations arise from a mathematical constraint on complex functions that represent physical processes.

For instance, the complex dielectric permittivity relates the electric field $\mathbf{E}$ and the dielectric displacement vector $\mathbf{D}$, which are real since they represent physical quantities. Therefore, it must satisfy a symmetry condition:

$$\epsilon(-\omega) = \epsilon^*(\omega).$$

Moreover, causality states that the polarizability must originate from an applied electric field. By applying these two concepts, it is possible to relate the real and imaginary parts of $\epsilon(\omega)$. Taking into account that dielectric permittivity is related to the refractive index, we understand that also for the real and imaginary parts of this quantity similar conditions are fulfilled. The mathematical procedure for obtaining these relations is complex and can be found in more specialized books; we just recall here the Kramers–Kronig relations connecting the real and imaginary parts of the dielectric permittivity,

$$\epsilon'_r(\omega) - 1 = \frac{2}{\pi}\wp \int_0^\infty \frac{\omega'\epsilon''_r(\omega')}{\omega'^2 - \omega^2}d\omega', \quad \text{with}$$

$$\epsilon''_r(\omega) = -\frac{2}{\pi}\wp \int_0^\infty \frac{\omega'\epsilon'_r(\omega')}{\omega'^2 - \omega^2}d\omega',$$

and the real and imaginary parts of the refractive index,

$$n'(\omega) - 1 = \frac{2}{\pi}\wp \int_0^\infty \frac{\omega'n''(\omega') - \omega n''(\omega)}{\omega'^2 - \omega^2}d\omega', \quad \text{with}$$

$$n''(\omega) = \frac{2}{\pi}\wp \int_0^\infty \frac{\omega'n'(\omega') - \omega n'(\omega)}{\omega'^2 - \omega^2}d\omega',$$

where $\wp$ stands for the principal part of the integral (divergent for $\omega = \omega'$).

These relations are very important, since they state that a dispersion for the real part corresponds to a dispersion for the imaginary part. For instance, by looking at Fig. 5.1, we see that a strong dispersion of the real part of the dielectric permittivity occurs near the absorption band, where the imaginary part changes rapidly.

These relations have been successfully used in the investigation of optical properties of solids, because they allow obtaining the real part of the optical response if the imaginary part is known in the whole frequency range and vice versa.

Color dispersion

The Lorentz approach explains very well the common dispersion effect that can be observed when a white light beam is refracted by a glass prism or when a rainbow appears in the sky after a thunderstorm. In Fig. IN5.1, the double refraction occurring at the air–glass and the glass–air interface in a glass prism is sketched for a light beam including a wide range of wavelengths in the visible spectrum.

Fig. IN5.1. Dispersion of a white light beam passing through a glass prism.

It is a common experience that a higher deviation is suffered by lower wavelength waves, i.e. the deviation angle is bigger for blue light than for red light. This can be explained if the refractive index increases by decreasing the wavelength (i.e. by increasing the frequency). In fact, the refraction angle is an increasing function of the ratio n_2'/n_1' when $n_2' > n_1'$ (first interface: air–glass) and of the ratio n_1'/n_2' when $n_1' > n_2'$ (second interface: glass–air). Since for a transparent medium $n'' \approx 0$, this means that, according to the Lorentz theory, we are far from the absorption region, namely in the normal dispersion region, where the refractive index increases with frequency.

In spite of its simplicity, this classical approach gives a correct picture of the optical behavior not only for the assembling of single atoms or molecules, but also for materials in the condensed phase such as dielectrics where the approximation of the electron tightly bound to the atomic nucleus through an elastic force is suitable.

This model can actually be extended to systems where electrons are not bounded, such as the electron gas of a metal considered under the free electron model, as will be shown in Section 5.6.

5.2. Quantum approach to optical properties

In order to have a rigorous treatment of the macroscopic optical properties of a material, it is necessary to link them to the correct quantum description of the interaction between the electromagnetic wave and the electrons of the material. As in the classical model, we consider a linearly polarized optical field given by

$$\mathbf{E} = \frac{E_0}{2}\hat{\mathbf{x}}(e^{-i\omega t} + e^{i\omega t}), \tag{5.17}$$

where the spatial dependence of the field is included in E_0 varying over lengths much bigger than the ones typical of the bounded electron. In fact, while the "optical" wavelengths are in the range of a few hundred nanometers, the bounded electron movement is in the range of nanometers, and therefore E_0 can be considered as a constant during the interaction. The electron displacement $\mathbf{r}$ from the atomic nucleus gives rise to an instantaneous electric dipole moment $\boldsymbol{\mu} = -e\mathbf{r}$, and thus the field given by Eq. (5.17) will interact with the electronic dipole, giving rise to a potential energy:

$$U(t) = -\boldsymbol{\mu} \cdot \mathbf{E} = ex\frac{E_0}{2}(e^{-i\omega t} + e^{i\omega t}). \tag{5.18}$$

If this quantity is much lower than the initial electron energy, it can be considered as a perturbation and we can apply the time-dependent perturbation theory (see Section 2.13) taking $\mathcal{H}'(t) = U(t)$. This allows obtaining the transition probability between a state m and a state k of the unperturbed Hamiltonian, which, according to Eq. (2.234), becomes

$$|a_k^{(1)}(t)|^2 = \frac{|ex_{km}E_0|^2}{\hbar^2}\frac{\sin^2\left[\frac{1}{2}(\omega_{km} - \omega)t\right]}{(\omega_{km} - \omega)^2}, \tag{5.19}$$

where

$$ex_{km} \equiv \mu_{km} = \int_0^V \phi_k^* ex\phi_m dV \tag{5.20}$$

is the matrix element of the dipole moment of the electron between the states ϕ_m and ϕ_k in the direction of the wave electric field.

Following the same steps as in Section 2.13, we can finally work out the transition probability per second between the states ϕ_m and ϕ_k [see Eq. (2.248)] as

$$W_{km}^{(i)} = \frac{1}{4\hbar^2}|\mu_{km}|^2|E_0|^2 g(\nu), \tag{5.21}$$

with $g(\nu)$ being the line shape that for each possible transition ($k \to m$) weights its probability to occur depending on the distance from the central resonance frequency $\nu_{km} = \omega_{km}/2\pi$, as discussed in the following.

The most important features of the induced process are:

(i) It has the highest probability to occur when the photon energy corresponds to the energy difference between the involved states:

$$\nu = \nu_{km} = \frac{\omega_{km}}{2\pi} = \frac{|E_k - E_m|}{h}. \qquad (5.22)$$

(ii) It can occur in either of the conditions $E_k > E_m$ and $E_k < E_m$. This is a very important remark, since it points out the possible occurrence of two effects due to this interaction that leads to the electronic transition from the state ϕ_m to the state ϕ_k: absorption of one photon when $E_k > E_m$ or emission of one photon when $E_k < E_m$. We will discuss in detail these aspects of the interaction.

(iii) The strength of the transition is determined by $|\mu_{km}|^2$, which is strongly dependent on the shape of the involved wave functions and on their dependence on the space coordinates. Since the wave functions are characterized by the quantum numbers n, l, m, the superposition integral (5.20) can be affected by their change from the state ϕ_m to the state ϕ_k. In particular, this leads to the well-known *selection rules* (i.e. constraints on the allowed changes of quantum numbers), according to which only specific transitions corresponding to nonzero values of $|\mu_{km}|^2$ are allowed.

Besides considering the main features of the light-induced electronic transition between different Hamiltonian eigenstates, it is useful to find a connection between the microscopic and the macroscopic quantities. This link allows achieving information on the electronic structure of materials through measurements performed on the optical properties. For this aim, it is useful to use a method similar to the one used in the classical approach, namely to get the atomic polarizability first so as to obtain eventually the dielectric permittivity.

Therefore, by considering a simple dipole interaction as highlighted above, we are interested in calculating the expectation value of the electronic dipole moment in the perturbed state:

$$\langle \mu \rangle = \int_{-\infty}^{+\infty} \psi^* e x \psi dV \equiv \int_{-\infty}^{+\infty} \psi^* \mu \psi dV. \qquad (5.23)$$

For this aim, we can use the results of Section 2.14 concerning the two-level system. In fact, we can calculate Eq. (5.23) for the transition occurring between two levels and after that extend the result to all the possible transitions just by summation over all the final states.

We start by considering the wave function ψ given by Eq. (2.250) as superposition of the eigenfunctions of the two states:

$$\langle \mu \rangle = \int_{-\infty}^{+\infty} \left[a_1^* \phi_1^* e^{\frac{iE_1 t}{\hbar}} + a_2^* \phi_2^* e^{\frac{iE_2 t}{\hbar}} \right] \mu \left[a_1 \phi_1 e^{-\frac{iE_1 t}{\hbar}} + a_2 \phi_2 e^{-\frac{iE_2 t}{\hbar}} \right] dV. \tag{5.24}$$

Since $\mu_{11} = \mu_{22} = 0$, and recalling the definition (2.255), $\omega_0 \equiv (E_2 - E_1)/\hbar$, we have

$$\langle \mu \rangle = a_2^* a_1 \mu_{21} e^{i\omega_0 t} + a_1^* a_2 \mu_{12} e^{-i\omega_0 t}. \tag{5.25}$$

Now we use Eqs. (2.263) and (2.264), providing the first-order approximation for $a_1(t)$ [≈ 1, see Eq. (2.261)] and $a_2(t)$, and take into account the reality condition $\mu_{21} = \mu_{12}$, $H'_{21} = H'_{12}$. In this way, we get

$$\langle \mu \rangle = a_2^* \mu_{21} e^{i\omega_0 t} + a_2 \mu_{12} e^{-i\omega_0 t}$$

$$= -\frac{\mu_{12} H'_{12}}{2\hbar} \left\{ \frac{e^{-i[\omega_0 + \omega]t} - 1}{\omega_0 + \omega} + \frac{e^{-i[\omega_0 - \omega]t} - 1}{\omega_0 - \omega} \right\} e^{i\omega_0 t}$$

$$+ \frac{\mu_{12} H'_{12}}{2\hbar} \left\{ \frac{e^{i[\omega_0 + \omega]t} - 1}{\omega_0 + \omega} + \frac{e^{i[\omega_0 - \omega]t} - 1}{\omega_0 - \omega} \right\} e^{-i\omega_0 t}. \tag{5.26}$$

Then

$$\langle \mu \rangle \approx -\frac{\mu_{12} H'_{12}}{2\hbar} \left\{ \frac{e^{-i\omega t}}{\omega_0 + \omega} + \frac{e^{-i\omega t}}{\omega_0 - \omega} \right\}$$

$$+ \frac{\mu_{12} H'_{12}}{2\hbar} \left\{ \frac{e^{i\omega t}}{\omega_0 + \omega} + \frac{e^{i\omega t}}{\omega_0 - \omega} \right\}. \tag{5.27}$$

Equation (5.27) is obtained by taking into account only the time-dependent term in phase with the field perturbation (by considering the losses, it will be shown below that the other terms become negligible in the steady state). This means keeping only the terms with the same time dependence as the optical field given by Eq. (5.17). Then we can write

$$\langle \mu \rangle = -\frac{\mu_{12} H'_{12}}{2\hbar} \left\{ \frac{1}{\omega_0 - \omega} + \frac{1}{\omega_0 + \omega} \right\} (e^{-i\omega t} + e^{i\omega t}), \tag{5.28}$$

and since $H'_{12} = -\mu_{12} E_0$,

$$\langle \mu \rangle = \frac{|\mu_{12}|^2}{2\hbar} \left\{ \frac{1}{\omega_0 - \omega} + \frac{1}{\omega_0 + \omega} \right\} E_0 (e^{-i\omega t} + e^{i\omega t}). \tag{5.29}$$

In order to extend this result to all the possible transitions from a generic state m, we make the substitutions $1 \to m$, $2 \to k$, and $\omega_0 \to \omega_j$, where ω_j is the resonance frequency of the transition $m \to k$. Finally, we sum over all the possible final states for transitions from the state ϕ_m:

$$\langle \mu \rangle = \left[\sum_j |\mu_{km}|^2 \frac{1}{\hbar} \left\{ \frac{1}{\omega_j - \omega} + \frac{1}{\omega_j + \omega} \right\} \right] \frac{E_0}{2} (e^{-i\omega t} + e^{i\omega t}).$$

$$(5.30)$$

The term in square brackets define the *atomic polarizability*:

$$\alpha(\omega) = \sum_j |\mu_{km}|^2 \frac{1}{\hbar} \left\{ \frac{1}{\omega_j - \omega} + \frac{1}{\omega_j + \omega} \right\} = \sum_j |\mu_{km}|^2 \frac{1}{\hbar} \left\{ \frac{2\omega_j}{\omega_j^2 - \omega^2} \right\},$$

$$(5.31)$$

which shows the electronic response to the field application. In order to underline how each transition contributes to the atomic polarizability, it is useful to introduce the *oscillator strength*,

$$f_j = \frac{2m}{\hbar^2} |x_{km}|^2 \hbar \omega_j,$$

$$(5.32)$$

which satisfies the Thomas–Reiche–Kuhn sum rule:

$$\sum_j f_j = 1.$$

$$(5.33)$$

In other words, the oscillator strength gives the relative weight of a specific transition in determining the atomic polarizability.

In this way, Eq. (5.31) can be written as

$$\alpha(\omega) = \frac{e^2}{m} \sum_j \frac{f_j}{\omega_j^2 - \omega^2}.$$

$$(5.34)$$

By considering a density of N atoms per cm^3, we can link the atomic polarizability to the macroscopic quantity describing the material polarization induced by the electric field.

For this aim, we take into account the definition of a polarization vector:

$$\mathbf{P} = \epsilon_0 \underline{\chi}(\omega) \mathbf{E},$$

$$(5.35)$$

with $\underline{\chi}(\omega)$ being the dielectric susceptibility tensor, which reduces to a scalar quantity for isotropic media. In the following, we will restrict ourselves to this case.

Since at atomic level the polarizability is defined as the electric dipole moment divided by the electric field and the polarization vector is the electric dipole moment per unit volume, we can simply write

$$\frac{F}{E} = N\alpha(\omega).$$
(5.36)

From Eq. (5.35), it follows that the dielectric susceptibility is the material polarizability per unit volume:

$$\chi(\omega) = \frac{N\alpha(\omega)}{\epsilon_0}.$$
(5.37)

On the other hand, from Eq. (1.5) it is straightforward to get (in the SI system)

$$\epsilon_r = 1 + \chi(\omega).$$
(5.38)

Therefore, the dielectric permittivity becomes

$$\epsilon_r(\omega) = 1 + \frac{N\alpha(\omega)}{\epsilon_0} = 1 + \frac{Ne^2}{\epsilon_0 m} \sum_j \frac{f_j}{\omega_j^2 - \omega^2}.$$
(5.39)

Interestingly, this expression connects a macroscopic material response to the microscopic properties represented by the allowed electronic transitions.

Looking at Eq. (5.39) as well as Eq. (5.34) and at the previous expressions, we observe an anomalous behavior, since the material response goes to infinity at each resonance frequency $\omega = \omega_j$. This result has no physical meaning and is a consequence of neglecting losses that were not considered in the previous treatment. We can introduce them as a phenomenological damping of the a_2 coefficient due to spontaneous emission to the low-energy state.

Equations (2.259) and (2.260) can be modified as follows:

$$a_1^{(1)} = 1, \quad \frac{da_2^{(1)}}{dt} + \frac{\Gamma}{2}a_2^{(1)} = -\frac{i}{\hbar}\mathcal{H}'_{21}\exp(i\omega_0 t),$$

introducing the damping term $(\Gamma/2)a_2^{(1)}$. The differential equation can be integrated under the condition of no perturbation $\mathcal{H}'_{21} = 0$ and gives the exponential decay to level 1:

$$a_2^{(1)} = a_2^{(1)}(0)e^{-\frac{\Gamma}{2}t},$$

where $\tau = 2/\Gamma$ is the characteristic lifetime of the excited level. In introducing the harmonic perturbation represented by the electromagnetic wave with field given in Eq. (5.17) and the electric dipole interaction, we have to integrate the following equation:

$$\frac{da_2^{(1)}}{dt} + \frac{\Gamma}{2}a_2^{(1)} = \frac{i}{2\hbar}H'_{21}[e^{i(\omega_0+\omega)t} + e^{i(\omega_0-\omega)t}].$$

The solution is found as the sum of the zero perturbation solution and a particular solution, and we have

$$a_2^{(1)} = \frac{H'_{21}}{2\hbar}\left[\frac{e^{i(\omega_0+\omega)t}}{\omega_0+\omega-i\frac{\Gamma}{2}} + \frac{e^{i(\omega_0-\omega)t}}{\omega_0-\omega-i\frac{\Gamma}{2}}\right] + a_2^{(1)}(0)e^{-\frac{\Gamma}{2}t},$$

where it is clear that at a *steady state of excitation* $(t \to \infty)$ the second term becomes negligible.

By using this expression in Eq. (5.26) keeping the in-phase terms, we have

$$\langle\mu\rangle = a_2^*\mu_{21}e^{i\omega_0 t} + a_2\mu_{12}e^{-i\omega_0 t}$$

$$= -\frac{\mu_{21}H'_{21}}{2\hbar}\left[\frac{e^{-i\omega t}}{\omega_0+\omega+i\frac{\Gamma}{2}} + \frac{e^{i\omega t}}{\omega_0-\omega+i\frac{\Gamma}{2}}\right.$$

$$\left. + \frac{e^{i\omega t}}{\omega_0+\omega-i\frac{\Gamma}{2}} + \frac{e^{-i\omega t}}{\omega_0-\omega-i\frac{\Gamma}{2}}\right].$$

This can be rewritten as

$$\langle\mu\rangle = \frac{|\mu_{21}|^2}{2\hbar}\left[\frac{e^{-i\omega t}}{\omega_0+\omega+i\frac{\Gamma}{2}} + \frac{e^{i\omega t}}{\omega_0-\omega+i\frac{\Gamma}{2}}\right.$$

$$\left. + \frac{e^{i\omega t}}{\omega_0+\omega-i\frac{\Gamma}{2}} + \frac{e^{-i\omega t}}{\omega_0-\omega-i\frac{\Gamma}{2}}\right]E_0$$

or

$$\langle\mu\rangle = \frac{|\mu_{21}|^2}{2\hbar}\left\{\left[\frac{1}{\omega_0+\omega+i\frac{\Gamma}{2}} + \frac{1}{\omega_0-\omega-i\frac{\Gamma}{2}}\right]E_0 e^{-i\omega t}\right.$$

$$\left. + \left[\frac{1}{\omega_0+\omega-i\frac{\Gamma}{2}} + \frac{1}{\omega_0-\omega+i\frac{\Gamma}{2}}\right]E_0 e^{i\omega t}\right\}.$$

Then, by writing the dipole moment as

$$\langle\mu\rangle = \frac{1}{2}\{\alpha(\omega)e^{-i\omega t} + \alpha^*(\omega)e^{i\omega t}\}E_0 = R\{\alpha(\omega)e^{-i\omega t}\}E_0,$$

we find the atomic polarizability as

$$\alpha(\omega) = \frac{|\mu_{21}|^2}{\hbar}\left[\frac{1}{\omega_0 + \omega + i\frac{\Gamma}{2}} + \frac{1}{\omega_0 - \omega - i\frac{\Gamma}{2}}\right] \approx \frac{|\mu_{21}|^2}{\hbar}\left[\frac{2\omega_0}{\omega_0^2 - \omega^2 - i\Gamma\omega}\right],$$

where we have neglected the term with Γ^2 since $\Gamma \ll \omega_0$.

Summation over all the possible transitions occurring from a single state m gives the atomic polarizability as

$$\alpha(\omega) = \sum_j |\mu_{km}|^2 \frac{1}{\hbar}\left[\frac{2\omega_j}{\omega_j^2 - \omega^2 - i\Gamma_j\omega}\right] = \frac{e^2}{m}\sum_j f_j\left[\frac{1}{\omega_j^2 - \omega^2 - i\Gamma_j\omega}\right],$$

according to the definition (5.32) of the oscillator strength f_j.

The corrected atomic polarizability allows getting a more realistic expression of the dielectric permittivity:

$$\epsilon_r(\omega) = 1 + \frac{N\alpha(\omega)}{\epsilon_0} = 1 + \frac{Ne^2}{\epsilon_0 m}\sum_j f_j\left[\frac{1}{\omega_j^2 - \omega^2 - i\Gamma_j\omega}\right], \tag{5.40}$$

$$\epsilon_r(\omega) = 1 + \frac{Ne^2}{\epsilon_0 m}\sum_j f_j\left\{\frac{\omega_j^2 - \omega^2}{\left(\omega_j^2 - \omega^2\right)^2 + \omega^2\Gamma_j^2} + \frac{i\Gamma_j\omega}{\left(\omega_j^2 - \omega^2\right)^2 + \omega^2\Gamma_j^2}\right\}. \tag{5.41}$$

That is to say, each resonant transition contributes to a real part of permittivity given by

$$\epsilon_r'(\omega) = 1 + \frac{Ne^2}{\epsilon_0 m}f_j\frac{\omega_j^2 - \omega^2}{\left(\omega_j^2 - \omega^2\right)^2 + \omega^2\Gamma_j^2} \tag{5.42}$$

and to an imaginary part given by

$$\epsilon_r''(\omega) = \frac{Ne^2}{\epsilon_0 m}f_j\frac{\Gamma_j\omega}{\left(\omega_j^2 - \omega^2\right)^2 + \omega^2\Gamma_j^2}, \tag{5.43}$$

quite similar to the classical expressions (5.7) and (5.8).

From Eqs. (5.38) and (5.41), we get the real part of the dielectric susceptibility,

$$\chi'(\omega) = \frac{Ne^2}{\epsilon_0 m} f_j \frac{\omega_j^2 - \omega^2}{(\omega_j^2 - \omega^2)^2 + \omega^2 \Gamma_j^2} \tag{5.44}$$

and from Eq. (5.42) we get the imaginary part,

$$\chi''(\omega) = \frac{Ne^2}{\epsilon_0 m} f_j \frac{\Gamma_j \omega}{\left(\omega_j^2 - \omega^2\right)^2 + \omega^2 \Gamma_j^2}. \tag{5.45}$$

This can be rewritten by using the definition (5.32) of oscillator strength and considering a single resonance at $\omega_0 \equiv \omega_j$. Then

$$\chi''(\omega) = \frac{Ne^2}{\epsilon_0 m} \frac{2m}{\hbar^2} |x_{km}|^2 \hbar \omega_0 \frac{\Gamma \omega}{\left(\omega_0^2 - \omega^2\right)^2 + \omega^2 \Gamma^2}$$

$$= \frac{2Ne^2}{\epsilon_0 \hbar} |x_{km}|^2 \frac{\Gamma \omega \omega_0}{\left(\omega_0^2 - \omega^2\right)^2 + \omega^2 \Gamma^2} \equiv \frac{2Ne^2}{\epsilon_0 \hbar} |x_{km}|^2 \, \tilde{g}(\omega). \tag{5.46}$$

We have defined

$$\tilde{g}(\omega) = \frac{\Gamma \omega \omega_0}{(\omega_0^2 - \omega^2)^2 + \omega^2 \Gamma^2}, \tag{5.47}$$

which, in the approximation $\omega \approx \omega_0$, becomes

$$\tilde{g}(\omega) \approx \frac{\Gamma \omega^2}{[(\omega_0 - \omega)(\omega_0 + \omega)]^2 + \omega^2 \Gamma^2} \approx \frac{\Gamma \omega^2}{4\omega^2 (\omega_0 - \omega)^2 + \omega^2 \Gamma^2}$$

$$= \frac{\Gamma}{4(\omega_0 - \omega)^2 + \Gamma^2} = \frac{1}{4} \frac{\Gamma}{(\omega_0 - \omega)^2 + (\Gamma/2)^2}. \tag{5.48}$$

The maximum value at $\omega = \omega_0$ is

$$\tilde{g}(\omega_0) = \frac{1}{4} \frac{\Gamma}{(\Gamma/2)^2} = \frac{1}{\Gamma}. \tag{5.49}$$

Then the following expression allows evaluating the frequencies corresponding to half maximum:

$$\tilde{g}(\omega_{1/2}) = \frac{1}{4} \frac{\Gamma}{\left(\omega_0 - \omega_{1/2}\right)^2 + (\Gamma/2)^2} = \frac{1}{2} \tilde{g}(\omega_0) = \frac{1}{2\Gamma}, \tag{5.50}$$

i.e.

$$(\Delta\omega)^2 \equiv (\omega_0 - \omega_{1/2})^2 = \left(\frac{\Gamma}{2}\right)^2 \quad \rightarrow \quad \Gamma = \pm 2\Delta\omega. \tag{5.51}$$

Therefore, Γ is the width of the curve at half maximum.

More often, the actual frequency $\nu = \omega/2\pi$ rather than ω is considered. Then we get the function

$$\tilde{g}(\nu) = \frac{1}{4} \frac{(\Delta\nu/2\pi)}{(\nu_0 - \nu)^2 + (\Delta\nu/2)^2}, \qquad (5.52)$$

where, according to Eq. (5.51). we set $\Delta\nu = \Gamma/2\pi$, the *full width* at half maximum.

Finally, we introduce the normalized counterpart $g(\nu)$ of this function through the condition

$$\int_{-\infty}^{+\infty} g(\nu)d\nu = 1, \qquad (5.53)$$

calling it the *line shape* of the electronic transition. It is easy to show that $g(\nu) = 4\tilde{g}(\nu)$, i.e.

$$g(\nu) = \frac{(\Delta\nu/2\pi)}{(\nu_0 - \nu)^2 + (\Delta\nu/2)^2}. \qquad (5.54)$$

This bell-like function is called "Lorentzian" and plays a very important role in the processes of interaction between light and matter, since it represents the probability density that a transition occurs at a specific frequency. It points out the important concept that transition occurs only when the photon energy is close to the energy difference between the involved two levels. $[g(\nu) \to 0$ when either $\nu \gg \nu_0$ or $\nu \ll \nu_0.]$

Going back to Eq. (5.46), we have

$$\chi''(\nu) = \frac{2Ne^2}{\epsilon_0\hbar}|x_{km}|^2 \frac{g(\nu)}{4} = \frac{N}{2\epsilon_0\hbar}|\mu_{km}|^2 g(\nu) \qquad (5.55)$$

by using the transition dipole moment $\mu_{km} = ex_{km}$.

In this way, we get a very compact expression underlining the important quantities affecting the possible transition: the electron density N, which determines the number of possible transitions; the dipole moment μ_{km}, which determines the "strength" of each transition; and the line shape $g(\nu)$, which determines the probability of having a transition at a specific frequency.

According to the previous definition and under the same approximation $\omega \approx \omega_0$, the real part of the dielectric susceptibility can be written as

$$\chi'(\omega) = \frac{2Ne^2}{\epsilon_0\hbar}|x_{km}|^2 \frac{\omega_0(\omega_0^2 - \omega^2)}{(\omega_0^2 - \omega^2)^2 + \omega^2\Gamma^2} \approx \frac{Ne^2}{\epsilon_0\hbar}|x_{km}|^2 \frac{\omega_0 - \omega}{(\omega_0 - \omega)^2 + (\Gamma/2)^2} \qquad (5.56)$$

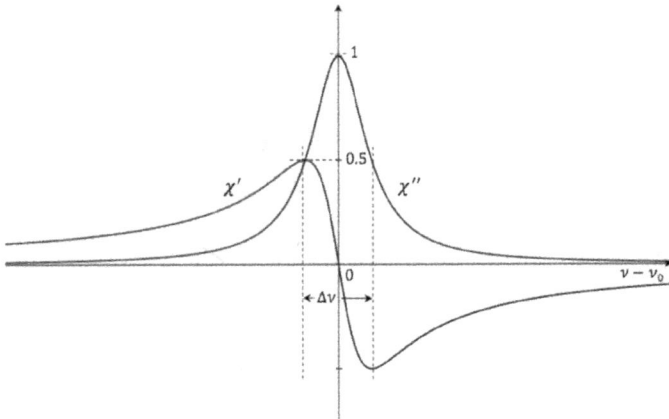

Fig. 5.2. The real and imaginary parts of the dielectric susceptibility versus $\nu - \nu_0$. The quantities are normalized to the maximum of the imaginary part: $\chi''(\nu_0)$.

or

$$\chi'(\nu) = \frac{Ne^2}{\epsilon_0 \hbar} |x_{km}|^2 \frac{1}{2\pi} \frac{\nu_0 - \nu}{(\nu_0 - \nu)^2 + (\Delta\nu/2)^2} = \frac{N}{\epsilon_0 \hbar} |\mu_{km}|^2 \frac{\nu_0 - \nu}{\Delta\nu} g(\nu).$$

(5.57)

The functional behavior of $\chi'(\nu)$ and $\chi''(\nu)$ is reported in Fig. 5.2 normalized to the maximum of the imaginary part $\chi''(\nu_0) = (N/2\epsilon_0\hbar)|\mu_{km}|^2$ $(1/\pi\Delta\nu)$. According to what was already highlighted, we see that $\chi''(\nu) \to 0$ for either $\nu \gg \nu_0$ or $\nu \ll \nu_0$ (i.e. far from resonance, when $|\nu_0 - \nu| \gg \Delta\nu$). The real part $\chi'(\nu) \to 0$ much more slowly, so that the tails of the real part related to one specific resonance may be nonnegligible at frequencies close to a different resonance.

Let us consider a monochromatic wave at frequency ν interacting with a material, and therefore close to a single resonance at frequency ν_0 and far enough from all the other transition frequencies of the medium. Under these conditions, while the resonance at ν_0 will contribute to $\chi'(\nu)$ and to $\chi''(\nu)$, the others may contribute only to the real part $\chi'(\nu)$ owing to the slowly decreasing tails of this function, since $\chi''(\nu) \approx 0$ for $|\nu_0 - \nu| \gg \Delta\nu$. Therefore, the summation appearing in Eq. (5.40) (appearing in the corresponding expression of the dielectric susceptibility) applies only to the real part, leading to a "nonresonant" contribution χ_{NR} whose frequency dependence can be neglected with respect to the resonant term (in the narrow range near the considered resonance). It must be added to the resonant $\chi_{\mathrm{R}}(\nu)$ originated by the response close to ν_0; as a consequence, we can write

$$\chi(\nu) = \chi_{\rm R}(\nu) + \chi_{\rm NR} = \chi'(\nu) + i\chi''(\nu) + \chi_{\rm NR}. \tag{5.58}$$

Therefore, we write

$$\epsilon'_r(\nu) = 1 + \chi'(\nu) + \chi_{\rm NR}, \tag{5.59}$$

$$\epsilon''_r(\nu) = \chi''(\nu). \tag{5.60}$$

Accordingly, we can write the complex refractive index as

$$\tilde{n}(\nu) = \sqrt{\epsilon'_r(\nu) + i\epsilon''_r(\nu)} = \sqrt{1 + \chi_{\rm NR} + \chi'(\nu) + i\chi''(\nu)} = \sqrt{\epsilon_{\rm NR} + \chi_{\rm R}(\nu)}. \tag{5.61}$$

In the approximation[2] $\epsilon_{\rm NR} \gg \chi_{\rm R}(\nu)$,

$$\tilde{n}(\nu) = \sqrt{\epsilon_{\rm NR}}\sqrt{1 - \frac{\chi_{\rm R}(\nu)}{\epsilon_{\rm NR}}} \approx \sqrt{\epsilon_{\rm NR}}\left[1 + \frac{\chi_{\rm R}(\nu)}{2\epsilon_{\rm NR}}\right]$$

$$= \sqrt{\epsilon_{\rm NR}}\left[1 + \frac{\chi'(\nu) + i\chi''(\nu)}{2\epsilon_{\rm NR}}\right]. \tag{5.62}$$

Here $\sqrt{\epsilon_{\rm NR}} = n_0$ is the usual refractive index of the medium far from resonance. Therefore, the real and imaginary parts of the refractive index become

$$n'(\nu) = n_0 + \frac{\chi'(\nu)}{2n_0}, \tag{5.63}$$

$$n''(\nu) = \frac{\chi''(\nu)}{2n_0}. \tag{5.64}$$

It is worth noting the direct proportionality between the imaginary part of the refractive index and that of the dielectric susceptibility.

It is straightforward to plot the functions $n'(\nu)$ and $n''(\nu)$ using Eqs. (5.63) and (5.64), as shown in Fig. 5.3.

Figure 5.3 is quite important, since it gives a clear explanation of the light dispersion effect. The real part (accounting for the propagation phenomena) is shown to increase with frequency except in a narrow region (width $\sim \Delta\nu$) close to a resonance frequency, where anomalous dispersion takes place (the real part of the refractive index decreases with frequency). In other words, when the material appears transparent to the incoming wave we should expect normal dispersion; on the contrary, in correspondence with absorption, we expect anomalous dispersion.

---

[2]Using a Taylor expansion for $x \ll 1$: $\sqrt{1+x} \approx 1 + (1/2)x$.

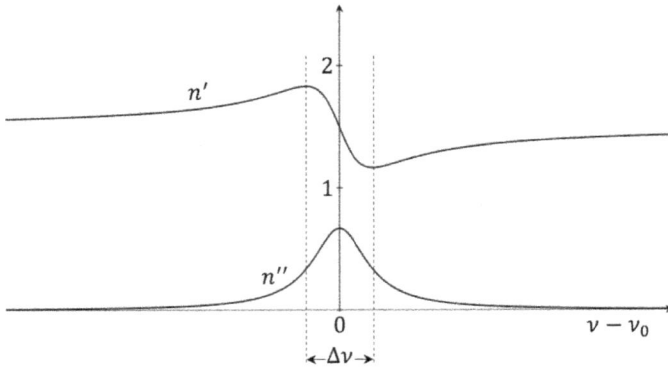

Fig. 5.3.   The real and imaginary parts of the refractive index versus $\nu - \nu_0$, in arbitrary units.

The actual value of the real part of the refractive index corresponding to a single resonance can be $n' < 1$ when $\nu > \nu_0$. This fact points out that the phase velocity c/n' has nothing to do with the velocity of the traveling signal. In fact, according to the theory of relativity, the wave signal cannot propagate faster than the vacuum light velocity c. A classical treatment of this problem has been given by L. Brillouin, who discussed in detail the speed of the light signal, which far from resonance can be identified with the *group velocity* as discussed in Section 2.2, but it is more difficult to determine at a frequency corresponding to the resonant absorption (see L. Brillouin, *Wave Propagation and Group Velocity* (Academic, 1960)).

In Eq. (1.45), we have seen the effect of the imaginary part of the refractive index: it reduces the field amplitude by a factor $e^{-n'' k_0 z}$, i.e. the wave intensity decreases by a factor $e^{-2n'' k_0 z}$, being proportional to the square modulus of the field:

$$I(z) = I_0 e^{-2n'' k_0 z} = I_0 e^{-\alpha z}, \qquad (5.65)$$

where I_0 is the wave intensity at $z = 0$. In Eq. (5.65), we have defined the absorption coefficient as

$\alpha = 2n'' k_0 = 2n''(\omega/c)$; using Eq. (5.64) it becomes

$$\alpha(\nu) = 2\frac{\chi''(\nu)}{2n_0}\frac{\omega}{c} = \frac{k_0}{n_0}\chi''(\nu). \qquad (5.66)$$

In this way, *Beer's law* [Eq. (5.65)] takes into account the energy losses of the electromagnetic wave. It shows that the exponential depletion of the intensity due to absorption is directly related to the imaginary part of the dielectric susceptibility $\chi''(\nu)$. Using Eq. (5.55), we can also write

$$\alpha(\nu) = \frac{k_0}{n_0} \frac{N}{2\epsilon_0 \hbar} |\mu_{km}|^2 g(\nu). \tag{5.67}$$

By defining the *absorption cross section*,

$$\sigma_{\text{abs}}(\nu) = \frac{k_0}{n_0} \frac{1}{2\epsilon_0 \hbar} |\mu_{km}|^2 g(\nu), \tag{5.68}$$

we can write the absorption coefficient as

$$\alpha(\nu) = \sigma_{\text{abs}}(\nu) N, \tag{5.69}$$

as the product of the concentration of absorber units N times $\sigma_{\text{abs}}(\nu)$, giving the contribution of each absorber in the process.

5.3. The Einstein coefficients

Up to now, we have focused the discussion on absorption processes, transitions from an energy level m to an upper one k, possibly characterized by an energy spread given by $g(\nu)$. However, when we consider two levels with energy E_2 and $E_1 (E_2 > E_1)$, an electronic population density N_2 and N_1 will be associated with them, and therefore, according to the results of the time-dependent perturbation theory (see Section 2.13), transitions can be induced by a harmonic perturbation both upward and downward. In the first case, a single transition corresponds to excitation of one electron from level 1 to level 2 with consequent *absorption* of one photon of energy $h\nu = E_2 - E_1$, while a single downward transition corresponds to the *stimulated emission* of one photon with the same energy as that of the incident photon, which on the other hand, must correspond to the energy level difference ($h\nu = E_2 - E_1$).

Besides these two processes, a third one may occur, involving the *spontaneous emission* of one photon when one electron performs a downward transition without being excited by other photons. This process is related to the electron tendency to fill empty states of lower energy if available. These three processes are sketched in Fig. 5.4.

A basic difference between the stimulated emission and the spontaneous emission is the following: the former is a "coherent effect," i.e. the emitted photon keeps the frequency and wave vector of the incident one in phase

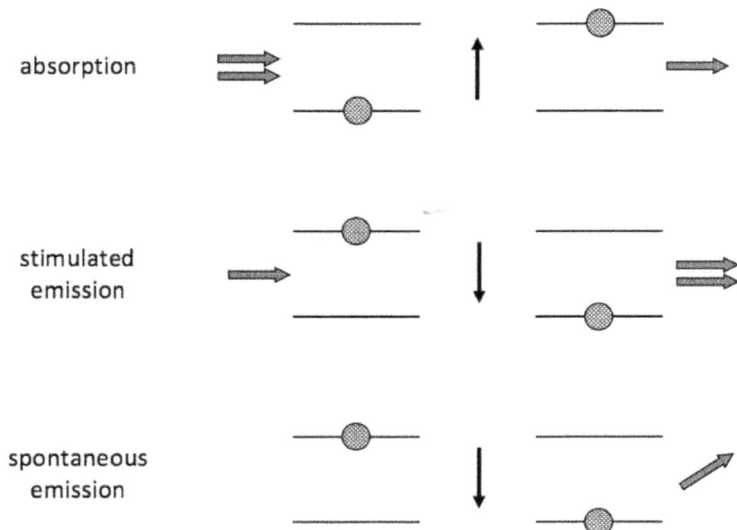

Fig. 5.4.   Absorption, stimulated emission and spontaneous emission by one electron in a two-level system.

with it, and the latter is an incoherent effect leading to emission of a photon in any undetermined direction with a frequency falling in the range determined by the line shape $g(\nu)$ of the considered transition.

We note that absorption has the same features of stimulated emission: both occur owing to incident radiation; the absorbed or emitted photons have the frequency determined by the incoming wave and the probability of the effect to occur depends on the frequency location within the range given by the line shape $g(\nu)$.

Of course, when light interacts with an atomic system, the overall effect that is macroscopically observed is the result of a statistical superposition of a large number of transitions due to the population densities N_1 and N_2.

Given an optical field with energy density per unit frequency $\rho(\nu)$, the induced transition rate (probability per unit time) from level 1 to level 2 (*absorption*) should be written as

$$W_{12}^{i} = B_{12}\rho(\nu), \tag{5.70}$$

where B_{12} is the coefficient characterizing the process. In the same way, the induced transition rate from level 2 to level 1 (*stimulated emission*) is given by

$$W_{21}^{i} = B_{21}\rho(\nu). \tag{5.71}$$

On the contrary, *spontaneous emission* does not require an incoming optical field to occur, and therefore its transition rate can be written as

$$W_{21}^{\text{sp}} = A. \tag{5.72}$$

The coefficients A and B are related to the material response and they must be valid for any distribution of the field energy, and therefore it is useful to employ the well-known expression given by Eq. (2.5) for the blackbody radiation:

$$\rho(\nu, T) = \frac{8\pi n_0^3}{c^3} \frac{h\nu^3}{e^{\frac{h\nu}{k_B T}} - 1},$$

with n_0 being the nonresonant real part of the refractive index [see Eq. (5.63)].

At equilibrium the electronic population of the two levels is "statistically" constant in time $(d/dt = 0)$. This means that the total number of transitions per unit time for absorption equals that for emission. This quantity is given by the level population multiplied by the transition rate, and therefore the steady state condition reads

$$N_1 W_{12}^{\text{i}} = N_2(W_{21}^{\text{i}} + W_{21}^{\text{sp}}), \quad \text{in the unit } (\text{s} \cdot \text{cm}^3)^{-1}. \tag{5.73}$$

Using Eqs. (5.70)–(5.72), we have

$$N_1 B_{12} \rho(\nu) = N_2(B_{21} \rho(\nu) + A), \tag{5.74}$$

$$N_1 B_{12} \frac{8\pi n_0^3}{c^3} \frac{h\nu^3}{e^{\frac{h\nu}{k_B T}} - 1} = N_2 \left(B_{21} \frac{8\pi n_0^3}{c^3} \frac{h\nu^3}{e^{\frac{h\nu}{k_B T}} - 1} + A \right). \tag{5.75}$$

According to Eq. (3.144), the population ratio given by the Boltzmann statistic is

$$\frac{N_2}{N_1} = \frac{g_2}{g_1} e^{-\frac{h\nu}{k_B T}}, \tag{5.76}$$

where g_2 and g_1 are the degeneracy of the quantum levels and $h\nu$ is the energy difference between the two levels.

Performing the calculation

$$\frac{8\pi n_0^3}{c^3} \frac{h\nu^3}{e^{\frac{h\nu}{k_B T}} - 1} = \frac{\frac{N_2}{N_1} A}{B_{12} - \frac{N_2}{N_1} B_{21}} \tag{5.77}$$

we finally get

$$\frac{8\pi n_0^3}{c^3} \frac{h\nu^3}{e^{\frac{h\nu}{k_B T}} - 1} = \frac{\frac{g_2}{g_1} A}{B_{12} e^{\frac{h\nu}{k_B T}} - \frac{g_2}{g_1} B_{21}}. \tag{5.78}$$

This equation is satisfied by simultaneous fulfillment of the conditions

$$B_{12} = \frac{g_2}{g_1} B_{21}, \tag{5.79}$$

$$\frac{A}{B_{21}} = \frac{8\pi n_0^3 h\nu^3}{c^3}. \tag{5.80}$$

These results allow writing the induced transition rates as

$$W_{21}^{\mathrm{i}} = B_{21}\rho(\nu) = A\frac{c^3}{8\pi n_0^3 h\nu^3}\rho(\nu), \tag{5.81}$$

$$W_{12}^{\mathrm{i}} = \frac{g_2}{g_1} A\frac{c^3}{8\pi n_0^3 h\nu^3}\rho(\nu). \tag{5.82}$$

In the case $g_1 = g_2$, we have $W_{21}^{\mathrm{i}} = W_{12}^{\mathrm{i}}$, confirming that absorption is just the reverse of stimulated emission.

An independent expression for the spontaneous emission coefficient A cannot be calculated here, since it requires a full quantum theory, i.e. quantization of both the atom and the optical field. However, we notice from Eq. (5.72) that A is the inverse of the time constant τ_{sp} characterizing the spontaneous emission from the excited level:

$$A = \frac{1}{\tau_{\mathrm{sp}}}. \tag{5.83}$$

In fact, multiplying the transition rate for spontaneous emission by the electronic population N_2, we have the variation in time of this population due to this process,

$$\frac{dN_2}{dt} = -AN_2, \tag{5.84}$$

which is easily solved to give

$$N_2(t) = N_2(0)e^{-At} \equiv N_2(0)e^{-\frac{t}{\tau_{\mathrm{sp}}}}, \tag{5.85}$$

showing an exponential decay of the upper-level population due to spontaneous emission with the characteristic time τ_{sp}.

Taking into account Eq. (5.83), we can rewrite Eq. (5.81) as

$$W_{21}^{\mathrm{i}} = B_{21}\rho(\nu) = \frac{c^3}{8\pi n_0^3 h\nu^3 \tau_{\mathrm{sp}}}\rho(\nu). \tag{5.86}$$

This expression relates to a wide radiation spectrum given by $\rho(\nu)$, much broader than the emission or absorption band of the considered transition.

We can analyze the opposite situation, where a monochromatic wave interacts with a two-level system characterized by a specific line shape $g(\nu)$. In this case, we have to take into account the probability that at a given frequency the transition occurs and therefore we have to substitute $\rho(\nu)$ in Eq. (5.86) with the product between the energy density at the given frequency ρ_ν and $g(\nu)$:

$$W_{21}^i = B_{21}\rho(\nu) = \frac{c^3}{8\pi n_0^3 h\nu^3 \tau_{\rm sp}}\rho_\nu g(\nu). \tag{5.87}$$

In order to have an expression more useful from the experimental point of view, we wish to have the light intensity I_ν rather than the radiation energy density.

The relationship between the light intensity and the corresponding energy density is easily found by considering a finite volume crossed by the incoming wave, as depicted in Fig. IN5.2.

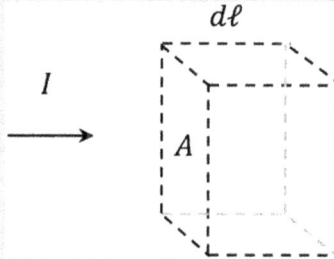

Fig. IN5.2. The volume crossed by an optical wave of intensity I.

The volume dV is given by a cross section A multiplied by the length $d\ell$ traveled in the time interval dt:

$$dV = Ad\ell = A\frac{c}{n_0}dt.$$

The wave energy in this volume is Pdt, with P being the incident power, and therefore the energy density is

$$\rho_\nu = \frac{Pdt}{A(c/n_0)dt} = \frac{P/A}{c/n_0} = \frac{I_\nu}{c/n_0} \quad \text{since} \quad I_\nu = \frac{P}{A},$$

corresponding to Eq. (5.88).

Additionally,

$$\frac{d\rho_\nu}{dt} = \frac{d}{dt}\frac{I_\nu}{\frac{c}{n_0}} = \frac{dI_\nu}{\frac{c}{n_0}dt} = \frac{dI_\nu}{dz} \quad \text{since} \quad dz = \frac{c}{n_0}dt.$$

Namely, the time variation of the energy density equals the space variation of the intensity.

We notice that

$$\rho_\nu = \frac{n_0}{c}I_\nu, \tag{5.88}$$

to get

$$W_{21}^i = B_{21}\rho(\nu) = \frac{c^2}{8\pi n_0^2 h\nu^3 \tau_{sp}}I_\nu g(\nu). \tag{5.89}$$

5.4. Light absorption, amplification and spontaneous emission in atomic systems

Let us consider a monochromatic wave at frequency ν traveling through a medium showing a two-level resonance at frequency $\nu_0 \approx \nu$. If spontaneous emission is neglected, the energy exchange between the optical field and the medium is due to absorption and stimulated emission. Obviously, the variation in time of the energy density of the wave is given by the imbalance between the losses of energy due to absorption and the gain due to stimulated emission. Each term is given by the number of transitions per unit time and volume multiplied by the energy content ($h\nu$) of each transition[3]:

$$\frac{d\rho_\nu}{dt} = \frac{dI}{dz} = [N_2 W_{21}^i - N_1 W_{12}^i]h\nu, \tag{5.90}$$

where z is the coordinate along the propagation direction.

Using Eqs. (5.82) and (5.89), we have

$$\frac{dI}{dz} = \left[N_2 - N_1\frac{g_2}{g_1}\right]\frac{c^2}{8\pi n_0^2 \nu^2 \tau_{sp}}Ig(\nu). \tag{5.91}$$

We define

$$\gamma(\nu) = \Delta N \frac{c^2}{8\pi n_0^2 \nu^2 \tau_{sp}}g(\nu), \tag{5.92}$$

---

[3]We drop the subscript ν in the intensity, implying that we are dealing with a monochromatic wave.

with

$$\Delta N = N_2 - N_1 \frac{g_2}{g_1}, \tag{5.93}$$

the population difference between the upper and the lower level (the unit is cm^{-3}). Then it is possible to write

$$\frac{dI}{dz} = \gamma(\nu)I. \tag{5.94}$$

In the approximation $\Delta N \approx$ const, the integration of Eq. (5.94) is straightforward:

$$I(z) = I_0 e^{\gamma z}, \tag{5.95}$$

where $I_0 = I(0)$.

First of all, we note that Eq. (5.95) is a generalization of Beer's law [Eq. (5.65)] of absorption. In fact, it corresponds to it when $\gamma < 0$. In this case, by setting $\alpha \equiv -\gamma > 0$, Eq. (5.95) reduces to Eq. (5.65) and α is the absorption coefficient.

However, the condition $\gamma > 0$ is also allowed and according to Eq. (5.95) this leads to growth in intensity, i.e. *amplification* of the traveling wave. For this reason, γ is called the *gain coefficient*.

By looking at the definition (5.92), we see that the only quantity affecting the sign of γ is the population difference ΔN, determining which is the dominant effect between absorption and stimulated emission. In fact, $\Delta N > 0$ leads to $\gamma > 0$ and $\Delta N < 0$ leads to $\gamma < 0$.

At thermal equilibrium, the ratio of the electronic population densities is given by Eq. (5.76), which together with Eq. (5.93) gives

$$\Delta N = N_2 \left[1 - e^{\frac{h\nu}{k_B T}} \right] = N_1 \frac{g_2}{g_1} \left[e^{-\frac{h\nu}{k_B T}} - 1 \right]. \tag{5.96}$$

This population difference is always negative for $h\nu > k_B T$, a condition holding for electronic transitions in the optical spectrum range. In fact, in this case $h\nu$ is of the order of a few eV, while at room temperature $(T = 300 \text{ K})$ one gets $k_B T \approx 0.026$ eV, and therefore $e^{\frac{h\nu}{k_B T}} \gg 1$, making $\Delta N < 0$ at thermal equilibrium.

This roughly means (exactly if $g_1 = g_2$) that at room temperature the condition $N_2 \ll N_1$ is always fulfilled for a system at thermal equilibrium and absorption prevails by far over stimulated emission. For this reason, in order to achieve the amplification condition it is necessary to provide energy to the atomic system and bring it out of thermodynamic equilibrium. In other words, only through a *pumping* mechanism will it be possible to reach the *population inversion* leading to $\Delta N > 0$.

Spontaneous emission and line shape broadening

By comparing Eq. (5.67) to Eq. (5.92), we can find a relationship between the square modulus of the transition dipole moment $|\mu_{km}|^2$ and the spontaneous emission lifetime τ_{sp}. First of all, we should generalize the expression (5.67) taking into account also the stimulated emission transitions from level 2 to level 1, therefore replacing N with $-\Delta N$,

$$\alpha(\nu) = -\frac{k_0}{n_0} \frac{\Delta N}{2\epsilon_0 \hbar} |\mu_{km}|^2 g(\nu), \qquad (5.97)$$

and compare it to Eq. (5.92), which, according to the definition $\alpha(\nu) = -\gamma(\nu)$, becomes

$$\alpha(\nu) = -\Delta N \frac{c^2}{8\pi n_0^2 \nu^2 \tau_{sp}} g(\nu). \qquad (5.98)$$

Then we obtain

$$\frac{1}{\tau_{sp}} = |\mu_{km}|^2 \frac{16\pi^3 n_0 \nu^3}{\epsilon_0 h c^3}, \qquad (5.99)$$

which can be independently calculated by a full quantum theory.

Since Eq. (5.97) has been worked out considering the one-dimensional case, a complete expression must consider the calculated value as $1/3$ of the average total transition dipole moment, which takes into account all the directions in space, and therefore a better value is

$$\frac{1}{\tau_{sp}} = |\mu_{km}|^2 \frac{16\pi^3 n_0 \nu^3}{3\epsilon_0 h c^3}. \qquad (5.100)$$

We need to remember that from the definition (5.83) τ_{sp} is the excited state lifetime owing to the spontaneous emission of photons from level 2 to level 1. Actually, the occurrence of spontaneous emission in such a system is quite a serious problem of quantum mechanics. In fact, in principle, any energy eigenstate should be a steady state, and therefore its lifetime should be infinite, in agreement with the uncertainty principle, which connects energy and time: $\Delta E \Delta \tau \geq \hbar$. On the contrary, the observation of spontaneous emission gives $\Delta \tau \sim \tau_{sp}$. Then we have

$$\Delta E \geq \frac{\hbar}{\tau_{sp}} \quad \rightarrow \quad h\Delta\nu \geq \frac{\hbar}{\tau_{sp}}.$$

In this way, we get the minimum linewidth of the transition, denominated the *intrinsic linewidth*:

$$\Delta\nu_i = \frac{1}{2\pi\tau_{sp}}. \qquad (5.101)$$

This relationship could be found from the treatment in Section 5.2, setting the decay time constant

$$\tau_{sp} = \frac{1}{\Gamma} \quad \text{and exploiting the result } \Gamma = 2\pi\Delta\nu. \tag{5.102}$$

As a matter of fact, the spectral linewidth, which is observed in experiments, is much broader than the intrinsic one; in other words, the linewidth appearing in the experimental line shape $g(\nu)$ gives

$$\Delta\nu \gg \Delta\nu_i = \frac{1}{2\pi\tau_{sp}} \quad \text{or} \quad 2\pi\tau_{sp}\Delta\nu \gg 1. \tag{5.103}$$

For instance, typical values in semiconductors are $\tau_{sp} \sim 10^{-10}$ s and $\Delta\nu \sim 10^{13}\text{s}^{-1}$, leading to $2\pi\tau_{sp}\Delta\nu \sim 6 \cdot 10^3$; while the intrinsic linewidth is $\Delta\nu_i \sim 1.6 \cdot 10^9\text{s}^{-1} \ll \Delta\nu$.

This means that the absorption or emission spectrum observed and given by $g(\nu)$ is much wider than the one originated by a single event of spontaneous emission; this is due to several physical mechanisms.

Some of them lead to *homogeneous broadening* when all the atoms appear to have the same transition energy. In this case, broadening can be induced by: collisions with other atoms or molecules, or with lattice vibrations in solids; transitions to levels different from the two considered; effects due to saturation (power broadening) (see below). The resulting line shape is the Lorentzian function given by Eq. (5.54):

$$g(\nu) = \frac{\left(\frac{\Delta\nu}{2\pi}\right)}{(\nu_0 - \nu)^2 + \left(\frac{\Delta\nu}{2}\right)^2}. \tag{5.104}$$

A different condition is the one occurring when atoms are distinguishable; in this case, the line shape broadening is actually due to the convolution of the frequency spread of the electronic transition of each atom. This leads to *inhomogeneous broadening*, where each frequency range within $g(\nu)$ corresponds to a different group of atoms. This effect can be the line broadening due to the Doppler effect at low pressure in gases or to impurities in crystals. It usually leads to a Gaussian frequency distribution slightly different from the previously mentioned Lorentzian and given by

$$g(\nu) = \sqrt{\left(\frac{4\ln 2}{\pi}\right)} \frac{1}{\Delta\nu} e^{-\left[4\ln 2\left(\frac{\nu_0 - \nu}{\Delta\nu}\right)^2\right]}. \tag{5.105}$$

The actual broadening mechanism of the considered transition is very important in determining the optical behavior under oscillation conditions in lasers and when the high intensity of the traveling wave gives rise to saturation.

The typical linewidth of an atomic or molecular system obtained by spectral analysis is in the range of 10^6–10^{10} s^{-1}, several orders of magnitude narrower than in condensed matter.

It is quite interesting to compare a Lorentzian to a Gaussian line shape. The maximum for a Lorentzian line shape given by Eq. (5.104) is

$$g_L(\nu_0) = \frac{2}{\pi \Delta \nu},$$

while the maximum for a Gaussian line shape given by Eq. (5.105) is

$$g_G(\nu_0) = \sqrt{\left(\frac{4\ln 2}{\pi}\right)} \frac{1}{\Delta \nu},$$

and thus, for the same full width $\Delta \nu$, the Gaussian curve is higher and decreases to zero faster, as shown in Fig. IN5.3.

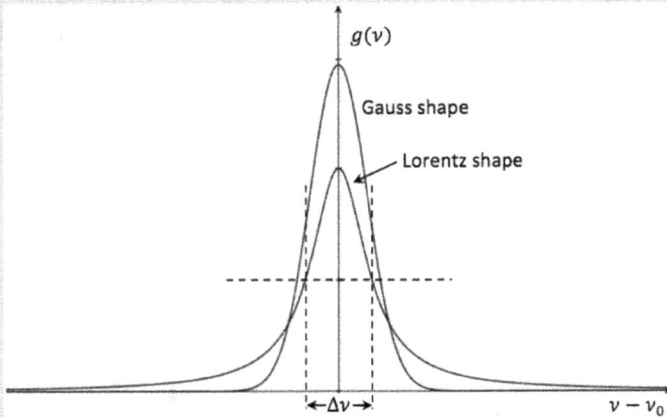

Fig. IN5.3. Comparison between Lorentzian and Gaussian line shapes with the same full width $\Delta \nu$.

Absorption

As already pointed out, Eq. (5.95) can be written as

$$I(z) = I_0 e^{-\alpha z}, \tag{5.106}$$

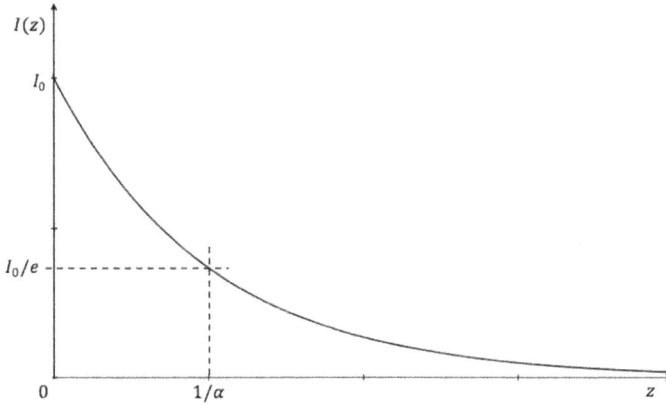

Fig. 5.5. Plot of Eq. (5.106) showing the intensity versus the propagation distance inside an absorbing medium.

by defining $\alpha = -\gamma > 0$ as the absorption coefficient. This means that light intensity exponentially decreases with traveling distance, as depicted in Fig. 5.5.

The meaning of α becomes clear from this figure since at $z_e = 1/\alpha$ the intensity is reduced to I_0/e, and therefore $1/\alpha$ and is a measure of the penetration depth of the wave into the absorbing medium. Under thermal equilibrium conditions, $N_1 \gg N_2$, and therefore, as in Eq. (5.69), we can write

$$\alpha(\nu) = N_1 \sigma_a(\nu), \tag{5.107}$$

where

$$\sigma_a(\nu) = \frac{c^2}{8\pi n_c^2 \nu^2 \tau_{\rm sp}} g(\nu) = \frac{\lambda_0^2}{8\pi n_0^2 \tau_{\rm sp}} g(\nu). \tag{5.108}$$

This expression highlights the role of the considered transition in inducing absorption at the chosen frequency through the absorption cross section $\sigma_a(\nu)$, separating it from the concentration of absorbers N_1. Equation (5.107) is also useful for evaluating the effective absorption depth given a concentration N_1 of absorbing electrons.

Let us suppose that a monochromatic wave at $\lambda_0 = 633\,\mathrm{nm}$ is impinging on a gas having a homogeneously broadened line transition with the maximum at $\nu_0 = c/\lambda_0$, with the following data: $\Delta\nu = 1.5 \cdot 10^9\mathrm{s}^{-1}$,

$\tau_{\rm sp} = 3 \cdot 10^{-7}$s, $n_0 \sim 1$. We obtain an absorption cross section:

$$\sigma_a(\nu_0) = \frac{\lambda_0^2}{8\pi n_0^2 \tau_{\rm sp}} g(\nu_0) = \frac{(633)^2 \cdot 10^{-14}}{8\pi \cdot 3 \cdot 10^{-7}} \frac{2}{\pi \cdot 1.5 \cdot 10^9} \sim 2.2 \cdot 10^{-13} \mathrm{cm}^{-2}.$$

A density of valence electrons $N_1 = 0.45 \cdot 10^{13}\,\mathrm{cm}^{-3}$ gives rise to an absorption coefficient $\alpha \sim 1\,\mathrm{cm}^{-1}$; this means that the intensity is reduced to I_0/e after traveling for a depth of 1 cm in the absorbing medium.

A very important effect that takes place at high intensities is *saturation*. By considering a simple two-level system, the occurrence of saturation is clearly understood if we take into account that Eq. (5.94) has been solved in the approximation $\Delta N \sim$ const.

Let us first consider only light absorption; then, for each absorbed photon and consequent excitation of one electron from level 1 to level 2, we have a population variation: $\delta(\Delta N) = (N_2 + 1) - (N_1 - 1) = \Delta N + 2$ (assuming $g_2 = g_1$). As initially $\Delta N < 0$, this means that each absorption transition decreases the absolute value of the population difference; according to Eq. (5.98), this means a decrease in the absorption coefficient. When the induced variation $\delta(\Delta N)$ is not negligible owing to the presence of many photons inducing many absorption transitions, it is evident that the approximation used to solve Eq. (5.94) is no longer valid and the exponential behavior given by Eq. (5.106) is no longer realized. In this case, the dependence of intensity on z must be calculated in a more realistic way, taking into account all the electronic levels involved in the process and solving the so-called *rate equations* for the electronic populations N_2 and N_1. The reader is referred to more technical texts for a complete treatment of the problem, and we will consider one example of it in the next chapter by discussing the population inversion process. The final result is that we can write the dependence of the population difference on the intensity as

$$\Delta N = \frac{\Delta N_0}{1 + (I/I_s(\nu))}, \tag{5.109}$$

defining ΔN_0 as the initial population difference, and the saturation intensity:

$$I_s(\nu) = \frac{4\pi n_0^2 h\nu^3}{(\tau/\tau_{\rm sp})c^2 g(\nu)}. \tag{5.110}$$

Here τ is the effective lifetime of the level 2 caused by all the radiative or nonradiative decay processes. It is usually shorter than τ_{sp}. According to Eq. (5.98), we can exploit Eq. (5.109) to write the effective absorption coefficient as

$$\alpha(\nu) = \frac{\alpha_0(\nu)}{1 + (I/I_s(\nu))}, \tag{5.111}$$

where $\alpha_0(\nu)$ stands for the low signal absorption coefficient corresponding to $\Delta N_0 \sim$ const.

As expected, Eq. (5.109) shows a decrease in ΔN as I increases, in such a way that $\Delta N \to 0$ as $I \to \infty$. The meaning of the parameter I_s is clear from Eq. (5.111), as the light intensity corresponding to a decrease in the absorption coefficient to a value $\alpha_0(\nu)/2$.

The saturation effect leads to *bleaching* of the material at the considered wave frequency; in fact, as absorption decreases, transmission increases and the material may reach good transparency at the resonance frequency if the impinging intensity is $I \gg I_s$. This effect can be understood by looking at Figs 5.6 and 5.7.

The transmission coefficient of a material slab of thickness L is simply defined by

$$T(I) = \frac{I(L)}{I_0}. \tag{5.112}$$

In the case of negligible scattering and after taking into account the reflection losses at the slab interfaces, $I(L)$ is the result of the intensity depletion

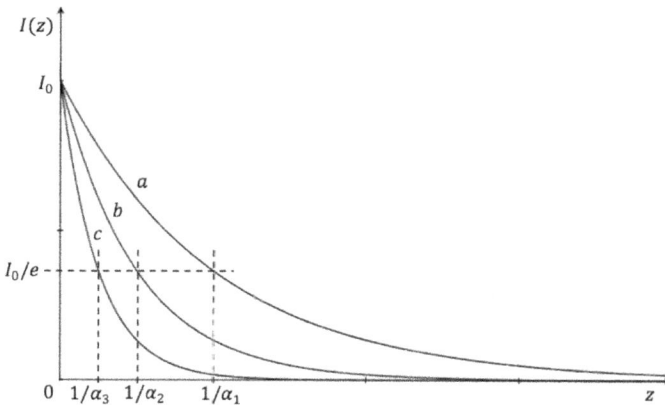

Fig. 5.6. Plot of Eq. (5.106) for different values α, showing that as saturation takes place the intensity decrease is slower: $\alpha_1 < \alpha_2 < \alpha_3$.

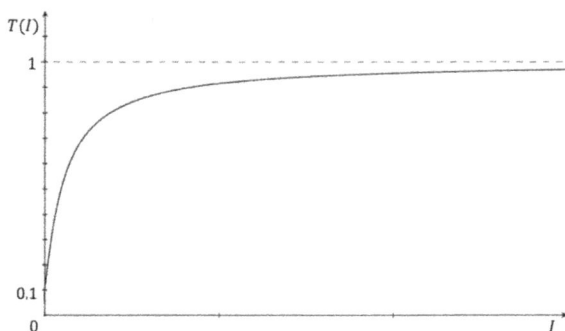

Fig. 5.7. Plot of transmission versus intensity for an absorbing slab, where $T(0) = 0.1$.

shown in Fig. 5.6. As the value of the absorption coefficient decreases by increasing I, the actual curve $I(z)$ moves to the one corresponding to lower α and the overall transmission coefficient increases as reported in Fig. 5.7, where $T(I)$ has been calculated using Eqs. (5.106) and (5.111).

This effect is extremely important for many applications and has to be taken into account when filtering of the incoming light is needed to avoid damage to a light detector.[4]

In general, the transmittivity T of any material of thickness L is affected not only by absorption, but also by light scattering and reflections at the interfaces of the medium. Considering I_0 as the light intensity impinging on the medium, a correct expression for the transmitted intensity would be

$$I(L) = I_0(1 - R_1)e^{-(\alpha+\alpha_s)L}(1 - R_2),$$

with α_s being the extinction coefficient for scattering and R_i the reflectivity at the two interfaces. Then the overall transmittivity is

$$T = (1 - R_1)(1 - R_2)e^{-(\alpha+\alpha_s)L}.$$

Gain and amplification

We now discuss the main consequences of Eq. (5.95) when $\gamma > 0$, leaving to Chapter 6 a more detailed discussion of the amplification process. As being

[4]For instance, glasses needed to observe intense light sources must be properly designed to work far from the saturation condition in order to avoid damage to the eyes.

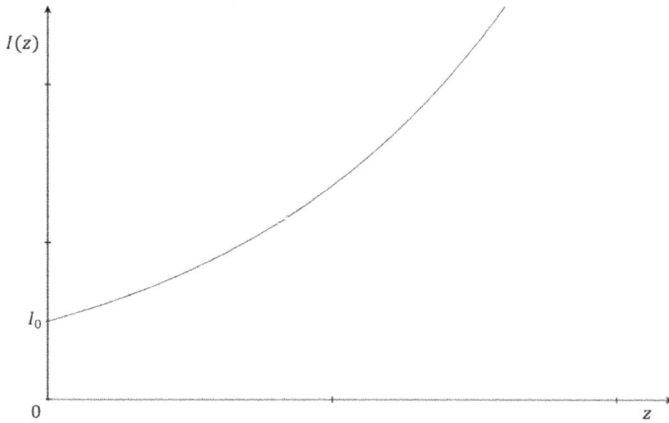

Fig. 5.8.   Plot of intensity versus propagation length according to Eq. (5.95) with $\gamma > 0$.

opposite to Fig. 5.5 we have now an exponential growth of the intensity versus the propagation length, as depicted in Fig. 5.8.

We can write Eq. (5.95) as

$$I(z) = I_0 G(\nu, z). \tag{5.113}$$

By defining the *gain factor* as

$$G(\nu, z) = e^{\gamma(\nu)z}. \tag{5.114}$$

It gives the relative increase in intensity at a location z:

$$G(\nu, z) = \frac{I(z)}{I_0} = \frac{I_0 + \Delta I(z)}{I_0} = 1 + \frac{\Delta I(z)}{I_0} \rightarrow \frac{\Delta I(z)}{I_0} = G(\nu, z) - 1. \tag{5.115}$$

The dependence of the gain factor on frequency is remarkable; in fact, owing to the exponential dependence on $\gamma(\nu)$ given by Eq. (5.114), the gain process may lead to narrowing of an initial broad line. This is actually the effect occurring on the onset of the laser emission, as will be discussed in the next chapter: the spontaneous emission can be amplified if light meets a system where population inversion has been achieved. In this case, at the beginning the intensity distribution is given by $g(\nu)$, but according to Eq. (5.114) it will have a different gain factor at each frequency, being maximum at $\nu = \nu_0$. The consequence is a spectral distribution with a linewidth narrower than the one due to the spontaneous emission. This process leads to the so-called amplified spontaneous emission (ASE), sometimes confused with laser emission.

The discussion that has been done about saturation of absorption can be repeated and becomes even more important for the gain coefficient γ. In this case, we start from an inverted population ($\Delta N > 0$) which is lowered by 2 for each emitted photon. Owing to amplification, the intensity may grow quickly and the population difference ΔN may significantly decrease according to Eq. (5.109). In this way, the gain coefficient decreases with intensity and, in the case of homogeneous broadening, can be written as

$$\gamma(\nu) = \frac{\gamma_0(\nu)}{1 + (I/I_s(\nu))}, \tag{5.116}$$

with $\gamma_0(\nu)$ being the *small signal* gain coefficient, corresponding to $\Delta N \sim$ const.

We should remember that the actual population inversion and the $\gamma(I)$ dependence must take into account the effective electronic levels involved in the process. We will consider that in the next chapter, discussing the pumping mechanism that can allow achieving the population inversion.

We note here that additional electronic states must be involved to get population inversion, besides the two-level system considered. In fact, we have to provide energy in order to bring the system to the nonequilibrium condition $\Delta N > 0$ since we start from $\Delta N < 0$. In case we would like to exploit absorption from level 1 to level 2 to do it, we would find a decreasing absorption as N_2 approaches N_1 until we get $\Delta N \sim 0$. In this condition $\alpha(\nu) \sim 0$ and at the same time $\gamma(\nu) \sim 0$, and therefore no further change of the population difference is possible. In other words, in playing with only two levels, it is not possible to reach the population inversion. We need the electrons to transit through one or more energy states in the way up or down from/to the ground state. For this reason, it is necessary the solve the rate equations governing the variation of electronic population in the different states involved in the process in order to find the actual conditions for amplification.

As for absorption, we could define a stimulated emission cross-section as

$$\sigma_{\text{em}}(\nu) = \frac{c^2}{8\pi n_0^2 \nu_0^2 \tau_{\text{sp}}} g(\nu), \quad \text{with} \quad \sigma_{\text{em}}(\nu) = \sigma_a(\nu), \tag{5.117}$$

such that

$$\gamma(\nu) = \Delta N \sigma_{\text{em}}(\nu). \tag{5.118}$$

However, owing to the level structure, which may be rather complex, a more general expression for the gain coefficient should take into account the actual degeneracies of the two considered levels. Then

$$\sigma_{em}(\nu) \neq \sigma_a(\nu). \tag{5.119}$$

As a consequence, the gain coefficient should be written as

$$\gamma(\nu) = N_2\sigma_{em}(\nu) - N_1\sigma_a(\nu). \tag{5.120}$$

5.5. Optical properties of molecular systems

The discussion in the previous sections has been focused on the light interaction with a two-level system involving valence electrons, where the energy difference between the two levels is in the range of a few eV; now we will be more specific, applying those concepts to different material systems. We may roughly consider two main categories: a system where the single molecule properties determine the overall optical response and a system where the properties arising from condensation, like in the solid state, strongly affect the optical behavior. In this section, we briefly discuss some features of molecular systems.

We have depicted the optical response of diatomic molecules in Section 3.6. In these systems, the concepts discussed in the previous sections of the present chapter can be applied to couples of energy levels that are usually represented by plotting the electronic energy versus the internuclear distance. A general picture of this type can be used for any molecular system both in a vapor and in a condensed state where the material response is strongly characterized by the single molecule electronic structure. The main difference in a condensed state will be the densities of allowed vibrational states leading to a quasicontinuum sequence of levels. In the case of separated molecules, we get (either in absorption or in spontaneous emission) a spectrum as the one shown in Fig. 3.13, while for a condensed system we obtain a broad absorption band as well as a broad emission band.

Figure 5.9 shows the electronic energy of a diatomic molecular system versus the distance between the atomic nuclei. Each curve $E(R)$ (associated with a specific value of the principal quantum number n) has a minimun corresponding to the equilibrium distance between the two atoms. For each curve, there are several vibrational states for increasing values of the vibrational quantum number ν and, for each of them, there are several rotational states for increasing values of the quantum number J which determines the molecule's angular momentum. Therefore, the concepts discussed in the previous section apply to couples of such energy levels according to the *selection rules*, which fix the allowed transitions (the ones

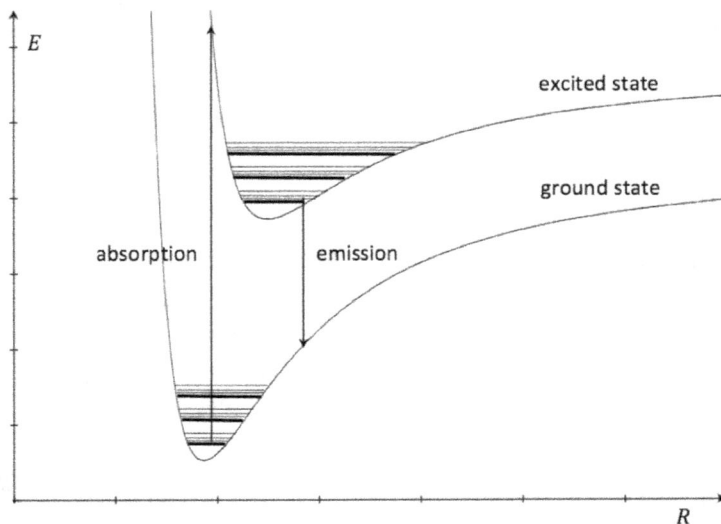

Fig. 5.9. Electronic energy versus intermolecular distance in diatomic molecules and vertical transitions according to the Born–Oppenheimer approximation.

where $|\mu_{km}|^2 \neq 0$, leading to the conditions $\Delta\nu = \pm 1$ and $\Delta J = \pm 1$). In particular, we are interested in transitions involving a change of the principal quantum number n, since they correspond to photons in or near the visible range (a few eV).

The absorption or emission transitions are represented in this scheme by vertical lines according to the Franck–Condon principle, otherwise denominated the Born–Oppenheimer approximation. This assumption is based on the high speed of the electronic transition, which does not allow the internuclear separation to change in this time range. Since after the transition the electronic energy does not correspond to the minimum of the potential energy curve, there will be a rearrangement of the internuclear distance and a consequent lowering of the electronic energy according to what is depicted in Fig. 5.9.

The absorption spectra of these systems are usually characterized by spectral lines, each of them having a fine structure corresponding to the rotovibrational transitions shown in Fig. 3.13.

We can describe the optical properties of molecules dissolved in condensed matter (it could be liquid or solid) in a similar way taking into account that the stronger molecular interaction due to the high molecular density gives rise to a quasicontinuum of states, even if we may consider the active molecules isolated from each other, since their concentration

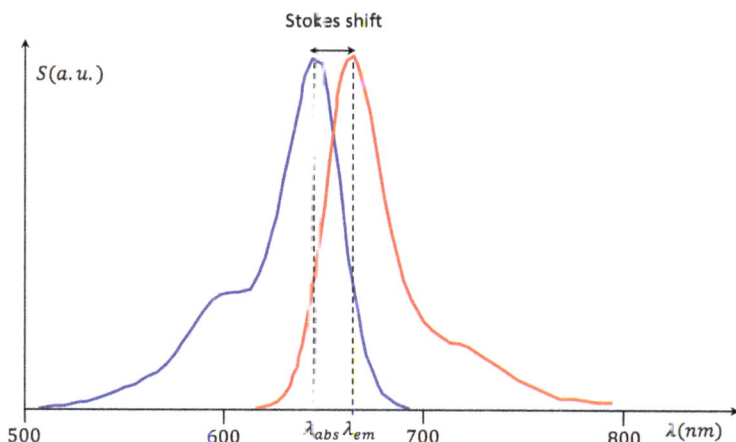

Fig. 5.10. Typical absorption (blue line) and emission (red line) spectra of solutions of molecular compounds. The Stokes shift is defined as the difference between the peak value λ_{em} of the emission band and the peak value λ_{abs} of the absorption band.

is usually in the range of 10^{-3}–10^{-4} with respect to the host molecular density. Therefore, the absorption/emission spectra in the case of white light excitation appear as a continuum band, as shown in the example of Fig. 5.10. Of course, the emission band is also dependent on the spectrum of the exciting beam, since it determines the excited levels involved in the process.

The spectra shown in Fig. 5.10 point out a very general feature, fulfilled also in the solid state — the so-called *Stokes shift* between the absorption and the emission band. It is the shift of the peak of the emission band toward higher wavelengths (redshift) with respect to the peak of the absorption band. This effect is a consequence of what is displayed in Fig. 5.9, where the length of the arrow corresponding to the energy difference in the emission is definitely lower than the one corresponding to the energy difference in the absorption.

5.6. Effects of band structure on the optical properties of materials

In crystalline solids, the electron energy $E(k)$ is determined by the band structure, which plays a fundamental role also in characterizing the optical properties of the material. The levels involved in optical transitions usually belong to different electronic bands — in other words, they concern

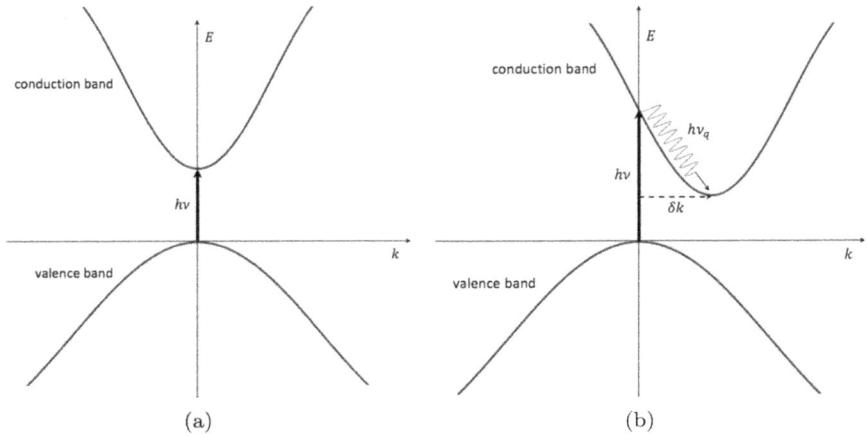

Fig. 5.11. Photon absorption due to interband transitions in a solid. (a) Direct transition: the momentum of the electron is conserved from the initial to the final state. (b) Indirect transition: the momentum conservation involves a lattice vibration of energy $h\nu_q$ and momentum $q = \delta k$, either absorbed or emitted in the transition.

interband transitions. In the case of absorption, a photon excites an electron from an occupied state of the valence band to an unoccupied state in the conduction band, as depicted in Fig. 5.11.

First of all, we note that in this picture optical transitions are represented by vertical lines. In fact, an energy difference in the range of a few eV corresponds to the photon wavelength in or near the visible range (say, $\lambda \sim 1000\text{--}10,000\text{Å}$), while the lattice parameter is $a \sim 1\text{--}5\text{Å}$. As a consequence, the photon wave vector

$$k_{\text{ph}} = \frac{2\pi}{\lambda} \ll \frac{\pi}{a} = k_{\text{Br}}, \qquad (5.121)$$

with k_{Br} being the electron wave vector on the edge of the Brillouin zone. This means that a photon gives a negligible additional momentum to the electron. For this reason, the change of electron momentum is negligible during the transition, which in this picture must be indicated by a vertical line. As can be observed in Fig. 5.11, the minimum photon energy necessary for having the transition must be $h\nu \geq E_G$.

Actually, we have to distinguish between direct and indirect transitions. In a *direct transition*, we have no change of the electron momentum between the initial and the final electronic state $[k_f = k_i;$ Fig. 5.11(a)]; in this case, the final state corresponds to the one reached by the electron with the absorption of the photon energy. On the contrary, in an *indirect transition*, a lattice vibration (*photon*) is involved ($k_f = k_i \pm q$); in fact, in this case a change of electron momentum is necessary in order to reach the empty state of minimum energy in the conduction band.

Since the number of transitions depends on the number of available states, it is remarkable that photons of a specific energy are more effective if the actual dependence on k of the band shape is weak, because in this case many different initial and final states can be coupled by the same photon energy. As a consequence, the band shape determines the absorption and emission spectrum. thus making the study of optical properties very important for their determination.

Looking at Fig. 5.12, we see that energy conservation requires the following condition for the photon energy:

$$hv = E_G + \frac{\hbar^2 k^2}{2m_e} + \frac{\hbar^2 k^2}{2m_h}, \tag{5.122}$$

where m_e is the effective mass of electron in the final state (conduction band) and m_h that of the hole in the initial state (valence band). In the case shown in the figure, the actual final state of the electron can be at an energy lower than the energy achieved by photon absorption, since the excited electron can relax quickly toward the empty state of the conduction

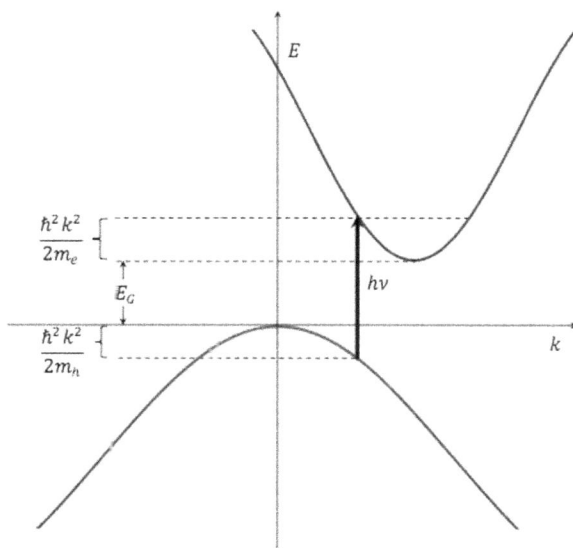

Fig. 5.12.   Energy balance in interband transitions.

band with minimum energy exciting lattice vibrations with wave vector
$q = k - k_f$; according to it, the energy of the final state depends on the
energy $h\nu_q$ of the excited phonon $E_f = h\nu - h\nu_q$.

Equation (5.122) can also be written as

$$h\nu = E_G + \frac{\hbar^2 k^2}{2m_r}, \qquad (5.123)$$

using the reduced mass

$$m_r = \frac{m_e m_h}{m_e + m_h}. \qquad (5.124)$$

The gain or absorption coefficient of such a system can be obtained following
the method used in Section 5.4, where the calculation of the transition
probability between the two involved levels must take account (for each
wave vector) of the *joint density of states* $\rho(k)$ between the valence and
the conduction band, and of the probability that level 1 is an occupied
state and level 2 is empty. In the following, we consider direct transitions.
According to it, for wave vectors in the range of $k \div k + dk$ the possible
absorption transitions per unit volume are

$$dN_1 = \frac{\rho(k)dk}{V}\{f_v(E_1)[1 - f_c(E_2)]\}, \qquad (5.125)$$

with $f_v(E_1)$ being the probability of having the state at energy E_1 filled
in the valence state and $1 - f_c(E_2)$ the probability of finding empty the
state at energy E_2 in the conduction band, where $f(E)$ is the Fermi–Dirac
distribution function.

If we consider also the possible downward transition between the same
levels, we have

$$dN_2 = \frac{\rho(k)dk}{V}\{f_c(E_2)[1 - f_v(E_1)]\}, \qquad (5.126)$$

since in this case we need to start from an occupied state to go down to an
empty state. Therefore, the actual population difference for each dk is

$$d(N_2 - N_1) = \frac{\rho(k)dk}{V}\{f_c(E_2)[1 - f_v(E_1)] - f_v(E_1)[1 - f_c(E_2)]\}$$

$$= \frac{\rho(k)dk}{V}\{f_c(E_2) - f_v(E_1)\}. \qquad (5.127)$$

According to Eq. (5.98), we can write the absorption coefficient as

$$d\alpha(\nu) = -\frac{\rho(k)dk}{V}\{f_c(E_2) - f_v(E_1)\}\frac{c^2}{8\pi n_0^2 \nu^2 \tau_{sp}}g(\nu), \qquad (5.128)$$

which must be integrated over all possible wave vectors:

$$\alpha(\nu) = -\int_0^\infty \frac{\rho(k)dk}{V}\{f_c(E_2) - f_v(E_1)\}\frac{c^2}{8\pi n_0^2 \nu^2 \tau_{sp}}g(\nu). \tag{5.129}$$

Here we can identify τ_{sp} as the electron–hole recombination time from the conduction to the valence band.

Using Eq. (5.123), we have

$$d\nu = \frac{1}{2\pi}\frac{\hbar}{m_r}kdk \tag{5.130}$$

and

$$k = \sqrt{(h\nu - E_g)\frac{2m_r}{\hbar^2}}. \tag{5.131}$$

By using the simplified expression of the density of states given by Eq. (4.12),

$$\frac{\rho(\nu)d\nu}{V} = \frac{k^2}{\pi^2}dk = \frac{k^2}{\pi^2}d\nu\frac{2\pi m_r}{\hbar}\frac{1}{k} = \frac{4\pi}{h^2}\sqrt{h\nu - E_G}(2m_r)^{3/2}\,d\nu. \tag{5.132}$$

Then Eq. (5.129) becomes

$$\alpha(\nu) = -\int_0^\infty \frac{4\pi}{h^2}\sqrt{h\nu - E_G}(2m_r)^{3/2}\{f_c(E_2)$$

$$- f_v(E_1)\}\frac{c^2}{8\pi n_0^2 \nu^2 \tau_{sp}}g(\nu)d\nu. \tag{5.133}$$

Integration can be performed by replacing $g(\nu)$ with a delta function $\delta(\nu - \nu_0)$ to get

$$\alpha(\nu_0) = -\sqrt{\left(\nu_0 - \frac{E_G}{h}\right)}\left(\frac{2m_r}{h}\right)^{3/2}$$

$$\times \{f_c(E_2) - f_v(E_1)\}\frac{c^2}{2n_0^2 \nu_0^2 \tau_{sp}}. \tag{5.134}$$

This expression provides the contribution of each single frequency $\nu_0 = (E_2 - E_1)/h$ to the absorption coefficient. The actual coefficient will depend on the frequency spectrum involved in the process.

It is interesting that the difference $f_c(E_2) - f_v(E_1)$ plays the same role as ΔN does in Eq. (5.98), determining the sign of this coefficient, which has a real value only if

$$\nu_0 > \frac{E_G}{h}. \tag{5.135}$$

This is the threshold frequency for the onset of the process.

According to Eq. (5.134), we have

$$absorption\ [\alpha(\nu_0) > 0] \quad if\ f_c(E_2) - f_v(E_1) < 0; \tag{5.136}$$

$$amplification\ [\gamma(\nu_0) = -\alpha(\nu_0) > 0] \quad if\ f_c(E_2) - f_v(E_1) > 0. \tag{5.137}$$

The condition (5.136) corresponds to thermal equilibrium, since the occupation probability is higher in the valence band than in the conduction band: $f_v(E_1) > f_c(E_2)$.

A typical frequency dependence of the absorption coefficient of semiconductors is shown in Fig. 5.13.

The realization of the opposite condition (inversion) can be done in an inhomogeneous semiconductor, such as in the p–n junction, and leads to gain and light amplification. It will be discussed in the next chapter.

5.7. Optical properties of metals, insulators and semiconductors

In condensed matter, we usually have a high density of electron energy levels that lead to a broad absorption spectrum when white light impinges

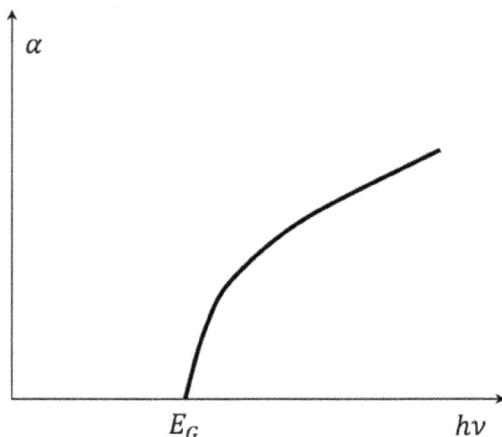

Fig. 5.13. Typical frequency dependence of the absorption coefficient in a semiconductor.

on the material. After excitation, electrons spontaneously relax to lower energy states, thus giving back to the electromagnetic field a portion of the absorbed light and fulfilling the Stokes shift rule, as explained Section 5.5: the emission spectrum will be redshifted with respect to the absorption spectrum, as shown in Fig. 5.10. This spontaneous emission process is usually denominated *luminescence*. When it occurs with a decay time of a few nanoseconds or faster, it is called *fluorescence*, while a much slower process is usually called *phosphorescence*.

Anyway, the absorption process has different characteristics, depending on the electron filling of the energy bands. Since it determines the conduction properties of the material, we expect that metals, insulators and semiconductors present different and peculiar optical properties.

Metals

In this case, we have a conduction band partially filled, and thus a high number of electrons can be considered a nearly free gas, as discussed in Chapter 4. We can start from the classical expression given by Eq. (5.1), dropping the restoring elastic force by considering the electron free:

$$-eE(t) - \beta\frac{dx}{dt} = m\frac{d^2x}{dt^2}. \tag{5.138}$$

Equation (5.138) can be compared to Eq. (4.24) to give $\beta = m/\tau$. Again we consider a plane wave $E(t) = E_0 e^{-i\omega t}$ and the in-phase electron oscillation $x(t) = x_0 e^{-i\omega t}$ to get

$$-\omega^2 x - i\frac{\omega}{\tau}x = -\frac{e}{m}E(t), \tag{5.139}$$

i.e.

$$x(t) = \left(-\frac{e}{m}\right)\frac{E(t)}{-\omega^2 - i(\omega/\tau)}. \tag{5.140}$$

Then, according to Eq. (5.6),

$$\epsilon_r(\omega) = 1 - \frac{Ne^2}{\epsilon_0 m}\frac{1}{\omega^2 + i(\omega/\tau)}. \tag{5.141}$$

By defining the *plasma frequency* ω_p such that

$$\omega_p^2 = \frac{Ne^2}{\epsilon_0 m}, \tag{5.142}$$

we have

$$\epsilon_r(\omega) = 1 - \frac{\omega_p^2}{\omega^2 + i(\omega/\tau)} = 1 - \frac{\omega_p^2/\omega^2}{1 + i(1/\omega\tau)}. \qquad (5.143)$$

In the approximation of low damping $\omega\tau \gg 1$,

$$\epsilon_r(\omega) = 1 - \frac{\omega_p^2}{\omega^2}. \qquad (5.144)$$

Since the refractive index is the square root of $\epsilon_r(\omega)$, Eq. (5.144) clearly shows the macroscopic optical behavior of a metal. In fact, if $\omega < \omega_p$ we have $\epsilon_r(\omega) < 0$; then the refractive index is a pure imaginary number, and the wave cannot propagate in the medium. By looking at the reflectivity at the interface of the material, we find that the electromagnetic energy is reflected rather than absorbed, as highlighted in Section 1.4, mentioning the high reflectivity of metal surfaces. In fact, at normal incidence we have

$$R = \left|\frac{\tilde{n}-1}{\tilde{n}+1}\right|^2 = \frac{n'+in''-1}{n'+in''+1} \cdot \frac{n'-in''-1}{n'-in''+1} = \frac{(n'-1)^2+n''^2}{(n'+1)^2+n''^2} \qquad (5.145)$$

and, if $n'' \gg n'$,

$$R \approx \frac{1+n''^2}{1+n''^2} = 1. \qquad (5.146)$$

At frequency $\omega > \omega_p$, Eq. (5.144) provides a real permittivity and a real refractive index. In this case, $n' \gg n''$, and the material reflectivity is low

$$R \approx \frac{(n'-1)^2}{(n'+1)^2}. \qquad (5.147)$$

The material is transparent, with a behavior similar to that of an insulator as long as the frequency is below the threshold for an interband transition.

At $\omega = \omega_p$, $\epsilon_r(\omega_p) = 0$ and we have *plasma oscillations*, i.e. collective oscillations of the electron gas absorbing the electromagnetic energy. It is easy to recognize these as *longitudinal oscillations*, where the wave vector is parallel to the electric field. In fact, if we allow the presence of localized charges in the medium, the approximated expression (1.13) of the operator $\nabla \times (\nabla \times \mathbf{E})$ does not hold anymore. Then the wave equation in lossless media (1.15) must be modified as

$$\nabla^2 \mathbf{E} - \nabla(\nabla \cdot \mathbf{E}) - \mu_0 \epsilon \frac{\partial^2 \mathbf{E}}{\partial t^2} = 0. \qquad (5.148)$$

By using the plane wave solution (1.22) for the field, we can write

$$-k^2 \mathbf{E} + \mathbf{k} \, (\mathbf{k} \cdot \mathbf{E}) = -\epsilon_r \frac{\omega^2}{c^2} \mathbf{E}. \qquad (5.149)$$

When $\epsilon_r(\omega_p) = 0$, this equation is fulfilled if $\mathbf{k} \parallel \mathbf{E}$ (longitudinal wave) so that the left term also reduces to zero.

The charge localization and the onset of plasma oscillations can be understood by thinking of a small displacement of the electrons on the metal surface due to the impinging optical wave. Such a displacement δx gives rise to a net positive charge in the same location due to the ions of the crystal lattice. This originates an opposite increase in negative charges on the other side of the conductor, thus producing an electrostatic field between the excess positive and negative charges. Considering a metallic slab with section A and thickness L of the two charged regions, we have the polarization (dipole moment per unit volume)

$$P = -\frac{(NeA\delta x)L}{AL} = -Ne\delta x,$$

which obviously corresponds to the surface charge density σ. The induced electrostatic field inside the slab is easily calculated as $-\sigma/\epsilon_0 = -P/\epsilon_0$ [the latter expression is also derived from Eq. (1.5) setting $\epsilon = 0$].

$$E = -\frac{P}{\epsilon_0} = \frac{Ne\delta x}{\epsilon_0}.$$

Therefore, an electron in the polarized region feels a force,

$$-eE = -\frac{Ne^2\delta x}{\epsilon_0},$$

and must follow the motion equation of a harmonic oscillator,

$$-\frac{Ne^2\delta x}{\epsilon_0} = m\frac{d^2(\delta x)}{dt^2},$$

where the oscillation frequency ω is found as

$$\frac{Ne^2}{m\epsilon_0} = \omega^2 = \omega_p^2,$$

which is exactly the plasma frequency defined in Eq. (5.142).

The values of the plasma frequency for most metals fall in the near-UV region of the electromagnetic spectrum, usually with $\nu_p > 10^{15} \text{s}^{-1}$, and the corresponding energies $(h\nu_p = \hbar\omega_p)$ are in the range of 3–10 eV.

A sketch of the frequency dependence of the metal reflectivity is shown in Fig. 5.14, for the case of negligible damping. We observe a minimum of the reflectivity near $\nu = \nu_p$. By increasing the energy of the incident photons, the excitation of the electrons to higher energy bands may be possible, and therefore the metal reflectivity can be modulated by such absorption effects.[5]

Since the color appearance of a metal is due to the reflection spectrum, it will be determined by the location of the plasma frequency with respect to the visible range and to the possible occurrence of interband transitions in this range. This explains, for instance, the different appearance of silver and aluminum. In fact, in the first case, the plasma frequency on the edge of the visible range at about 4 eV, and therefore the reflection in the violet is lower than in aluminum, whose plasma frequency falls at 6.3 eV, farther from the visible range. On the other hand, the reddish color of copper is due to the absorption caused by interband transitions at a wavelength shorter than 560 nm.

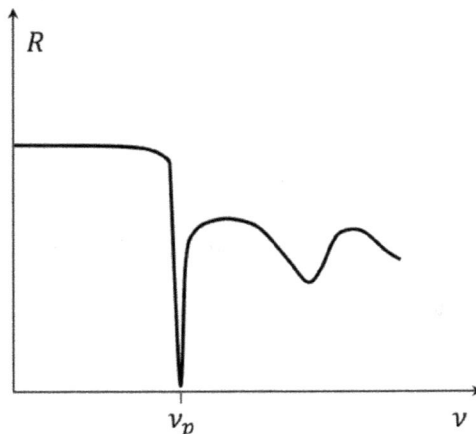

Fig. 5.14. Sketch of the reflectivity of a metal versus the frequency of the impinging electromagnetic wave.

---

[5]See for instance the reflectivity of silver in E.A. Taft *et al.*, *Phys. Rev.*, **121**, 1100 (1961).

Plasma oscillations are also quantized, and we call the quantum energy related to such an oscillation a plasmon.

Surface plasmons

At a dielectric–metal interface, the conditions for excitation and propagation of electrogmagnetic surface waves can be realized. These waves are confined to the surface and are exponentially evanescent in a direction perpendicular to it. They are originated by the coupling between the field oscillations and the electron plasma of the conductor.

The most simple geometry sustaining surface plasmons is given by a flat interface between a dielectric nonabsorbing medium half space ($z > 0$) with a positive real dielectric constant and an adjacent conducting half space ($z < 0$) described by a dielectric function with a negative real part — a condition fulfilled in metals at frequencies below the bulk plasmon frequency ω_p.

The surface plasmon (SP) excitation can be detected as a minimum in the reflection spectrum, since at those wavelengths the electromagnetic energy is not reflected but absorbed by the SP excitation, as shown in Fig. IN5.4.

The study of SPs started before 1970, but only in the past two decades have they been deeply investigated and become the basis of a research stream in nano-optics denominated *plasmonics*.

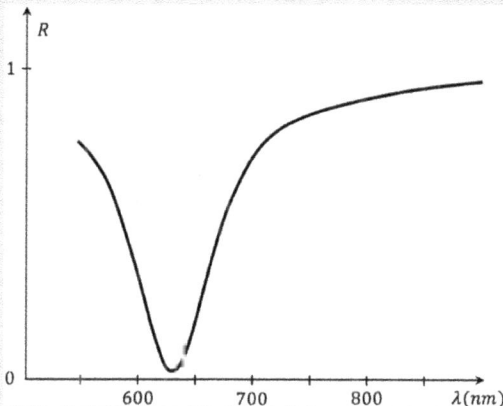

Fig. IN5.4. Typical plasmonic resonance leading to a minimum of the reflectivity at a metal–dielectric interface.

Insulators

In insulators, the band gap is pretty large and at room temperature the number of conduction electrons is negligible, and therefore absorption can be induced only by photons with an energy above the visible range (ultraviolet). For this reason, these materials are transparent to the visible light. For instance, diamond has a band gap $E_G \approx 5.5\,\mathrm{eV}$ and thus it absorbs photons with wavelength $\lambda < 230\,\mathrm{nm}$. The minimum energy and the corresponding photon wavelength necessary for getting light absorption is denominated by the *absorption edge*. In Table 5.1, this quantity is reported for some insulators.

Actually, many insulators are not transparent in the whole visible range in spite of the large band gap. In these cases, the color is not due to interband transitions, but to transitions occurring between levels of a molecular system such as the one described in Section 5.5. The optical absorption in these cases is due to defects or dopants embedded in the host crystal lattice, behaving as molecules dispersed in the crystal with concentrations much lower with respect to the crystal atomic density. A typical example is that of ruby, which is made of aluminum oxide (Al_2O_3) including at most 1% of chromium (Cr^{+3}) impurities (substituting Al^{+3} ions), which provide the electronic levels inducing the absorption of blue light, which leads to the characteristic red color of the crystal.

As a matter of fact, several optical materials in the solid state, such as the ones used as laser materials, are made of crystals including dopants or defects in the lattice sites, thus providing the electronic levels, which correspond to optical transitions. Among those we can think of the

Table 5.1. The absorption edge of some insulators given as minimum absorption energy and the corresponding photon wavelength at room temperature. [Data from: (1) J.F.H. Custers and F.A. Raal, *Nature* **179**, 268 (1957); W. Saslow, T.K. Bergstresser and M.L. Cohen, *Phys. Rev. Lett.* **16**, 354 (1966); (2) R.H. French, *J. Am. Ceram. Soc.* **73**, 477 (1990); (3) S.S. Nekrashevich and V.A. Gritsenko, *Phys. Solid State* **56**, 207 (2014); (4) R.A. Heaton and C.C. Lin, *Phys. Rev.* **B22**, 3629 (1980)].

Material	Absorption edge (eV)	λ (nm)
Diamond (C_4)	5.5[1]	225
Alumina (Al_2O_3)	8.8[2]	141
α-Quartz (SiO_2)	8.4[3]	148
CaF_2	10.0[4]	124

Nd: YAG crystal, where an ittrium aluminum garnet structure is doped with neodimium (Nd) atoms to provide the level system used to get light amplification in the near-infrared (see Chapter 6).

Noncrystalline amorphous materials are a very important class of insulators, and among them glasses are the most important. They are either transparent or colored, depending on the chemical composition due to the possible inclusion of impurities. For what concerns transparent glasses, which are needed to fabricate a huge number of widely used optical devices, the most important optical parameter is the refractive index and its dispersion, namely its dependence on the light wavelength.

This macroscopic parameter is of paramount relevance in the fabrication of optical devices such as lenses for any kind of application, from eyeglasses to precision objectives for fine instruments.

The refractive index of the most-used transparent glass (BK7) in the central part of the visible spectrum is $n \sim 1.52$ and decreases with wavelength according to the plot of Fig. 5.15. This is a low-dispersion glass, since the refractive index variation in the visible region is less than 0.02. A different example is given by superflint glass SF10, reported in the same figure; it presents a higher refractive index and higher dispersion in the visible range.

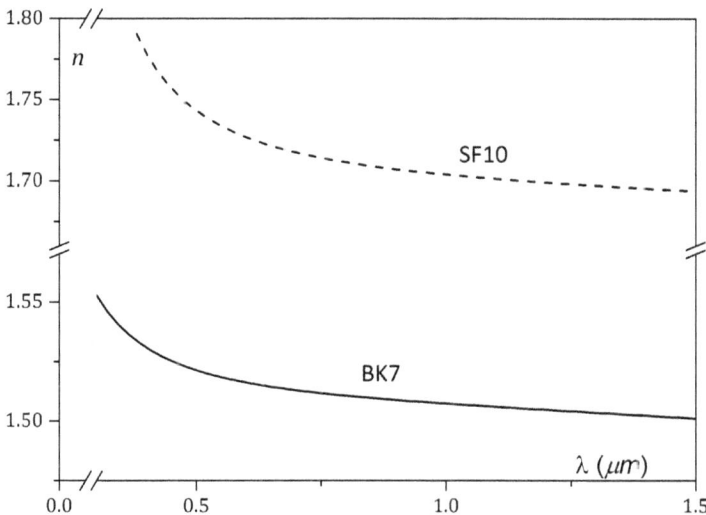

Fig. 5.15. The refractive index of glass BK7 and SF10 versus wavelength [data from: M.N. Polyanskiy, "Refractive index database," https://refractiveindex.info].

Dispersion formulae

The attempt at explaining the dispersion behavior in transparent media is older than the theory presented in Section 5.1. Using the elastic solid theory of light, Augustin-Louis Cauchy in 1836 proposed an empirical relation for the dispersion of the refractive index:

$$n(\lambda) = A + \frac{B}{\lambda^2} + \frac{C}{\lambda^4},$$

where the constants A, B, C have to be determined from experimental data taken at widely separated wavelengths. Actually, this simple formula is very satisfactory for many transparent materials as long as the considered wavelength is far enough from any material resonance, i.e. far from the anomalous dispersion region. In other words, the Cauchy dispersion formula can be used in the normal dispersion region, where the refractive index decreases slowly by increasing the wavelength.

A more accurate expression was subsequently calculated by Sellmeier, who considered the vibrations of atoms and molecules to be the main cause of light dispersion. He got a formula that applies to transparent media, but it keeps its validity even close to absorption bands located at wavelengths λ_m:

$$n^2(\lambda) = 1 + \sum_m \frac{D_m \lambda^2}{\lambda^2 - \lambda_m^2},$$

where D_m are parameters to be determined from experimental observations. Actually, Sellmeier's formula can be derived from Eq. (5.9) under suitable approximations. Usually, a summation limited to three terms gives a very satisfactory fit with experimental data.

Semiconductors

The band gap of semiconductors is narrower than that of insulators, being around or below 1 eV, and therefore the absorption edge occurs in the visible or near-infrared region. For instance, in gallium phosphide (GaP) $E_G = 2.27\,\text{eV}$ at room temperature; then absorption starts at $\lambda < 545\,\text{nm}$ in the green region of the optical spectrum and it is responsible for the orange color of this crystal, which corresponds to the transmitted wavelengths (under white light illumination). For silicon (Si) $E_G = 1.12\,\text{eV}$ at room temperature, and therefore it absorbs all the wavelengths in the optical range, being opaque to visible light.

Actually, the absorption of semiconductors is strongly affected by the rather complex band structure and by the possible occurrence of indirect transitions. In fact, the onset of an interband transition is possible at photon energies higher than

$$h\nu = E_G - h\nu_q \tag{5.150}$$

if the phonon energy $h\nu_q$ is absorbed. Therefore, light absorbance may occur at wavelengths slightly longer than the ones corresponding to the band edge. Then, while in direct band gap semiconductors the threshold energy for interband transition is E_G, in indirect band gap semiconductors it is given by Eq. (5.150).

The experimental absorption coefficient of Si and Ge is shown in Fig. 5.16.

The absorption of semiconductors may also be affected by impurities. We have seen in Chapter 4 how donors create new hydrogen-like levels close to the bottom of the conduction band and acceptors create levels close to the top of the valence band. Therefore, at a photon energy lower than the band edge we may have a contribution to absorption from electrons at these energy levels created by impurities (either donor-like or acceptor-like). A similar effect can be induced even in undoped semiconductors or in insulators when an electron excited to the conduction band creates a hole in the valence band. The couple electron–hole can be bounded by

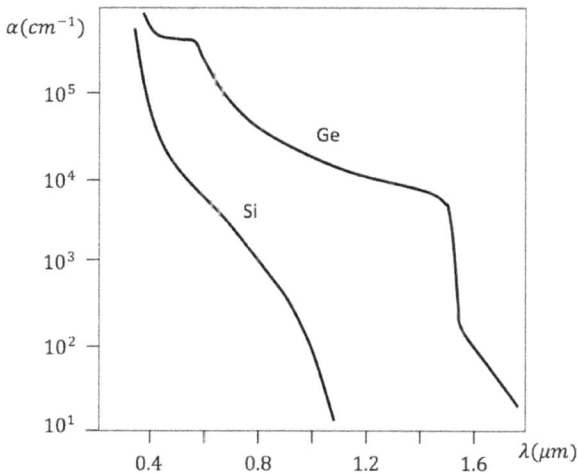

Fig. 5.16.   The measured absorption coefficient of Si and Ge versus wavelength.

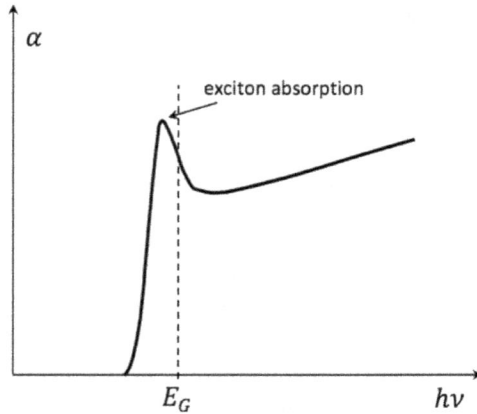

Fig. 5.17.   The typical shape of the absorption coefficient of a semiconductor with a peak due to the exciton absorption.

an electrostatic force and gives rise to a hydrogen-like system called the *exciton*. This system is similar to the one originated by impurities in semiconductors, but in this case the masses of the interacting charges are of the same order of magnitude, while in the case of impurities the ion mass is much higher than that of the electron. This results in the exciton usually not being localized, but traveling in the crystal.

However, the hydrogen-like system gives rise to energy levels below the conduction band, and the consequence is absorption at an energy slightly lower than the band edge. This is shown by a peak in the absorption coefficient occurring at a photon energy slightly below the energy gap[6] E_G, as depicted in Fig. 5.17.

5.8. Photoconductivity

When light is incident on a poorly conducting material, such as an insulator or a semiconductor, and the photon energy is sufficient to excite an electron from the valence to the conduction band, the number of free carriers is increased owing to the increased number of electrons in the conduction band and holes in the valence band. In this way, the material conductivity given by Eq. (4.165) is increased by a quantity

$$\Delta\sigma = (\Delta n \, e\mu_e + \Delta p \, e\mu_h), \tag{5.151}$$

---

[6]See for instance: M.D. Sturge, *Phys. Rev.* **127**, 768 (1962).

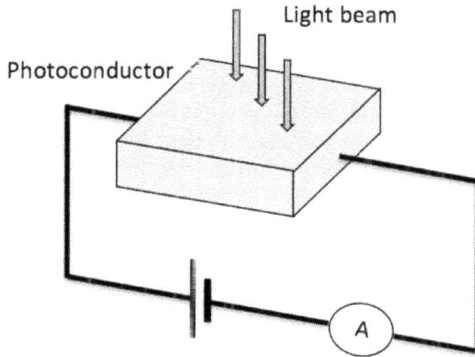

Fig. 5.18.   Scheme of detection of the photoconductivity effect.

where $\Delta n = \Delta p$ since the free carriers are generated as pairs in the photoexcitation process. Then

$$\Delta\sigma = \Delta n\, e(\mu_e + \mu_h). \tag{5.152}$$

From Eq. (5.152), we expect a bigger effect in materials where the mobilities are high. By applying a voltage to the material under these conditions as sketched in Fig. 5.18, the effect of *photoconductivity* is detected through the measurement of a *photocurrent*.

The photoexcited electron–hole pairs will contribute to the photocurrent until they recombine with each other or are trapped by localized defects. Of course, the observation of photoconductivity is possible in a material with a very low concentration of electrons thermally excited in the conduction band and a long lifetime τ_{ph} for the photoexcited carriers. In order to get a steady state photocurrent, a rate of generation of electron–hole pairs G must be provided to compensate for the recombination effects:

$$G = \frac{\Delta n}{\tau_{ph}}. \tag{5.153}$$

Since the number of excited electrons is proportional to the number of incident photons, we expect a linear dependence of the induced photocurrent on the photon flux as long as saturation effects or other competitive phenomena are negligible.

We can take into account that only a portion of the incident photons generates free carriers since not all the photons are absorbed. We define this fraction $\eta(< 1)$ as *quantum efficiency*, being the probability for an incident photon to generate an electron–hole pair. If Φ is the photon flux (number of photons per second), the generation rate is

$$G = \eta \frac{\Phi}{V}, \tag{5.154}$$

with V being the irradiated volume. From Eq. (5.153), we get the photoelectron concentration,

$$\Delta n = \eta \frac{\Phi \tau_{\text{ph}}}{V} \tag{5.155}$$

and from (5.152)

$$\Delta \sigma = \eta \frac{\tau_{\text{ph}}}{V} e(\mu_e + \mu_h)\Phi. \tag{5.156}$$

This means that the current density obtained by applying an electric field to the crystal edges is proportional to the photon flux:

$$J = \Delta \sigma E = \eta \frac{\tau_{\text{ph}}}{V} e(\mu_e + \mu_h)E\Phi. \tag{5.157}$$

Considering a crystal length ℓ with cross section $A, V = A \cdot \ell$ and the induced photocurrent is

$$i_{\text{ph}} = JA = \eta \frac{\tau_{\text{ph}}}{\ell} e(\mu_e + \mu_h)E\Phi. \tag{5.158}$$

Taking $\mu_h \ll \mu_e$ and writing the electron drift velocity as $v_e = \mu_e E$, we get

$$i_{\text{ph}} = \eta \frac{\tau_{\text{ph}}}{\ell} e v_e \Phi = \eta \frac{\tau_{\text{ph}}}{\tau_{\text{tr}}} e \Phi = \eta \frac{\tau_{\text{ph}}}{\tau_{\text{tr}}} e \frac{A}{h\nu} I_\nu, \tag{5.159}$$

where we have defined the transit time $\tau_{\text{tr}} = \ell/v_e$ of the photogenerated electron in the crystal and used the impinging light intensity at frequency ν, with $I_\nu = \Phi h\nu/A$.

Equation (5.159) shows the linear dependence of the induced photocurrent on the light intensity, and it is confirmed by many experimental data obtained in different materials. In this expression, we see the role played by the ratio $\tau_{\text{ph}}/\tau_{\text{tr}}$ in determining the value of the photocurrent. This term shows the importance of geometry, since a small value of the photoconductor size ℓ leads to a short transit time that increases the photocurrent provided that the other parameters are not changed.

We also have to take into account that Eq. (5.159) holds for each frequency of the impinging light beam; in particular, the quantum efficiency depends on frequency, and therefore we expect that each photoconductor is characterized by a spectral response showing the range of wavelengths where the conversion of photons into conduction electrons is effective.

This electron–hole pair generation through photon absorption is the basic process for detecting optical radiation using semiconductors. However, the simple photoconductor described above can be much improved with the aim of realizing an efficient detector of optical radiation, using different architectures.

Semiconductor photodiodes exploit the photoexcited electron–hole pairs to convert photon energy into electronic energy, i.e. a light signal into an electric signal, but they are based on the properties of the p–n junction in order to get a higher sensitivity and a shorter response time; p–i–n and avalanche photodiodes are typical examples of such devices. Sensitivity is measured as induced photocurrent per incident optical power and can be of the order of ampere/watt(A/W), with a response time as short as 1 ns or less. As outlined above, the sensitivity is dependent on the wavelength of the incident wave; however, over the absorption edge of the material it can cover a broad spectral range although with different efficiency, as in the example shown in Fig. 5.19 concerning a silicon avalanche photodiode.

In the description given above, the generated free carrier's are randomly distributed in the photoconductor and effective current arises if an electric

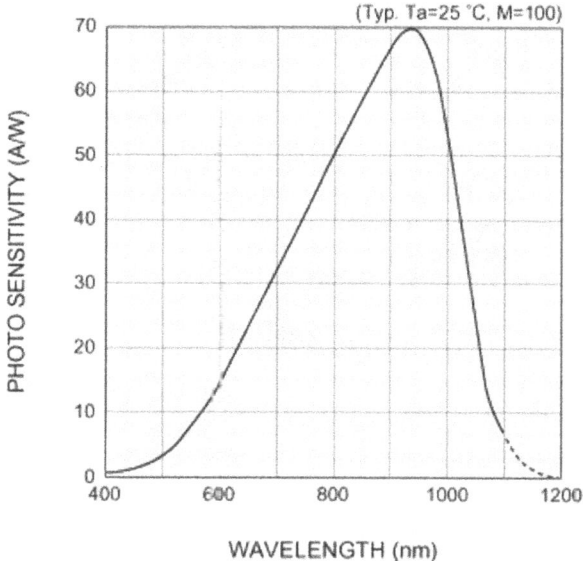

Fig. 5.19. The spectral sensitivity of a silicon photodiode (courtesy of Hamamatsu; http://www.hamamatsu.com/jp/en/product/category/3100/4003/4110/index.html-S 8890).

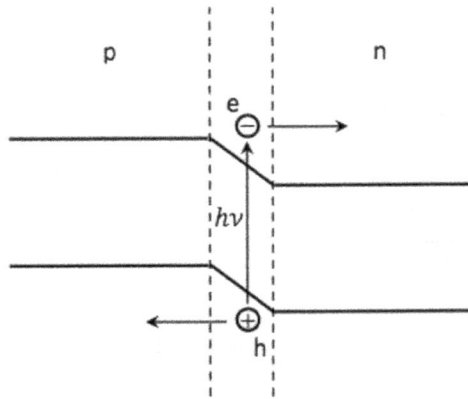

Fig. 5.20.   The photovoltaic effect in the p–n junction.

field is applied to the crystal edges. In some cases, depending on the structure of the system, the photogenerated electrons and holes move in opposite directions, and as a consequence an excess of negative charges may appear on one side while an excess of positive charges is present on the other side. In this way, an electric voltage difference is established between the two opposite sides and we speak of the *photovoltaic effect*.

A typical example of a photovoltaic device is a p–n junction, where a potential difference is present at equilibrium in the depletion region. Light-induced carriers generated near or inside the depletion region can migrate (electrons go to the n region while holes go to the p region), thus producing an additional voltage difference between the p and n sides of the junction. Solar cells exploit this mechanism to generate electrical energy. Figure 5.20 gives a sketch of this process.

5.9. Effects of light polarization

Light polarization plays a fundamental role in light–matter interaction: it may affect the absorption coefficient when the photon energy is high enough to induce optical transitions or may affect the propagation velocity and direction in transparent and anisotropic media. Moreover, the variety of polarization states coupled to the variety of structural symmetries of the materials leads to a number of relevant different types of interactions that cannot be described in an introductory text. Actually, some of these configurations are still the subject of deep investigation owing to the interest in applications. Among them, we should note the wide field of displays where light polarization effects play a fundamental role in all the devices based on liquid crystals.

Here, we briefly discuss only the basic phenomena involving light polarization in the interaction with materials, leaving any detailed study to more specialized texts.

Absorption

The importance of light polarization is highlighted by Eq. (5.18), accounting for the electric dipole interaction with the electromagnetic field. By considering a more general condition of a wave traveling in the z direction ($E_z = 0$) with any polarization in the xy plane, we have

$$\mathcal{H}' = -\boldsymbol{\mu} \cdot \mathbf{E} = -(\mu_x E_x + \mu_y E_y) = -\mu E \cos\theta, \tag{5.160}$$

and therefore the amount of perturbation depends on the mutual direction of the electric field with respect to the dipole moment of the system. From Eq. (5.160), we see that if the material system has different and nonzero values of the dipole components ($\mu_x \neq \mu_y \neq 0$), an impinging light wave with linear polarization induces a perturbation that depends on the polarization direction. For instance, by taking a light polarization along x we have $E = E_x$ and $\mathcal{H}'_x = -\mu_x E$, while for polarization along y we have $E = E_y$ and $\mathcal{H}'_y = -\mu_y E$. Since according to Eq. (5.67) the absorption coefficient is proportional to $|\mu|^2$, we expect different absorption for the two different polarizations. This effect is called *linear dichroism* and is strongly dependent on the structure of the material and in particular on the symmetry of the system (on both the micro and the macroscale).

Very important is also *circular dichroism*, i.e. different absorption for right and left circular polarization, a property that appears in some materials showing a helical arrangement of the molecules.

These few remarks allow understanding that the investigation of the polarization characteristics of the light absorbed/emitted by a material give information on its symmetry on a macroscopic scale.

Birefringence

The most relevant effect related to wave polarization concerns light propagation in transparent media. We have described in Section 3.13 the effects of material anisotropy on the wave propagation, showing that in this case we may have a splitting of a wave into two waves traveling with different velocities, i.e. the material shows two different refractive indices dependent on the polarization and propagation direction of light. This *birefringence* becomes more evident when it causes a spatial separation of the two rays, but it results in a phase shift between the two traveling waves without any geometrical separation in the case of normal incidence or very thin materials.

The anisotropy leading to such an effect is typical of most crystalline solids (except for the cubic phases) and of liquid crystals. On the contrary, gas, common liquids and amorphous materials are usually isotropic.

On the other hand, this effect is very important since its exploitation allows realization of optical devices such as polarizers and wave plates, needed to obtain any type of polarization (linear, circular, elliptical) starting from unpolarized light; devices that are also used to analyze the polarization state of light.

Polarizers and wave plates

A polarizer based on the birefringence of a material is usually realized by gluing together two right angle prisms made of the material, using glue with a refractive index in between the ordinary and the extraordinary one of the birefringent material. In this way, it is possible to realize a condition of total reflection at the interface between the first prism and the glue for only one of the two traveling waves (say, the s component), while the other one is transmitted (such as the p component). This is shown in Fig. IN5.5.

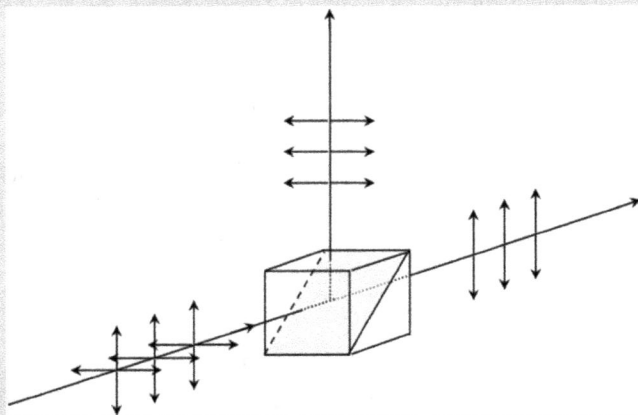

Fig. IN5.5. A polarizing beam splitter.

In this example, an unpolarized beam is incident on the devices and two linearly polarized beams are transmitted in directions perpendicular to each other. Looking at the transmittivity of the device for the two polarizations, one gets high transmission of the p wave and low transmission of the s wave.

Of course, the same device can be used to analyze the polarization state of the incoming radiation.

A linear polarizer can also be obtained by using a dichroic material where one of the polarizations is completely absorbed, and thus the transmitted beam consists of a linearly polarized wave, as sketched in Fig. IN.5.6. In this case, the polarization direction of the transmitted beam corresponds to the direction of minimum absorption.

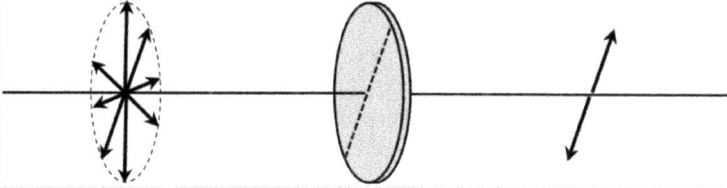

Fig. IN5.6 A dichroic polarizer.

When the polarizer is used to analyze the light polarization, the intensity transmitted is easily worked out by taking into account that in the transversal **xy** plane we have two main axes corresponding to the two linearly polarized waves traveling in the medium with the polarizations orthogonal to each other. Let us suppose a linearly polarized wave incident on the polarizer with an electric field with amplitude E_0 oscillating at an angle α with respect to one of those directions (i.e. **x** as the *optic axis* of the material). Then it will be split into two waves polarized along **x** and along **y** with amplitudes $E_x = E_0 \cos \alpha$ and $E_y = E_0 \sin \alpha$, respectively. If only the x component is transmitted by the device, the light intensity measured by a detector after the polarizer will be

$$I = I_0 \cos^2 \alpha,$$

because the intensity is proportional to the square of the electric field. This is the well-known *Malus law*, useful for the analysis of light polarization.

On the contrary, if the two polarizations are transmitted in the same direction (this happens in the case of normal incidence when refraction does not produce any deviation), according to Eqs. (1.89)–(1.92) the total electric field of the output wave has a polarization dependent on the phase delay ϕ between the ordinary and the extraordinary wave. When a plane wave of wavelength λ_0 travels through a distance d in a medium with refractive index n_i, it gets an additional phase, $\phi_i = (2\pi/\lambda_0)n_i d$; then the phase delay between the two components at the output of the anisotropic slab of the transparent material is

$$\phi = \left(\frac{2\pi}{\lambda_0}\right)(n_e - n_o)d,$$

where n_e and n_o are the refractive indices for the two waves.
Therefore, given a fixed value of the refractive index anisotropy $\Delta n = n_e - n_o$, the polarization state of the output wave is determined by the slab thickness d.

In this way, the so-called wave plates are realized by adjusting the product $(n_e - n_o)d$ in order to have a *half-wave plate*, when the phase delay is π, or a *quarter-wave plate*, when the phase delay is $\pi/2$. The first one can be used to rotate the polarization of the incoming wave by keeping the linear state, while the second one is used to generate circular polarization when starting with a linearly polarized wave, as shown in Fig. IN5.7.

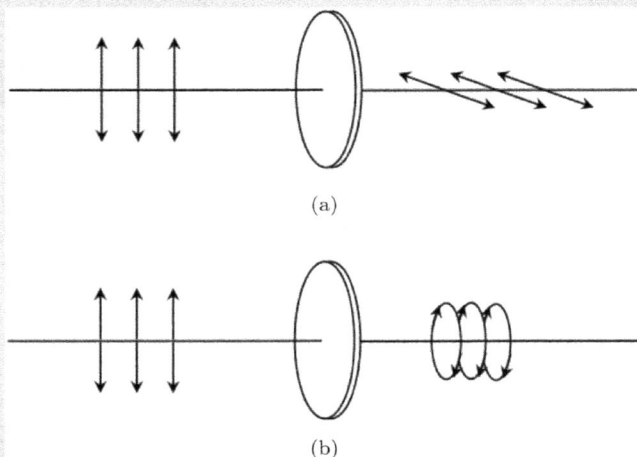

(a)

(b)

Fig. IN5.7. A half-wave plate (a) and a quarter-wave plate (b).

5.10. Temperature dependence of the optical properties

It is clear that the optical properties of materials are strongly affected by the thermal state of the system, i.e. all the effects described above can be modified by changes of temperature.

From the macroscopic point of view, this can be summarized by the temperature dependence of the complex refractive index $\tilde{n} = \tilde{n}(T) = n'(T) + in''(T)$, since both the real and the imaginary part of the refractive index are temperature-dependent.

Many different phenomena may be responsible for changes of $\tilde{n}$ depending on the state of aggregation and on structural properties of the considered material; however, they all are intrinsically due to variations of the kinetic energy of particles of the system. In fluids, the increased velocity of the molecules leads to stronger interaction with the surrounding molecules owing to the shorter time between collisions. In solids, the main effect is due to the ions which can oscillate around their equilibrium position, and the amplitude of vibrations increases with temperature. Actually, these vibrations are quantized (*phonons*) and strongly affect the electronic properties of the material owing to scattering effects or transitions between different electronic states including absorption or emission of phonons.

Looking at Eqs. (5.42) and (5.43), we see that temperature should affect the material response through the density $N(T)$ and the energy losses summarized in $\Gamma(T)$. In general, we have a substantial effect at light frequency near a transition resonance; in fact, a temperature rise increases $\Gamma(T)$ because the energy decay becomes faster owing to the increased number of collisions between atoms or molecules in gas and fluids or scattering events with lattice vibrations in solids. Since the linewidth of the optical transition is $\Delta\nu = \Gamma/2\pi$, the first effect will be a broadening of its line shape.

Moreover, the effect of temperature on electron density becomes important when approaching a phase transition, usually leading to a sudden jump of this parameter. Its change affects the absolute value of the absorption coefficient. The expected behavior of the absorption line shape of an atomic transition with increasing temperature is shown in Fig. 5.21.

When dealing with crystalline solids, we have to take into account the effects due to the energy band structure. In this case, the temperature dependence of energy gap $E_G = E_G(T)$ strongly affects the optical response.

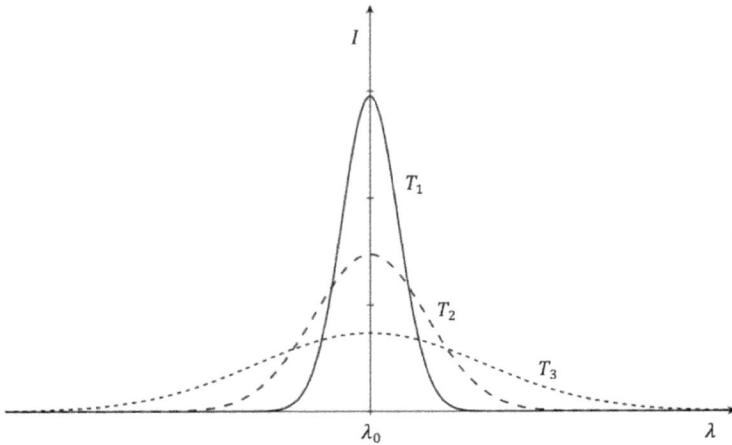

Fig. 5.21. Expected line broadening versus temperature in atomic transitions. The figure shows the intensity of a spectroscopic signal in arbitrary units versus wavelength for three values of temperature, $T_3 > T_2 > T_1$.

A phenomenological expression of this dependence is given by the Varshni equation[7]:

$$E_G(T) = E_0 - \frac{\alpha T^2}{T + \beta},\qquad(5.161)$$

where α and β are fitting parameters to be found from the experimental data. In Fig. 5.22, the quantity E_G versus temperature is reported for some semiconductors.

The decrease of the energy band gap as the temperature increases is a consequence of a modification of the electron–lattice interaction, due to the increase in amplitude of the lattice vibrations.

The most evident effect of this E_G narrowing is the reduced threshold energy for the photon inducing the interband transition, shifting the absorption band to longer wavelengths.

An additional effect of temperature is also due to the electron statistics, so that the occupation probability of higher energy levels increases with temperature according to the Fermi–Dirac distribution.

Considering transparent materials, namely when the light wavelength is far from any resonance, the temperature dependence is limited to the real part of the refractive index. Its change is strongly dependent on the

---

[7]Y.P. Varshni, *Physica* **34**, 149 (1967).

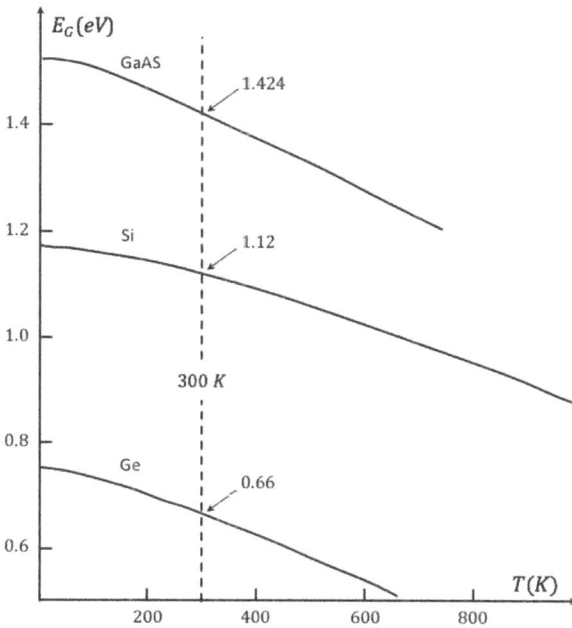

Fig. 5.22. Plot of $E_G(T)$ versus T for some semiconductors [data from: E.F. Schubert, *Light-Emitting Diodes*, 2nd ed. (Cambridge University Press, 2006)].

material structure, and therefore only a few general features will be given here, while a detailed analysis requires a specific discussion for each different type of material, which is out of the scope of this introductory text.

For a solid material, the *thermo-optical coefficient* dn/dT is affected by thermal expansion, which decreases the mass density, thus reducing the refractive index $(dn/dT < 0)$, and by an increase in polarizability with volume expansion, which gives a positive contribution to $n(dn/dT > 0)$. The overall variation of the refractive index results from a balance of these opposite contributions.

The thermo-optical coefficient is a very important parameter, since it defines the range of possible applications for an optical material. For this reason, the measurement of dn/dT is made for all optical glasses. Figure 5.23 reports dn/dT versus T for some optical glasses and shows that it can change value and sign depending on the chemical composition, being in the range of 10^{-6}K^{-1}.

In a fluid material, temperature also affects molecular orientation, and therefore the thermo-optical coefficient is the sum of one term accounting

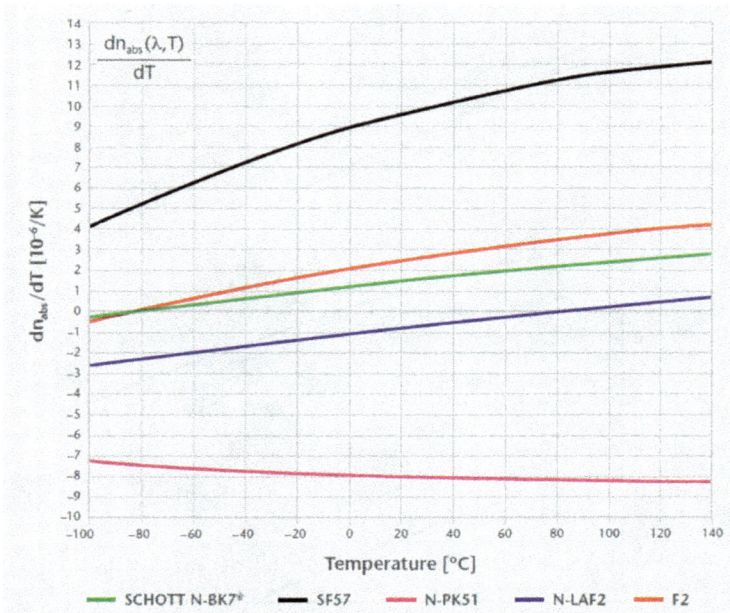

Fig. 5.23. The thermo-optical coefficient of different optical glasses versus T measured at wavelength $\lambda = 587\,\text{nm}$ (courtesy of Schott; www.schott.com).

for density changes and a second term accounting for order parameter changes. For this reason, the overall effect is higher than in the solid state. An example is given in Table 5.2, where the refractive index of water versus temperature is reported, measured at $\lambda = 589.0\,\text{nm}$ and nearly atmospheric pressure (0.1 MPa).

This table shows a thermo-optical coefficient in the range of 10^{-4}–10^{-3}K^{-1}, i.e. 2–3 orders of magnitude higher than in the optical glasses reported in Fig. 5.23. Moreover, we observe the effect of phase transition on the refractive index: the big change we observe at $100°\text{C}$ is due to the transition from the liquid to the vapor state.

Another interesting case concerns liquid crystals, where the order parameter plays a leading role in determining the material properties and contributes to making large the thermo-optical coefficients. Actually, liquid crystals are a good example of the behavior of anisotropic materials; their peculiarity is the opposite signs of the thermo-optical coefficients of the ordinary and extraordinary refractive indices when approaching the phase transition temperature to the isotropic phase. In fact, the index ellipsoid of

Table 5.2. The refractive index of water (H_2O) versus T measured at wavelength $\lambda = 589.0\,\text{nm}$ and pressure of $0.1\,\text{MPa}$ [from: P. Schiebener and J. Straub, *J. Phys. Chem. Ref. Data.* **19**, 677 (1990)]. The calculated thermo-optical coefficient is shown for steps of $10°\text{C}$.

$T(°\text{C})$	n	$\Delta n/\Delta T(10^{-4°}\text{C}^{-1})$
0	1.33432	—
10	1.33408	−2.4
20	1.33336	−7.2
30	1.33230	−10.6
40	1.33095	−13.5
50	1.32937	−15.8
60	1.32757	−18.0
70	1.32559	−19.8
80	1.32342	−21.7
90	1.32109	−23.3
100	1.00019	−3209.0

a uniaxial material reduces to a sphere in the isotropic state with a radius whose length is in between the lengths of the ellipsoid axes. This means that, in positive anisotropic materials where $n_e > n_o$, the ordinary refractive index increases by increasing the temperature near the phase transition to the isotropic state, while the extraordinary index decreases; this leads to having $dn_e/dT < 0$ and $dn_0/dT < 0$. In Fig. 5.24, the dependence on temperature of the refractive indices of the nematic liquid crystal 5CB is shown. This is the typical behavior in nematic liquid crystals.

Data reported in Fig. 5.24 allow calculating the corresponding thermo-optical coefficients reported in Fig. 5.25 for the extraordinary and ordinary refractive indices. We note the very large value of these coefficients — in the range of 10^{-3}–10^{-2}K^{-1}.

5.11. Additional remarks on light–matter interaction

We have discussed in the previous sections the basic mechanisms of the light–matter interactions that may show more complex features when additional parameters are taken into account. We will mention here only a few aspects which can lead the reader to more specialized books.

First of all, we recall that we have considered the material parameters not dependent on the field amplitude, so that the constitutive relations (1.5)–(1.7) are *linear equations*. This approximation is no longer fulfilled

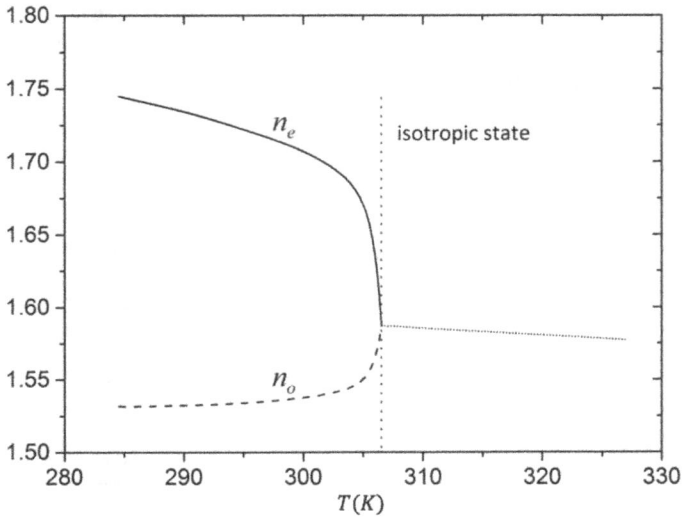

Fig. 5.24.   Dependence of the extraordinary (n_e) and ordinary (n_o) refractive indices versus T in the nematic liquid crystal 5CB at $\lambda = 589.0$ nm. The dashed line corresponds to the phase transition temperature to the isotropic state [data from: J. Li, doctoral thesis (University of Central Florida, 2005); http://stars.library.ucf.edu/etd].

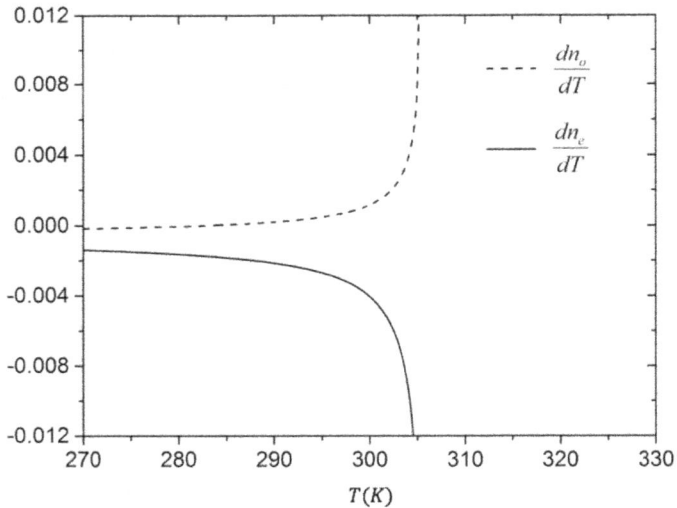

Fig. 5.25.   Dependence of the thermo-optical coefficients dn_e/dT and dn_o/dT versus T in the nematic liquid crystal 5CB at $\lambda = 589.0$ nm [calculated from data reported in: J. Li, doctoral thesis (University of Central Florida, 2005); http://stars.library.ucf.edu/etd].

when fields are high enough to modify the electronic distribution or molecular orientation in the medium. In this case, the general expressions related to the electric field become

$$\mathbf{D} = \underline{\epsilon}(\mathbf{E})\mathbf{E}, \quad \mathbf{P} = \underline{\chi}(\mathbf{E})\,\mathbf{E}. \tag{5.162}$$

The classical approach to solving the wave propagation problem has been proposed[8] by Nicholas Bloembergen, using a series expansion of the polarization vector under the electric dipole approximation:

$$\mathbf{P} = \underline{\chi}^{(1)}\mathbf{E} + \underline{\chi}^{(2)}\mathbf{E}\cdot\mathbf{E} + \underline{\chi}^{(3)}\mathbf{E}\cdot\mathbf{E}\cdot\mathbf{E} + \cdots. \tag{5.163}$$

The first term gives rise to the usual linear polarization $\mathbf{P}^L$ of the medium and the other terms correspond to the *nonlinear polarization* $\mathbf{P}^{NL}$, responsible for nonlinear effects:

$$\mathbf{P} = \mathbf{P}^L + \mathbf{P}^{NL}. \tag{5.164}$$

The series expansion (5.163) implies that the coefficients of the expansion become smaller and smaller, which explains why we need higher and higher field values to observe nonlinear effects of higher order.[9]

The use of such an expression in the wave equation allows describing a huge number of new phenomena occurring as a consequence of the nonlinear dependence (5.162), such as second and third harmonic generation or the intensity-dependent refractive index. This is actually a branch of science called *nonlinear optics*, which originated after the discovery of laser and deals with a large class of optical phenomena now considered for several applications.

In this case, the medium response to the electromagnetic wave is not linear, because it depends on the field amplitude itself. In a similar way, even in the linear optics regime, it is possible to affect the response of the material using a static or low-frequency electric field that induced modification in the optical properties, thus affecting light propagation. These types of phenomena fall into the frame of *electro-optics*. The classical linear electro-optic effect is the change of the refractive index proportional to the applied electric field; however, the effect of the electric field application strongly depends on the material properties and on its aggregation state. In fluids,

---

[8]N. Bloembergen, *Nonlinear Optics*, 4th ed. (World Scientific, 1996).
[9]The first observations of nonlinear optical effects required a high local field, i.e. high light intensities. However, during the more than 50 years of investigations, nonlinear effects induced by fields of moderate amplitude have been discovered in several materials.

the field-induced molecular orientation can play a strong role in affecting the optical properties. In liquid crystals, this effect is able to induce a change in the real part of the refractive index that is orders of magnitude higher than in the other materials and in a reversible way; this is due to the collective behavior consequent to the long-range orientational correlation in these materials. This property makes these materials very successful for display applications such as TV and computer screens.

Light interacting with matter may also induce changes in the structure of the material, modifications that can be either reversible or not and can also find a number of applications.

For instance, it is possible to get *light-induced phase transitions* when the aggregation state of the material changes after photon absorption. Using this effect, it is possible to record information on a material surface, thus realizing *optical data storage*, a technology used to record music and movies on optical disks (CD and DVD).

Another possible phenomenon consequent to light absorption is the molecular dissociation occurring when the energy of the absorbed photon overcomes the molecular bond energy. Exploiting this effect, it is possible to get *light-induced ablation* of molecular layers with high accuracy — a method widely used, for instance, in eye surgery.

Photochemistry

Among the variety of phenomena that can occur when light hits a material, the ones related to structural changes are of paramount relevance because of their impact on technology. They actually compose the branch of science denominated *photochemistry*. Its relevance can be easily understood by thinking of the process that allowed the existence of photography before the digital era. The film roll in the camera undergoes a phototransformation when hit by photons, allowing the subsequent developing procedure to get the picture. The basis of a photochemical process is still the absorption of photons with electron excitation, but this process is followed by energy transfer, which leads to chemical reactions or molecular rearrangement.

A very important photochemical effect is *photoisomerization*, which corresponds to a change of the shape of the molecule consequent to

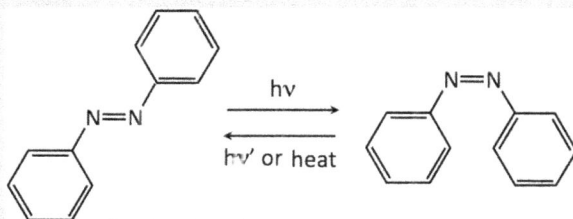

Fig. IN5.8. Photoisomerization of azobenzene.

photon absorption. This process is illustrated in the Fig. IN5.8. Shown is the azobenzene molecule, made by two benzene rings linked through a double bond of nitrogen (N). The more stable elongated *trans* form is transformed into the *cis* form when a photon in the UV (300–400 nm) is absorbed. The molecule can be converted back into the *trans* form by absorption of a photon of lower energy or by heat. Since the *cis* form is not stable, it will relax to the *trans* form after a characteristic time.

This phototransformation may have important consequences for the optical properties of the material where it is induced. For instance, if the azobenzene moiety is part of a molecule in the liquid crystal state, the photoisomerization process can make unstable the mesomorphic state so that one observes a light-induced phase transition from the liquid-crystalline state to the isotropic state.

Another example is the light-induced reorientation in polymers containing azocompounds, consequent to the *trans–cis–trans* cyclic transformation occurring in the photoisomerization process. In fact, if in the initial state the molecules are randomly oriented, after many cycles the most probable orientation will be the one corresponding to lower absorption, usually the one where the long axis of the *trans* isomer is perpendicular to the incoming light polarization, thus obtaining a uniform alignment of the polymer molecules.

Another very important photochemical effect is the one leading to *photopolymerization*. This is one of the most popular methods used to get a polymer starting from a fluid mixture containing the proper monomer. In this case, the mixture includes a photoactive compound able to initiate the polymerization, usually denominated the *photoinitiator*. In free radical photopolymerization (FRP), the photoinitiator (PI) after absorption becomes excited (PI*) and generates a radical $R^\bullet$ that reacts with radical monomers to develop the polymer chain, following the reaction summarized below (see Section 3.12):

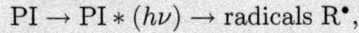

$$PI \to PI * (h\nu) \to \text{radicals } R^\bullet,$$

$R^\bullet$ + radical monomer $\to$ polymer.

The initial mixture is usually in the liquid phase, while the final polymer is a soft solid.

Chapter 6

Laser Electronics

6.1. Quantum electronics

The possibility of coherent amplification of radiation at optical frequencies suggests the realization of devices such as optical amplifiers and optical oscillators. The study of the conditions necessary for realizing them and the study of their physical properties present a kind of parallelism with the usual electronics that deals with electromagnetic fields at much lower frequency and with signals traveling as current through conductive wires. In our case, the signals are traveling on light waves based on quantum energy exchanges with electrons in the material. For these reasons, the study of these optical systems is conventionally called *quantum electronics*.

The scheme of an optical amplifier is very simple and it is shown in Fig. 6.1.

In order to achieve the condition for amplification, we need to provide a minimum power to the material through the pumping mechanism to get the population inversion producing a positive gain coefficient γ at the desired wavelength; in this case, the medium is called an *active material*, since it is able to provide energy to the input light beam, so that the output light beam becomes amplified. Therefore, a key issue is the efficiency of the pumping mechanism for the chosen energy levels involved in the stimulated emission. The choice will also be affected by the light wavelength to be amplified.

In dealing with conventional electronics, it is well known that wave oscillations can be obtained from an amplifying system if appropriate feedback is realized in order to bring back into the amplifier a portion of the output signal; such a condition is realized by connecting the output and

PUMPING

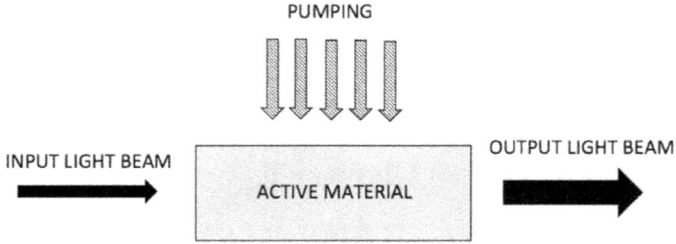

INPUT LIGHT BEAM   OUTPUT LIGHT BEAM

ACTIVE MATERIAL

Fig. 6.1.   Scheme of an optical amplifier.

PUMPING

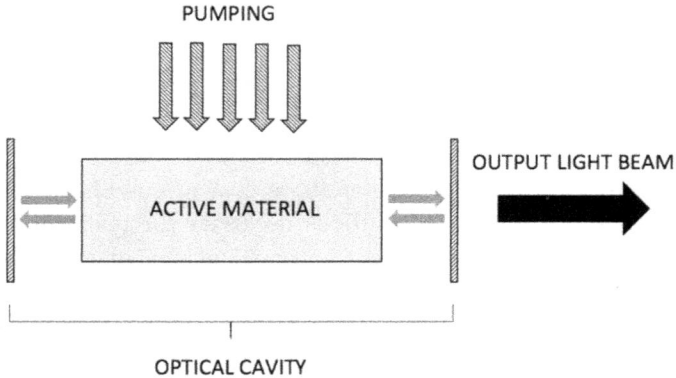

OUTPUT LIGHT BEAM

ACTIVE MATERIAL

OPTICAL CAVITY

Fig. 6.2.   Scheme of the optical oscillator called *laser*.

input of the amplifier through suitable circuit elements and wires. When optical waves are concerned, a couple of mirrors is the simplest solution to get feedback: a fraction of the output light beam is reflected back into the active material. In this way, it is possible to achieve the oscillation conditions, i.e. the onset of steady state wave oscillations, without a need for any input light beam. The scheme of an optical oscillator is shown in Fig. 6.2.

The active material is placed inside an *optical cavity* given (in the simplest design) by two plane mirrors. There is no input light beam, but when the oscillation conditions are fulfilled a coherent light beam is obtained through the partially reflective output mirror. Then an optical oscillator is realized which is usually called *laser*, an acronym for "light amplification by stimulated emission of radiation." Oscillation means that after one round trip in the cavity the amplitude of the wave recovers its initial value so that a steady state wave travels forth and back in the cavity and a portion of it comes out through one partially transmitting mirror.

An obvious question should arise: Where does the radiation to be amplified come from, since there is no input light beam? The answer is: optical noise, i.e. spontaneous emission occurring with the characteristic time τ_{sp} after excitation of the electrons to the upper level concerned with the process of radiation emission.

In this chapter, we will discuss the basic methods for obtaining the population inversion needed to get light amplification and the properties of radiation in an optical cavity, and finally the conditions for obtaining laser emission. We will also describe the properties of laser radiation and give some example of laser devices.

6.2. Population inversion and rate equations

In Section 5.4, we have discussed the main aspects related to the gain coefficient γ which arises between two levels when stimulated emission overcomes absorption, i.e. when a population inversion is established such that the electronic density at the high energy level is bigger than the one at the low energy level ($N_2 > N_1$). We have pointed out that the involvement of one or more additional energy levels is necessary for achieving this condition, owing to the saturation effect in a two-level excitation. In general, two pumping schemes are considered: *three-level* and *four-level* systems. The conditions for achieving $\Delta N > 0$ can be found by solving the *rate equations*, i.e. the coupled equations for the time evolution of the electronic population of the involved levels.

The two schemes are shown in Fig. 6.3. Both are characterized by excitation through a suitable pumping mechanism to an energy state higher

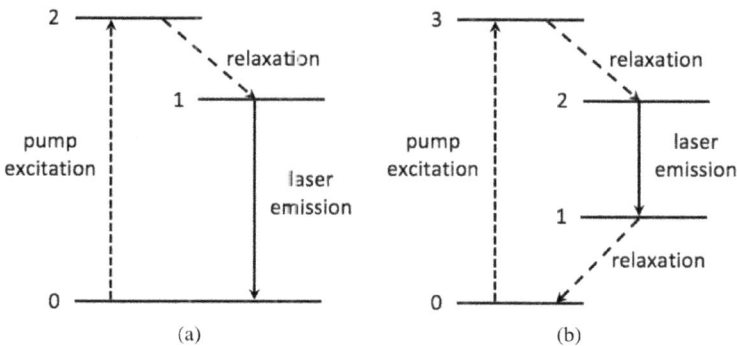

Fig. 6.3. Pumping schemes for light amplification: (a) three-level system, (b) four-level system.

than the upper level considered for the optical gain transition. A subsequent
fast decay of the electrons to this level is then exploited to achieve the
required population inversion with respect to a third level at lower energy.
The basic difference between the two schemes is that for the three-level
system the lower level is the ground state, while for the four-level system
it is an excited state at a much higher energy ($E_1 \gg k_B T$), but with a
short decay time to the ground state. In this case, the thermal equilibrium
population of level 1 is negligible with respect to the ground state and, if its
lifetime is short compared to the lifetime of level 2, the former will always
be empty with respect to the latter and the population inversion can be
approximated as

$$\Delta N = \left(N_2 - \frac{g_2}{g_1} N_1 \right)^{4 \text{ levels}} \sim N_2^{4 \text{ levels}} \ll N_0, \tag{6.1}$$

with N_0 being the population of the ground state. On the contrary, in the
three-level system it is necessary to excite to the upper level a number
of electrons higher than $N_0/2$ to get inversion, and therefore the four-
level system is expected to require lower power to get the same population
inversion. In fact, in this case we have

$$N_2^{3 \text{ levels}} = \frac{N_0}{2} + \frac{\Delta N}{2}, \quad N_1^{3 \text{ levels}} = \frac{N_0}{2} - \frac{\Delta N}{2}. \tag{6.2}$$

Since $N_0 \gg \Delta N$, we get

$$\frac{N_2^{3 \text{ levels}}}{N_2^{4 \text{ levels}}} \approx \frac{N_0}{2\Delta N}. \tag{6.3}$$

If τ_2 is the overall lifetime of level 2, the minimum pumping power necessary
for maintaining the population difference ΔN is given by the total energy
associated with the stimulated emission transition divided by τ_2. The energy
is the product between the electron density N_2, the pumped volume V
and the energy $h\nu$ associated with a single transition; then the minimum
pumping power is $P = N_2 V h\nu / \tau_2$, and hence for the two different systems
we have

$$P^{3 \text{ levels}} = \frac{N_0 V h\nu}{2\tau_2}, \tag{6.4}$$

$$P^{4 \text{ levels}} = \frac{\Delta N V h\nu}{\tau_2}. \tag{6.5}$$

$\Delta N \ll N_0$, so $P^{4 \text{ levels}} \ll P^{3 \text{ levels}}$.

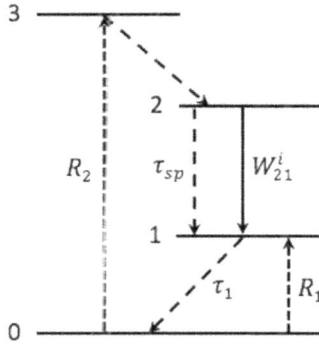

Fig. 6.4. The four-level system: energy levels and transition rates.

We now discuss the dynamic of the four-level system in order to get a more general expression for the gain coefficient useful for highlighting all the parameters important in the process, which have to be taken into account in the practical realization of a light amplifier. Looking at Fig. 6.4, we define: R_2 as the pumping rate to level 2, i.e. the electron density rise in time provided by the pumping mechanism; R_1 as the undesired pumping rate to level 1 which may occur; τ_2 as the lifetime of level 2, a combination of the spontaneous emission time τ_{sp} to level 1 and the decay time τ_{20} of energy from level 2 to the ground state not involving level 1 ($1/\tau_2 = 1/\tau_{\mathrm{sp}}+1/\tau_{20}$); and τ_1 as the decay time from level 1 to the ground state due to any possible effect (spontaneous emission, nonradiative thermal relaxation, etc.).

With regard to the time relaxation from level 2, we note that release of energy through other channels than spontaneous emission to level 1, summarized in the lifetime τ_{20}, can occur, for example, by energy transfer between atoms (in gas during collision) or by excitation of lattice vibrations (phonons, in solids). We may write a relation for the probability per unit time of decay of energy from level 2 as the sum of the probability of spontaneous emission and that of energy release through other channels:

$$A = A_{21} + A_{20}, \quad \text{i.e.} \quad 1/\tau_2 = 1/\tau_{\mathrm{sp}} + 1/\tau_{20}. \tag{6.6}$$

Usually, any transition different from the direct one from level 2 to 1 is a loss of energy for what concerns the amplification process, and therefore it is quite important to consider the quantum efficiency or *quantum yield* of the process:

$$\phi = \frac{A_{21}}{A} = \frac{\tau_2}{\tau_{\mathrm{sp}}}. \tag{6.7}$$

That should be as high as possible, being in the ideal case $\phi \to 1$.

According to this scheme and looking at Eq. (5.90), we write the corresponding rate equations:

$$\frac{dN_2}{dt} = R_2 - \frac{N_2}{\tau_2} - [N_2 W_{21}^i - N_1 W_{12}^i] = R_2 - \frac{N_2}{\tau_2} - \Delta N W_{21}^i, \quad (6.8)$$

$$\frac{dN_1}{dt} = R_1 - \frac{N_1}{\tau_1} + \frac{N_2}{\tau_{sp}} + \Delta N W_{21}^i. \quad (6.9)$$

These equations can be solved for the steady state condition $dN_2/dt = dN_1/dt = 0$, and they give

$$\Delta N = \frac{R_2 \tau_2 - (R_1 + (\tau_2/\tau_{sp})R_2)\tau_1(g_2/g_1)}{1 + [\tau_2 + (1 - (\tau_2/\tau_{sp}))\tau_1(g_2/g_1)]W_{21}^i}. \quad (6.10)$$

The result (6.10) is easily found by using the Cramer method to solve the system given by Eqs. (6.8) and (6.9) in the steady state condition, taking into account that $\Delta N = N_2 - N_1 g_2/g_1$:

$$\frac{g_2}{g_1} W_{21}^i N_1 - \left(W_{21}^i + \frac{1}{\tau_2}\right) N_2 = -R_2,$$

$$-\left(\frac{g_2}{g_1} W_{21}^i + \frac{1}{\tau_1}\right) N_1 + \left(W_{21}^i + \frac{1}{\tau_{sp}}\right) N_2 = -R_1.$$

Then we obtain

$$N_1 = \frac{\begin{vmatrix} -R_2 & -\left(W_{21}^i + (1/\tau_2)\right) \\ -R_1 & \left(W_{21}^i + (1/\tau_{sp})\right) \end{vmatrix}}{D}$$

$$= \frac{-R_2 \left(W_{21}^i + (1/\tau_{sp})\right) - R_1 \left(W_{21}^i + (1/\tau_2)\right)}{D}$$

$$= \frac{-(R_2 + R_1)W_{21}^i - (R_2/\tau_{sp}) - (R_1/\tau_2)}{D},$$

$$N_2 = \frac{\begin{vmatrix} (g_2/g_1)W_{21}^i & -R_2 \\ -\left((g_2/g_1)W_{21}^i + (1/\tau_1)\right) & -R_1 \end{vmatrix}}{D}$$

$$= \frac{-R_1(g_2/g_1)W_{21}^i - R_2\left((g_2/g_1)W_{21}^i + (1/\tau_1)\right)}{D}$$

$$= \frac{-(R_2 + R_1)(g_2/g_1)W_{21}^i - (R_2/\tau_1)}{D},$$

where

$$D = \begin{vmatrix} (g_2/g_1)W_{21}^i & -(W_{21}^i + (1/\tau_2)) \\ -((g_2/g_1)W_{21}^i + (1/\tau_1)) & (W_{21}^i + (1/\tau_{\mathrm{sp}})) \end{vmatrix}$$

$$= \frac{g_2}{g_1}W_{21}^i \left(W_{21}^i + \frac{1}{\tau_{\mathrm{sp}}}\right) - \left(\frac{g_2}{g_1}W_{21}^i + \frac{1}{\tau_1}\right)\left(W_{21}^i + \frac{1}{\tau_2}\right)$$

$$= \frac{1}{\tau_{\mathrm{sp}}}\frac{g_2}{g_1}W_{21}^i - \frac{W_{21}^i}{\tau_1} - \frac{1}{\tau_1\tau_2} - \frac{1}{\tau_2}\frac{g_2}{g_1}W_{21}^i$$

$$= -\frac{1}{\tau_1\tau_2}\left\{1 + \left[\tau_2 + \left(1 - \frac{\tau_2}{\tau_{\mathrm{sp}}}\right)\tau_1\frac{g_2}{g_1}\right]W_{21}^i\right\}.$$

Now it is possible to write:

$$\Delta N = N_2 - N_1\frac{g_2}{g_1}$$

$$= \frac{-(R_2 + R_1)(g_2/g_1)W_{21}^i - (R_2/\tau_1) + (R_2 + R_1)(g_2/g_1)W_{21}^i}{+(R_2/\tau_{\mathrm{sp}})(g_2/g_1) + (R_1/\tau_2)(g_2/g_1)}{-(1/\tau_1\tau_2)\left\{1 + [\tau_2 + (1 - (\tau_2/\tau_{\mathrm{sp}}))\tau_1(g_2/g_1)]W_{21}^i\right\}}$$

$$= \frac{R_2\tau_2 - (R_2(\tau_2/\tau_{\mathrm{sp}}) + R_1)\tau_1(g_2/g_1)}{1 + [\tau_2 + (1 - (\tau_2/\tau_{\mathrm{sp}}))\tau_1(g_2/g_1)]W_{21}^i},$$

corresponding to Eq. (6.10).

The low intensity inversion is obtained by considering $I_\nu \to 0$ and then setting $W_{21}^i = 0$ in Eq. (6.10). In this way, we get

$$\Delta N_0 = R_2\tau_2 - \left(R_1 + \frac{\tau_2}{\tau_{\mathrm{sp}}}R_2\right)\tau_1\frac{g_2}{g_1}. \tag{6.11}$$

This equation points out the dependence of the achievable maximum inversion ΔN_0 on the pumping rate and on material parameters, i.e. relaxation rates and level degeneracies.

In ideal cases, where $\tau_2 = \tau_{\mathrm{sp}}$ and $g_2 = g_1$, we have

$$\Delta N_0 \approx R_2\tau_2 - (R_1 + R_2)\tau_1, \tag{6.12}$$

where the advantage of a long lifetime for the upper state and a short one for the lower state is highlighted.

As a consequence, Eq. (6.10) can be written as

$$\Delta N = \frac{\Delta N_0}{1 + [\tau_2 + (1 - (\tau_2/\tau_{\rm sp}))\,\tau_1\,(g_2/g_1)]\,W_{21}^i} = \frac{\Delta N_0}{1 + (I/I_s)}, \qquad (6.13)$$

which is the expression (5.109) accounting for the saturation effect on the population inversion with the definition

$$I_s = \frac{8\pi n_0^2 h\nu^3 \tau_{\rm sp}}{c^2\,[\tau_2 + (1 - (\tau_2/\tau_{\rm sp}))\,\tau_1\,(g_2/g_1)]\,g(\nu)}. \qquad (6.14)$$

In the ideal case of $\tau_2 \approx \tau_{\rm sp}$, it reduces to

$$I_s = \frac{8\pi n_0^2 h\nu^3 \tau_{\rm sp}}{c^2 \tau_2 g(\nu)}, \qquad (6.15)$$

often written as

$$I_s = \frac{h\nu}{\tau_2 \sigma(\nu)}, \qquad (6.16)$$

by using the stimulated emission cross section, defined as

$$\sigma(\nu) = \frac{c^2}{8\pi n_0^2 \nu^2 \tau_{\rm sp}} g(\nu) \qquad (6.17)$$

such that

$$W_{21}^i = \frac{\sigma(\nu)I}{h\nu}, \qquad (6.18)$$

and Eq. (5.92) can be written as

$$\gamma(\nu) = \Delta N \sigma(\nu). \qquad (6.19)$$

That is to say,

$$\gamma(\nu) = \frac{\sigma(\nu)\Delta N_0}{1 + [\tau_2 + (1 - (\tau_2/\tau_{\rm sp}))\tau_1(g_2/g_1)]\,(\sigma(\nu)I/h\nu)} = \frac{\gamma_0(\nu)}{1 + (I/I_s)}. \qquad (6.20)$$

In deriving Eq. (6.20), we have assumed all atoms to be equivalent, i.e. a homogeneously broadened line shape, and then each part of the gain curve saturates to the same degree.

When the line shape is inhomogeneously broadened, saturation will occur only for the atoms whose line shape includes the frequency of the traveling wave. This leads to a spectral "hole" in the gain spectrum when the increasing gain saturates in a narrow frequency range. Details of this treatment (to be found in specialized books) lead to a saturated gain expression given by

$$\gamma(\nu) = \frac{\gamma_0(\nu)}{\sqrt{1 + (I/I_s)}}. \tag{6.21}$$

Equations (6.20) and (6.21) show the saturation effect in gain media, which leads to a decrease of the gain coefficient at increasing intensities, the maximum gain being the small signal value $\gamma_0(\nu) = \sigma(\nu)\Delta N_0$. It depends on the material characteristics included in $\sigma(\nu)$ and on the population inversion ΔN_0.

The three- or four-level pumping mechanism applies to atomic or molecular systems; they can be either in the vapor phase or embedded in host materials such as dye molecules in liquids or impurities in crystals.

Significant differences occur when the considered transition takes place between two different energy bands in crystals. We have found in Section 5.6 the expressions (5.133) and (5.134) for the absorption or gain coefficient in the case of direct gap materials. Here, we briefly discuss how the population inversion condition can be achieved.

In this case, we write the gain coefficient at the specific frequency ν_0 as

$$\gamma(\nu_0) = \sqrt{\nu_0 - \frac{E_G}{h}} \left(\frac{2m_r}{h}\right)^{3/2} \{f_c(E_2) - f_v(E_1)\} \frac{c^2}{2n_0^2\nu_0^2\tau_{\rm sp}}, \tag{6.22}$$

and therefore the amplification condition requires that $f_c(E_2) > f_v(E_1)$. This condition is sketched in Fig. 6.5, where we need to consider in each band a Fermi–Dirac distribution $f_v(E)$ and $f_c(E)$ with the corresponding pseudo-Fermi energy E_{Fv} and E_{Fc}.

The inversion condition gives

$$\frac{1}{e^{\frac{E_2 - E_{Fc}}{k_B T}} + 1} - \frac{1}{e^{\frac{E_1 - E_{Fv}}{k_B T}} + 1} > 0, \tag{6.23}$$

and therefore

$$e^{\frac{E_1 - E_{Fv}}{k_B T}} - e^{\frac{E_2 - E_{Fc}}{k_B T}} > 0. \tag{6.24}$$

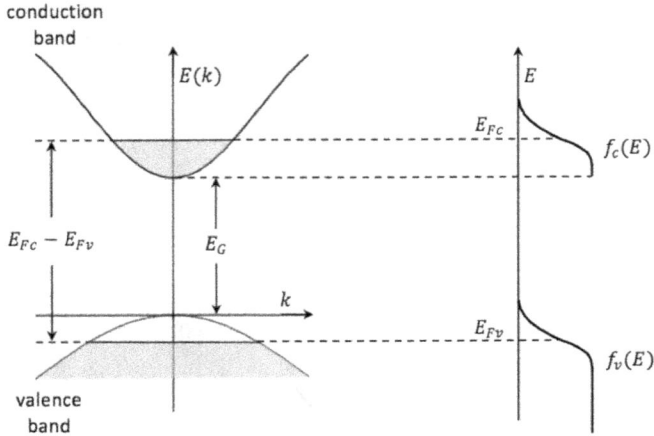

Fig. 6.5.   (a) Population inversion between conduction and valence bands; (b) Fermi–Dirac distributions in the two bands.

Then

$$E_1 - E_{Fv} > E_2 - E_{Fc} \tag{6.25}$$

or

$$E_{Fc} - E_{Fv} > E_2 - E_1. \tag{6.26}$$

In Fig. 6.6, it is shown how this condition can be realized in a p–n junction. We know that at equilibrium the Fermi level in the p and n semiconductors is the same.

Since the equilibrium condition requires that $E_{Fc} - E_{Fv} = 0$, population inversion is not possible. However, by applying a direct voltage to the p–n junction, the condition (6.26) may occur in the depletion region, which becomes the *active region* since in a narrow frequency range the following condition is fulfilled:

$$E_{Fc} - E_{Fv} > h\nu. \tag{6.27}$$

In this case, pumping is realized through electron *injection*, by keeping a minimum value of current I able to balance the spontaneous electron–hole recombination occurring with a characteristic time τ. With n_c being the concentration of injected electrons, through a section S in the volume V, we have

$$I = JS = n_c e v S = n_c e \frac{\ell}{\tau} S = n_c \frac{eV}{\tau}, \tag{6.28}$$

(a)

(b)

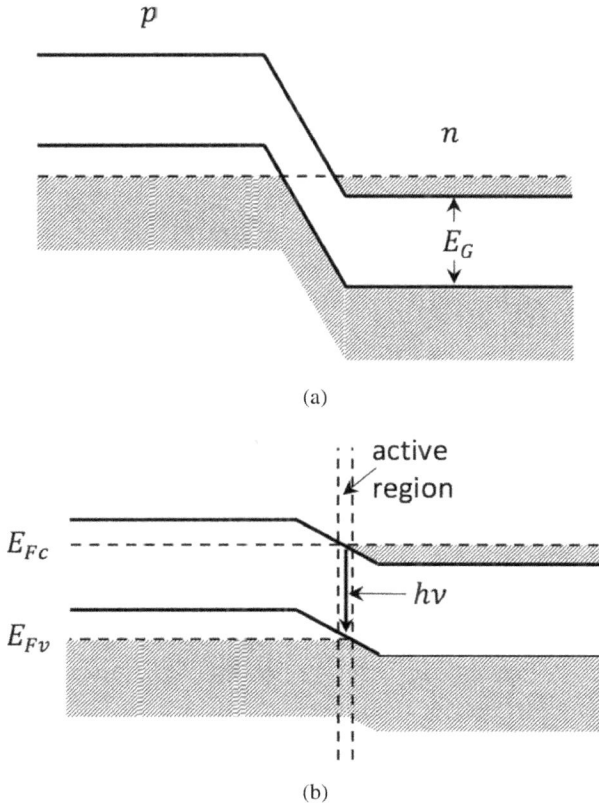

Fig. 6.6. (a) The *p–n* junction at equilibrium with the same Fermi level E_F, on the two sides; (b) the *p–n* junction under direct polarization and the resulting pseudo-Fermi levels E_{Fc} and E_{Fv}.

where ℓ is the length crossed by one electron during the time τ (at a speed v). Taking into account Eq. (4.129), we can write

$$n_c = \frac{I\tau}{e\mathcal{V}} = \frac{1}{V} \int_0^\infty f_c(E,T) g_c(E) dE$$

$$= \left(\frac{1}{2\pi^2 \hbar^3}\right) (2m_e)^{3/2} \int_0^\infty \frac{\sqrt{E}}{e^{\frac{E-E_{Fc}}{k_B T}} + 1} dE. \tag{6.29}$$

This expression links the injection current to the effective value of E_{Fc}. It can be numerically solved to get $E_{Fc}(I)$, thus providing the *threshold* current which leads to fulfillment of the condition (6.27).

6.3. Optical resonators

When one is dealing with optical resonators, the first issue to be addressed concerns the mode density of optical radiation in an optical cavity. If we consider a closed volume as a cube with side L, we can quickly evaluate the number of possible stationary waves. In fact, in order to be kept in a finite volume the electromagnetic waves must be reflected by the limiting walls, made by partially or totally reflecting mirrors. As long as the dimension of the cavity is large with respect to the light wavelength, we must take into account the spatial periodicity of the wave at the cavity walls:

$$e^{i\mathbf{k}\cdot\mathbf{r}} = e^{i(k_x x + k_y y + k_z z)} = e^{i(k_x[x+L]+k_y[y+L]+k_z[z+L])}. \tag{6.30}$$

This condition gives

$$k_x = q\frac{2\pi}{L}, \quad k_y = s\frac{2\pi}{L}, \quad k_z = t\frac{2\pi}{L}, \tag{6.31}$$

where q, s and t are integers.

As a consequence, we have a volume $(2\pi/L)^3$ per mode in $\mathbf{k}$-space. Therefore, the number of modes N_k with the wave vector from 0 to k is obtained by dividing the corresponding volume $(4/3)\pi k^3$ by the mode volume $(2\pi/L)^3$ and multiplying by 2, owing to the possible polarizations for each value of $\mathbf{k}$. We get

$$N_k = \frac{k^3 L^3}{3\pi^2}. \tag{6.32}$$

By using the frequency ν,

$$N_k = \frac{8\pi\nu^3 n^3 L^3}{3c^3}. \tag{6.33}$$

Therefore, the number of modes per unit frequency and unit volume is

$$p(\nu) = \frac{1}{L^3}\frac{dN_k}{d\nu} = \frac{8\pi\nu^2 n^3}{c^3}. \tag{6.34}$$

It is easy to verify that for a wavelength of $\lambda = 550\,\text{nm}$ at the center of the visible spectrum $p(\nu) = 8.3 \cdot 10^9$ modes/cm^3, that is a huge number. However, the idea of using open cavities gives the opportunity of a strong selection of modes allowing us to consider only the ones able to oscillate many times between opposite mirrors. Of course, there are several ways of creating resonators in a single direction (open cavity) using plane or curved mirrors, such as the two examples shown in Fig. 6.7.

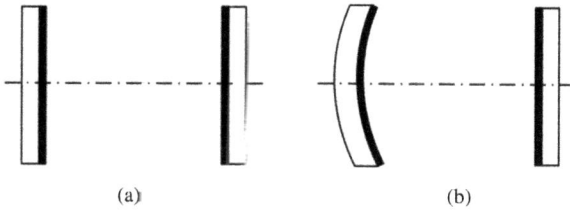

(a) (b)

Fig. 6.7. Examples of optical cavities realized with mirrors.

The cavity is called *stable* if the propagation direction of the wave traveling back and forth between the mirrors stays close to the optic axis of the system, i.e. even after many reflections the wave does not travel out of the cavity. The stability condition for a two-mirror cavity is easily found by using the *ray-tracing method* and is given by the simple expression

$$0 \leq g_1 g_2 \leq 1, \tag{6.35}$$

where

$$g_i = 1 - \frac{d}{R_i}, \tag{6.36}$$

with d being the mirrors' distance and R_i their curvature radius. It provides the stability diagram shown in Fig. 6.8, useful for establishing at a glance the cavity stability. It is interesting to identify in the diagram the points corresponding to the plane parallel mirror resonator (A) or to the symmetrical confocal mirror resonator[1] (B). The basic open optical cavity that we will analyze in the next section is the one with plane parallel mirrors, called a Fabry–Perot cavity.

The *ray-tracing* method is a geometrical optics approach based on the concept of a *ray*, defined as a line drawn along the propagation direction of the optical wave; in other words, a line parallel to the vector **k**. The changes of a ray due to an optical system are evaluated defining the distance r of the ray from the optic axis and its derivative r' (with respect to the linear coordinate on the axis) at the entrance and at the exit of the optical system. In this way, the effect of each optical system

---

[1] A symmetrical, confocal resonator is made up of two spherical mirrors with the same curvature radius R; since the mirror's focal length f is given by $f = R/2$, the mirror displacement is $d = R$ in order to make the focus of the two mirrors coincident at the same point.

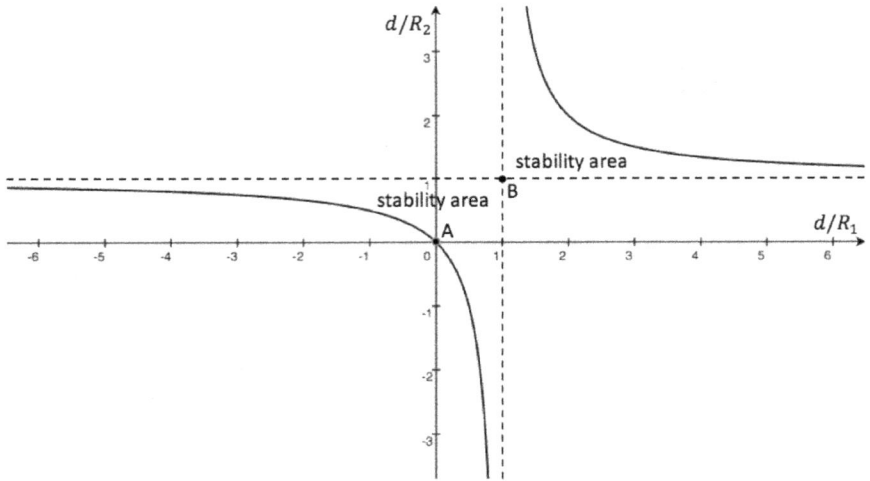

Fig. 6.8.   Stability diagram for a two-mirror optical cavity.

on the direction of a ray can be described by a 2×2 matrix. For instance, let us consider just the travel of a ray through two planes located at a distance d from each other. Looking at Fig. IN6.1, in the approximation of paraxial rays (a small angle with respect to the optic axis), it is easy to find that

$$r_2 = r_1 + dr'_1,$$
$$r'_2 = 0 + r'_1,$$

and therefore the corresponding matrix relation will be

$$\begin{bmatrix} r_2 \\ r'_2 \end{bmatrix} = \begin{bmatrix} 1 & d \\ 0 & 1 \end{bmatrix} \begin{bmatrix} r_1 \\ r'_1 \end{bmatrix}.$$

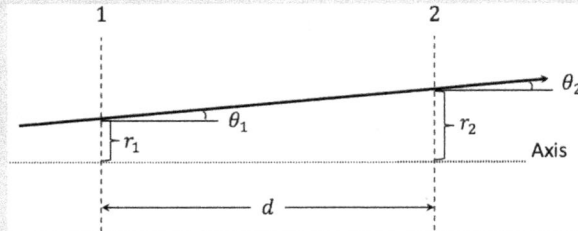

Fig. IN6.1. Ray-tracing in a homogeneous material of length d.

The same method can be applied to lenses, mirrors and their combination to obtain the corresponding matrix. For instance, we have

$$\begin{bmatrix} 1 & 0 \\ -\dfrac{1}{f} & 1 \end{bmatrix},$$

for a thin lens with focal length f, and

$$\begin{bmatrix} 1 & 0 \\ -\dfrac{2}{R} & 1 \end{bmatrix}$$

for a spherical mirror with curvature radius R.

In this way, any combination of basic elements reduces to a 2×2 matrix resulting from the product of the matrix corresponding to each optical element.

The matrix corresponding to a cavity made by two mirrors is generally indicated as

$$\begin{bmatrix} A & B \\ C & D \end{bmatrix}$$

and its solution leads to the stability condition (6.35). Details can be found in basic texts of quantum electronics.

Gaussian beams

Another aspect that cannot be neglected when one is dealing with optical resonators is the finite dimension of the cavity mirrors. As a consequence, the description of the oscillating electromagnetic wave as an ideal plane wave becomes very unsatisfactory owing to the finite dimension of the wave front. A more suitable description comes from a solution to the wave equation that takes into account a dependence on the transversal coordinates. If we consider the z axis as the main propagation direction (for example, corresponding to the optic axis of the resonator), we can still look for a quasi-plane wave solution where the field amplitude is dependent on the coordinates. Thus, for linear polarization we write

$$E = A(x, y, z)[\exp i(\mathbf{k} \cdot \mathbf{r} - \omega t)] = E_0 \psi(x, y, z)[\exp i(\mathbf{k} \cdot \mathbf{r} - \omega t)].$$
$$(6.37)$$

In the slowly varying envelope approximation (SVEA), the variations of the ψ function are small along a distance of a wavelength and it is possible to neglect its second derivative with respect to z. In this case, the *Gaussian beam* solutions are obtained for the wave equation. They are given by

$$A(x,y,z) = E_0 H_m \left[\frac{\sqrt{2}x}{w(z)} \right] H_n \left[\frac{\sqrt{2}y}{w(z)} \right] \frac{w_0}{w(z)} \exp \left[-\frac{x^2 + y^2}{w^2(z)} \right]$$

$$\times \exp \left\{ -i \left[kz - (1+m+n) \tan^{-1} \left(\frac{z}{z_0} \right) \right] \right\}$$

$$\times \exp \left[-i \frac{k(x^2 + y^2)}{2R(z)} \right], \tag{6.38}$$

where $H_n[u]$ are the nth-order Hermite polynomials:

$$H_n[u] = (-1)^n e^{u^2} \frac{d^n e^{-u^2}}{du^n}. \tag{6.39}$$

In order to understand the meaning of this expression, let us consider the lowest, order solution, i.e. the TEM$_{00}$ mode, the actual Gaussian mode, since the x–y dependence of the amplitude is represented by a Gaussian function.

Setting $n = m = 0$ in Eqs. (6.38) and (6.39), we have

$$A(x,y,z) = E_0 \frac{w_0}{w(z)} \exp \left[-\frac{x^2 + y^2}{w^2(z)} \right]$$

$$\times \exp \left\{ -i \left[kz - \tan^{-1} \left(\frac{z}{z_0} \right) \right] \right\}$$

$$\times \exp \left[-i \frac{k(x^2 + y^2)}{2R(z)} \right], \tag{6.40}$$

where

$$w^2(z) = w_0^2 \left[1 + \frac{z^2}{z_0^2} \right], \tag{6.41}$$

$$R(z) = z \left[1 + \frac{z_0^2}{z^2} \right], \tag{6.42}$$

$$z_0 = \frac{\pi w_0^2}{\lambda} = \frac{k w_0^2}{2}. \tag{6.43}$$

The first term of Eq. (6.40) gives the amplitude variation with the coordinates:

$$A(x,y,z) = E_0 \frac{w_0}{w(z)} \exp\left[-\frac{(x^2+y^2)}{w^2(z)}\right]$$

$$= E_0 \frac{w_0}{w(z)} \exp\left[-\frac{r^2}{w^2(z)}\right] \equiv A(r,z). \qquad (6.44)$$

The cylindrical symmetry allows reducing the transversal coordinate to the radial coordinate $r = \sqrt{x^2+y^2}$, which is the distance from the beam axis. For any z at $r = 0$, the field has the maximum amplitude:

$$A(0,z) = E_0 \frac{w_0}{w(z)}. \qquad (6.45)$$

Here $w(z)$ is a measure of the beamwidth called the spot size; in fact, when $r = w$ the field amplitude is[2] $A(0,z)/e$. From Eq. (6.41), we see that w increases with z and its minimum is[3] $w_0 = w(0)$ at $z = 0$, where the amplitude has its absolute maximum:

$$A(0,0) = E_0. \qquad (6.46)$$

On the other hand, the meaning of z_0 is found from Eq. (6.41) as the location where the spot size has grown by a factor of $\sqrt{2}$ with respect to its minimum w_0:

$$w^2(z_0) = w_0^2 \left[1 + \frac{z_0^2}{z_0^2}\right] = 2w_0^2 \rightarrow w(z_0) = \sqrt{2}w_0. \qquad (6.47)$$

It is worth noting the asymptotic linear rise of w for $z \gg z_0$:

$$w(z \gg z_0) = \frac{w_0 z}{z_0} = \frac{\lambda z}{\pi w_0}. \qquad (6.48)$$

Accordingly, the beam aperture angle θ can be easily evaluated by looking at Fig. 6.9, where the properties of TEM_{00} are summarized:

$$\tan\left(\frac{\theta}{2}\right) \approx \frac{\theta}{2} = \frac{dw}{dz} = \frac{\lambda}{\pi w_0} \rightarrow \theta = \frac{2\lambda}{\pi w_0}. \qquad (6.49)$$

---

[2]The width of a Gaussian function is conventionally taken as the value of the variable where the function is reduced by a factor of $1/e$ with respect to its maximum.

[3]In practical systems, w_0 is a parameter determined by geometrical factors, for instance by the mirrors' size and curvature in a laser cavity.

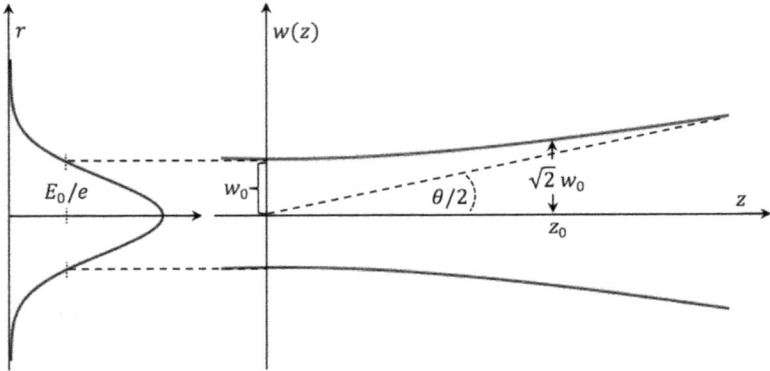

Fig. 6.9.   Amplitude distribution of the TEM$_{00}$ mode versus z.

This is the minimum divergence of a Gaussian beam whose minimum spot size is w_0.

Looking at Eq. (6.40), we see also two phase terms:

$$\exp\left\{-i\left[kz - \tan^{-1}\left(\frac{z}{z_0}\right)\right]\right\}, \tag{6.50}$$

$$\exp\left[-i\frac{k\left(x^2 + y^2\right)}{2R(z)}\right]. \tag{6.51}$$

The first one represents the longitudinal phase, the second one the radial phase. The first is the same as for a plane wave with a small correction consequence of the limited shape of the beam. The second is typical of a curved wave front; indeed, it can be shown that $R(z)$ is the curvature radius of the wave front changing during propagation. In fact, according to Eq. (6.42), the wave front is a plane when $z = 0$, since in this case $R(0) = \infty$; while when $z < 0$, $R < 0$ and the beam converges to the minimum spot size, from which diverges being $R > 0$ when $z > 0$.

Higher-order modes keep similar features, with the different amplitude distribution in space as shown in Fig. 6.10.

According to this discussion, the actual beam traveling inside an optical resonator can be the superposition of different Gaussian modes, being in the best case a single TEM$_{00}$ mode.

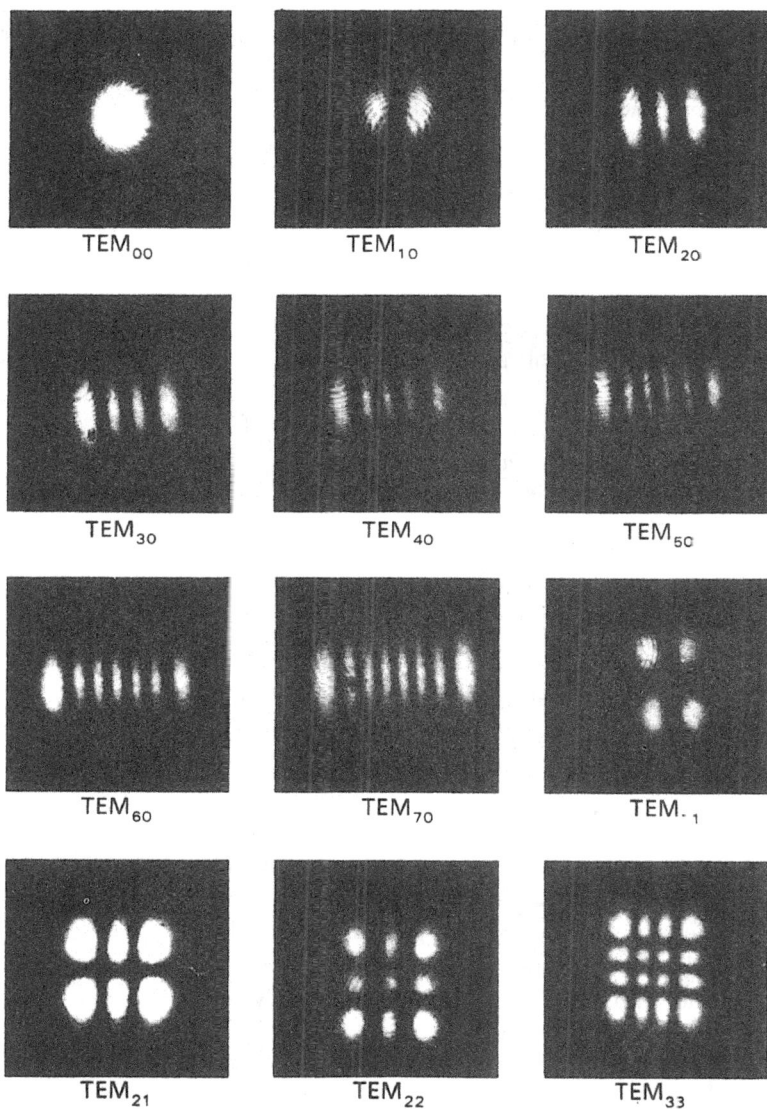

Fig. 6.10. Intensity distribution in the transversal plane of Gaussian beams for increasing mode indices [from: H. Kogelnik *et al.*, *Appl. Opt.* **5**, 1550 (1966)].

It is important to recall the power and intensity characteristics of a Gaussian beam. A given power should stay constant during propagation, and therefore we can evaluate it from the intensity equation (1.62)

$$I_0 = \frac{1}{2} c \varepsilon_0 n_1 E_0^2,$$

using $\mathcal{A}(x, y, z)$ in place of E_0. Here n_1 is the refractive index of the propagation medium. By integrating it on the transversal surface crossed by the wave front,

$$P = \frac{1}{2} \iint c \varepsilon_0 n_1 \mathcal{A}^2(x, y, z) d\sigma$$

$$= \frac{1}{2} c \varepsilon_0 n_1 E_0^2 \frac{w_0^2}{w^2(z)} \int_0^{2\pi} d\phi \int_0^\infty \exp \left[-\frac{2r^2}{w^2(z)} \right] r dr.$$

Since

$$\int_0^{2\pi} d\phi \int_0^\infty \exp \left[-\frac{2r^2}{w^2(z)} \right] r dr = \frac{\pi w^2(z)}{2},$$

we have

$$P = \frac{1}{4} c \varepsilon_0 n_1 E_0^2 \pi w_0^2 = I_0 \frac{\pi w_0^2}{2} \rightarrow I_0 = \frac{2P}{\pi w_0^2}.$$

6.4. Oscillation conditions

We discuss here the oscillation conditions originated by a Fabry–Perot resonator. The system is sketched in Fig. 6.11, where the space between two plane mirrors is filled by the gain medium. The distance between the mirrors is l, the amplitude of the incoming wave is E_i, and we want to evaluate the amplitude E_t of the transmitted wave. We define t_i and r_i

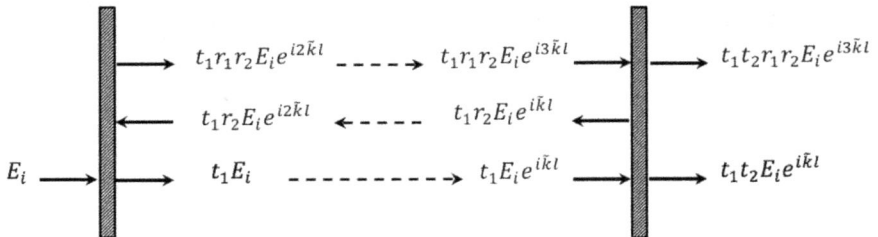

Fig. 6.11.   Calculation of the field transmitted by a Fabry–Perot resonator.

as the transmission and reflection coefficients for the wave amplitude of the two mirrors, and $\tilde{k}$ as the complex wave vector of the wave inside the medium, which takes into account both gains and losses due to dispersion, diffraction and absorption.

First of all, we must note that after a single trip in the cavity the phase factor of the wave will be multiplied by a factor $e^{i\tilde{k}l}$.

From the figure, we evaluate the transmitted field as

$$E_t = t_1 t_2 E_i e^{i\tilde{k}l}[1 + r_1 r_2 e^{i2\tilde{k}l} + (r_1 r_2)^2 e^{i4\tilde{k}l} + \cdots]. \tag{6.52}$$

By considering a high number of reflections ($\rightarrow \infty$; parallel mirrors), the geometrical series[4] can be summed to give

$$E_t = E_i \left[\frac{t_1 t_2 e^{i\tilde{k}l}}{1 - r_1 r_2 e^{i2\tilde{k}l}} \right], \tag{6.53}$$

where the complex wave vector can be written as

$$\tilde{k} = k - i\frac{\gamma - \beta}{2}, \tag{6.54}$$

taking into account the absorption coefficient in Eq. (5.66), and separating the gain contribution from any other losses included in β.

By including Eq. (6.54) in Eq. (6.53), we have

$$E_t = E_i \left[\frac{t_1 t_2 e^{ikl} e^{\left(\frac{\gamma-\beta}{2}\right)l}}{1 - r_1 r_2 e^{2ikl} e^{(\gamma-\beta)l}} \right], \tag{6.55}$$

with an active medium $\gamma > 0$: moreover, if its value is high enough, the denominator becomes small so that we can obtain the *oscillation condition*, which corresponds to an output field $E_t \neq 0$ even with $E_i \rightarrow 0$.

This is possible if

$$1 - r_1 r_2 e^{2ikl} e^{(\gamma-\beta)l} = 0, \tag{6.56}$$

giving the two conditions for the oscillating wave

$$r_1 r_2 e^{(\gamma-\beta)i} = 1 \quad \text{for the amplitude;} \tag{6.57}$$

$$e^{2ik_\cdot} = 1 \quad \text{for the phase.} \tag{6.58}$$

---

[4]Setting $x = r_1 r_2 e^{i2\tilde{k}l}$, the expression inside the square brackets is the geometrical series $S = \sum_0^\infty x^n = 1/(1-x)$.

The first condition provides the minimum (*threshold*) gain necessary for achieving the wave oscillation:

$$\gamma_{\text{th}} = \beta - \frac{1}{l}\ln(r_1 r_2). \qquad (6.59)$$

Since $\gamma = \sigma(\nu)\Delta N$, it corresponds to a threshold for the population inversion:

$$\Delta N_{\text{th}} = \frac{\beta - (1/l)\ln(r_1 r_2)}{\sigma(\nu)}, \qquad (6.60)$$

where

$$\sigma(\nu) = \frac{c^2}{8\pi n_0^2 \nu^2 \tau_{\text{sp}}} g(\nu).$$

The meaning of Eq. (6.59) is very clear: in order to keep oscillations, gain must balance losses, which are due to the limited reflectivity of mirrors ($r_1 r_2 < 1$) and to any other effect included in β (diffraction, scattering, absorption). Of course, higher losses require a higher threshold for the population inversion as given by Eq. (6.60).

The phase condition (6.58) provides the frequency of the waves that can produce stable oscillations:

$$kl = m\pi, \quad m = 1, 2, 3, \ldots. \qquad (6.61)$$

Since $k = n_0\omega/c$ or $k = n_0 2\pi\nu/c$, we get the possible oscillation frequencies, i.e. the cavity *longitudinal modes*:

$$\nu_m = m\frac{c}{2n_0 l}, \quad m = 1, 2, 3, \ldots. \qquad (6.62)$$

Here we notice that the possible oscillation frequencies are separated by $c/2n_0 l$, and therefore a longer cavity corresponds to closer longitudinal modes. Actually, the oscillation condition (6.58) should be slightly modified by taking into account the Gaussian shape of the beam, thus considering the longitudinal phase factor given by Eq. (6.50). This leads to the so-called *frequency-pulling effect* causing a small shift of the laser modes as compared to the passive cavity modes. This happens when the peak of the emission resonance does not coincide with a cavity mode; however, this change is often very small and so can be neglected.

We must emphasize that the conditions (6.59) and (6.62) must be fulfilled at the same time, which means that among the possible modes given by Eq. (6.62), only the ones realizing Eq. (6.59) will oscillate in the cavity. In the steady state, the oscillating mode keeps constant its amplitude after each round trip, and therefore the actual gain cannot exceed the threshold value γ_{th}. In fact, if $\gamma > \gamma_{th}$ we are not in the steady state and the energy in the cavity increases; however, owing to gain saturation, a lowering of the gain is induced, stabilizing it to a value close to γ_{th}.

6.5. Properties of laser radiation

We can summarize the onset of lasing as follows. Through energy *pumping* of an *active medium* in a *resonant cavity*, it is possible to get population inversion between two specific electronic levels with a characteristic line shape $g(\nu)$. After excitation, some electrons spontaneously decay to the lower energy state, thus emitting photons in the range of frequency given by $g(\nu)$. These photons produced by spontaneous emission can stimulate the emission of other photons from the same level, thus producing gain for the optical radiation. If the energy provided in pumping is high enough to achieve the threshold inversion for at least one cavity mode, we get laser oscillation and emission from the output mirror of a beam at the frequency of the oscillating mode and direction perpendicular to the mirror planes. In this section, we analyze the characteristics of such a laser beam.

Output power

We will consider the four-level system to analyze the emitted power. Since in this case $\Delta N \approx N_2$, we can write the power emitted due to stimulated emission as

$$P_e = (\Delta N_{th} V_m) h\nu_0 W_{21}^i, \tag{6.63}$$

with ν_0 being the frequency of the oscillating mode and V_m its volume in the active medium. Now we consider Eq. (6.13) in the approximation $\tau_2 \approx \tau_{sp}$ in such a way that at threshold we can write

$$\Delta N_{th} = \frac{\Delta N_0}{1 + \tau_2 W_{21}^i} \rightarrow W_{21}^i = \frac{1}{\tau_2}\left[\frac{\Delta N_0}{\Delta N_{th}} - 1\right]$$

$$= \frac{1}{\tau_2}\left[\frac{\gamma_0 l}{\beta l - \ln(r_1 r_2)} - 1\right], \tag{6.64}$$

and therefore

$$P_e = (\Delta N_{th} V_m)\frac{h\nu_0}{\tau_2}\left[\frac{\Delta N_0}{\Delta N_{th}} - 1\right], \tag{6.65}$$

which, neglecting the second term on the right-hand size of Eq. (6.12), can also be written as

$$P_e = (\Delta N_{\text{th}} V_m) \frac{h\nu_0}{\tau_2} \left[\frac{R}{R_{\text{th}}} - 1 \right], \tag{6.66}$$

where R represents the pumping rate above threshold and R_{th} is its value at threshold.

This expression shows the linear dependence of the emitted power on the pumping rate R. It is useful to compare P_e to the power emitted below threshold due to spontaneous emission in the oscillation mode. Since spontaneous emission is distributed over the whole line shape $g(\nu)$, the total power below threshold is spread into n_m modes in the volume V_m and falling into the line shape. Thus, we can write

$$n_m = p(\nu) V_m \Delta \nu, \tag{6.67}$$

with $p(\nu)$ being the number of modes per unit volume and unit frequency given by Eq. (6.34). Hence, the spontaneous emission power per mode is given by

$$P_b = (\Delta N_0 V_m) \frac{h\nu_0}{\tau_2} \frac{1}{n_m}. \tag{6.68}$$

Since the emitted power increases with ΔN_0 by increasing the pumping rate, by comparing Eq. (6.68) to Eq. (6.65) we see that the rise in the power emitted into the single mode below threshold is n_m times slower than above threshold. On the other hand, with values of wavelengths typical of the optical spectrum we have n_m in the range of $10^8 - 10^{10}$, which explains the sudden change of the slope of the emitted power above threshold. This plot is generally used as the fingerprint of the onset of laser oscillation and its typical behavior is shown in Fig. 6.12.

It is worth noting that above threshold the power output rise versus the pumping rate is due to the corresponding rise in the total number of emitters and not due to an increase in gain, since it is clamped at the threshold value γ_{th}. On the other hand, we expect saturation when all the available atoms in the volume are excited, obtaining the maximum absorption of the pumping energy.

However, in a practical laser system we must take into account that only a fraction of the total emitted power is exploited — that coming out of one semitransparent mirror. Then it is very important to optimize the reflectivity of such a mirror to get the maximum output from the

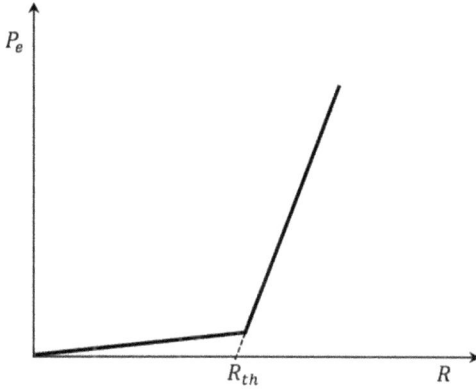

Fig. 6.12. The power emitted into a single laser oscillation mode below and above threshold versus the pumping rate.

laser source. For this aim, we can write the oscillation condition (6.57) for amplitude as

$$e^{\gamma_{th}l}(1-L) = 1, \tag{6.69}$$

by defining the fraction of energy losses for a single pass in the cavity as

$$L = 1 - r_1 r_2 e^{-\beta l}. \tag{6.70}$$

In the approximation $L \ll 1$, the conditon (6.69) reduces to[5]

$$\gamma_{th}l = L, \tag{6.71}$$

and using Eq. (6.20),

$$L = \frac{\gamma_0 l}{1 + (I/I_s)} = \frac{g_0}{1 + (I/I_s)}, \tag{6.72}$$

where $g_0 = \gamma_0 l$ is the single-pass gain.

Then we can write

$$I = I_s \left(\frac{g_0}{L} - 1\right). \tag{6.73}$$

With power being proportional to intensity, we can write

$$P = P_s \left(\frac{g_0}{L} - 1\right). \tag{6.74}$$

---

[5] $e^{\gamma_{th}l}(1-L) = 1 \rightarrow \ln[e^{\gamma_{th}l}] + \ln[1-L] = 0 \rightarrow \gamma_{th}l + \ln[1-L] = 0$, if $L \ll 1 \rightarrow \ln[1-L] \approx -L$.

The "useful" power is that transmitted by the output mirror, and therefore it is useful to separate the internal losses L_i from the ones due to the mirror transmittance T:

$$L = L_i + T. \tag{6.75}$$

Then the output power becomes

$$P_{\text{out}} = TP = P_s \left(\frac{g_0}{L_i + T} - 1 \right) T. \tag{6.76}$$

This expression allows optimizing the coefficient T in order to get the maximum output power given the internal losses L_i and the single-pass gain g_0. This can be done by setting $\partial P_{\text{out}}/\partial T = 0$,

$$\frac{\partial P_{\text{out}}}{\partial T} = \frac{\partial}{\partial T} \left[P_s \left(\frac{g_0}{L_i + T} - 1 \right) T \right] = 0, \tag{6.77}$$

which can easily be calculated to give the optimum transmittivity,[6]

$$T_{\text{opt}} = -L_i + \sqrt{g_0 L_i}, \tag{6.78}$$

leading to the optimum output power:

$$P_{\text{opt}} = P_s (\sqrt{g_0} - \sqrt{L_i})^2. \tag{6.79}$$

The typical behavior given by Eq. (6.76) is shown in Fig. 6.13. It leads to a maximum given by Eq. (6.79) dependent on the actual values of g_0 and L_i.

Emission spectrum

The emission spectrum of a laser is strongly affected by the characteristics of the line shape $g(\nu)$ — particularly by its homogeneous or inhomogeneous origin, as described in Section 5.4. This concept is explained by Figs. 6.14 and 6.15, showing the onset of laser oscillation occurring when the conditions (6.57) and (6.58) are fulfilled. Therefore, emission will take place for the resonant modes of the cavity achieving the threshold gain. Considering first a homogeneous broadening system, Fig. 6.14 reports the gain function $\gamma(\nu)$ versus the frequency ν for different pumping rate R_2. Since $\gamma(\nu)$ is proportional to $\Delta N \sim R_2 \tau_2$, gain increases with R_2. In the figure, the dashed horizontal line represents the threshold value γ_{th}.

---

[6] $\frac{\partial P_{\text{out}}}{\partial T} = \frac{\partial}{\partial T} \left[P_s \left(\frac{g_0}{L_i+T} - 1 \right) T \right] = T \frac{\partial}{\partial T} \left[P_s \left(\frac{g_0}{L_i+T} - 1 \right) \right] + P_s \left(\frac{g_0}{L_i+T} - 1 \right) \frac{\partial T}{\partial T} = 0;$

then $T \frac{-g_0}{(L_i+T)^2} + \frac{g_0}{L_i+T} - 1 = 0 \rightarrow -g_0 T + g_0(L_i+T) - (L_i+T)^2 \rightarrow g_0 L_i - (L_i+T)^2 = 0.$

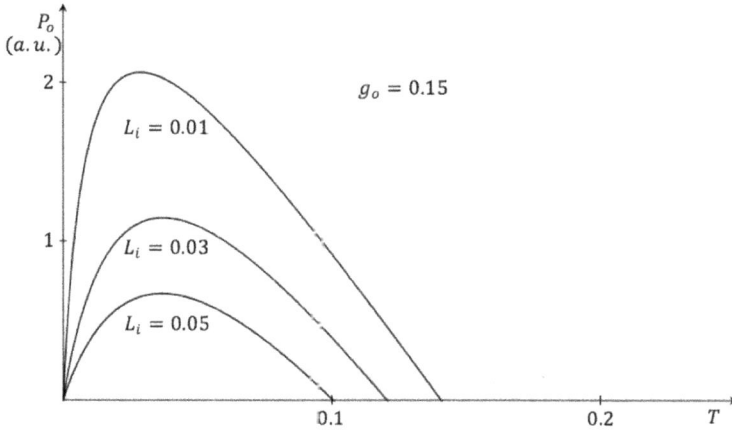

Fig. 6.13. Laser output power versus the output mirror transmittivity for different values of g_0 and L_i.

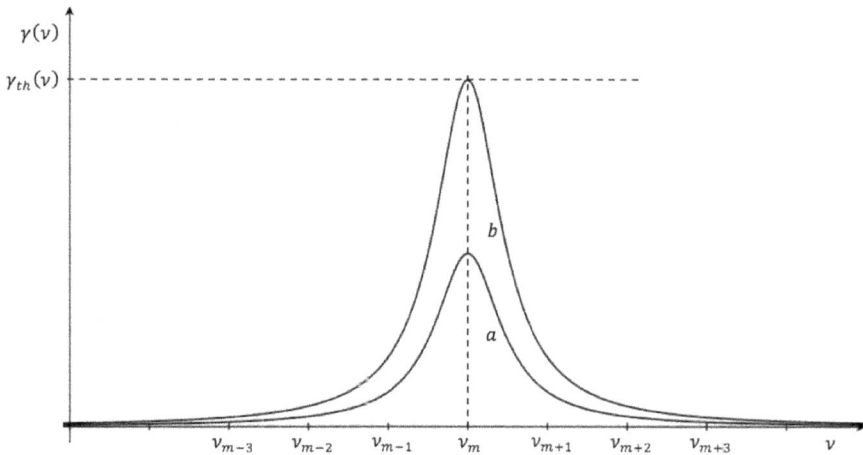

Fig. 6.14. Plot of $\gamma(\nu)$ versus ν for an increasing pumping rate ($b > a$) for a homogeneously broadened system. On the horizontal coordinate, the values of the cavity longitudinal modes are indicated. The horizontal dashed line corresponds to the threshold value γ_{th}. The vertical dashed line represents the cavity mode ν_m achieving the oscillation condition.

As soon as the oscillation condition is achieved by a cavity mode, owing to the homogeneous character of line shape broadening all the excited electrons can undergo stimulated emission at that frequency and we get the ideal *single-mode emission*. A further increase in the pumping rate

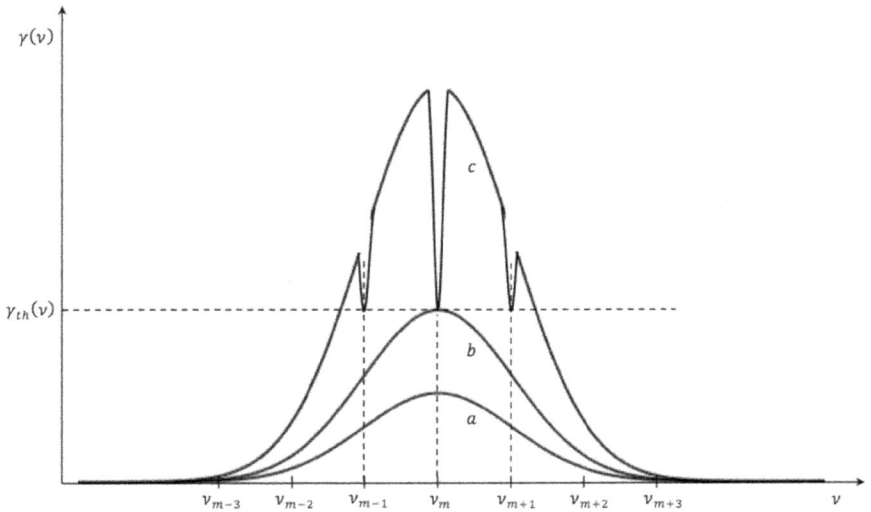

Fig. 6.15.   Plot of $\gamma(\nu)$ versus ν for an increasing pumping rate for an inhomogeneously broadened system. On the horizontal coordinate, the values of the frequency of the cavity longitudinal modes are indicated. The horizontal dashed line corresponds to the threshold value γ_{th}. The vertical dashed lines represent the cavity modes achieving the oscillation condition. The curves a, b and c correspond to increasing pumping rate. As soon as the threshold gain is achieved, the gain of the oscillating modes is clamped at this value.

does not modify the emission spectrum, since the gain is clamped at the threshold value and all the stored energy is given to the oscillating mode. As already outlined, an increase in the pumping power provides a linear increase in the output power owing to the consequent increase in excited atoms or molecules in the mode volume until saturation occurs.

The case of the inhomogeneously broadened system is shown in Fig. 6.15. In this case, each emitted frequency corresponds to different atoms/molecules and therefore as soon as one cavity mode achieves the threshold condition its gain is clamped at the threshold value, but the gain of other modes can increase further if the pumping rate is increased and they can eventually reach the oscillating condition. In other words, when the first mode oscillates, not all the stored energy is channeled into the stimulated emission at this frequency and other modes can achieve the oscillation condition when increasing the pumping energy. In this case, the gain curve $\gamma(\nu)$ is clamped only at the frequencies of the oscillating modes and can rise if additional pumping energy is provided. As a consequence we can have *multimode emission*.

The single-mode emission in a homogeneously broadened system can lead to an extremely narrow emission line. In order to make clear this point, we should evaluate the spectral width of the cavity mode. This is done by exploiting the definition of the quality factor Q of the resonator:

$$Q = \frac{\nu_0}{\delta\nu}, \tag{6.80}$$

with ν_0 being the frequency of the resonant mode and $\delta\nu$ its spectral width. That is to say, a high quality factor means a narrow linewidth $\delta\nu$. It is useful to connect it to the cavity losses using the definition

$$Q = 2\pi\nu_0 \frac{\mathcal{E}}{-(d\mathcal{E}/dt)}, \tag{6.81}$$

where $\mathcal{E}$ is the energy stored in the cavity at the resonance frequency ν_0 and $-d\mathcal{E}/dt$ is the power decay at the same frequency, resulting in an exponential decay of energy in the cavity in the absence of any input:

$$\mathcal{E}(t) = \mathcal{E}(0)e^{-(t/t_c)}. \tag{6.82}$$

Then

$$-\frac{d\mathcal{E}(t)}{dt} = \frac{\mathcal{E}(0)e^{-(t/t_c)}}{t_c} = \frac{\mathcal{E}(t)}{t_c}. \tag{6.83}$$

Therefore, Eq. (6.81) can be written as

$$Q = 2\pi\nu_0 t_c, \tag{6.84}$$

and using Eq. (6.80),

$$\delta\nu = \frac{1}{2\pi t_c}. \tag{6.85}$$

This expression shows the link between the mode linewidth and the energy decay time t_c of the cavity (*photon lifetime in the cavity*): the slower the energy decay from the cavity, the narrower the resonance linewidth (and the higher the cavity quality factor Q).

Of course, the photon lifetime t_c is linked to the fraction of the cavity energy losses per pass L given by (Eq. 6.70) or to the *round trip* losses given by

$$L_{\mathrm{rt}} = 1 - R_1 R_2 e^{-2\beta l}, \tag{6.86}$$

with $R_1 = (r_1)^2$ and $R_2 = (r_2)^2$ being the mirrors' reflectivity.

In the approximation $L_{rt} \ll 1$, we get[7] the single-pass fractional losses as

$$L = \beta l - \ln(\sqrt{R_1 R_2}). \tag{6.87}$$

In this way, we can write the fractional energy loss per second,

$$\frac{-(d\mathcal{E}/dt)}{\mathcal{E}} = \frac{L}{(nl/c)}, \tag{6.88}$$

since the right-hand-side term is the ratio between the single-pass fractional losses and the time nl/c necessary for a single pass of the radiation in the cavity. Then

$$\frac{d\mathcal{E}}{dt} = -\frac{cL}{nl}\mathcal{E}, \tag{6.89}$$

and comparing it to Eq. (6.83),

$$t_c = \frac{nl}{cL}. \tag{6.90}$$

In this way, Eq. (6.85) becomes

$$\delta\nu = \frac{cL}{2\pi nl} = \frac{c[\beta l - \ln(\sqrt{R_1 R_2})]}{2\pi nl}, \tag{6.91}$$

showing the direct connections between losses in the cavity and the oscillating mode linewidth.

In evaluating $\delta\nu$ using Eq. (6.91), we are considering just the passive optical resonator. In other words, we have not taken into account the presence of a gain medium in the cavity and the achievement of the lasing threshold for the considered longitudinal mode. A more detailed analysis that takes into account the noise (spontaneous emission) in a laser oscillator allows finding an expression linking the actual emission linewidth to the power emitted into the radiation field, which can be summarized as

$$\delta\nu_{\text{laser}} = K\frac{(\delta\nu)^2}{P_e}, \tag{6.92}$$

and a similar expression below threshold, with K being a parameter including the transition frequency and the populations of the levels involved in the transition. Since above the oscillation threshold we find a huge

---

[7]In fact using $x = R_1 R_2 e^{-2\beta l} \approx 1$, we have $-\ln(x) = \ln(1/x) \approx 1 - x$, and taking into account that $\ln(x)^2 = 2\ln(x)$, we can write $L_{rt} = 1 - R_1 R_2 e^{-2\beta l} = -\ln(R_1 R_2 e^{-2\beta l}) = -2\ln(r_1 r_2 e^{-\beta l}) \approx 2(1 - r_1 r_2 e^{-\beta l}) = 2L$; then $L = -\ln(r_1 r_2 e^{-\beta l}) = \beta l - \ln(\sqrt{R_1 R_2})$.

increase in the emitted power P_e, the lasing action leads to a further reduction of the emission linewidth. Details on this effect can be found in more specialized books on quantum electronics.

As a final remark, we should underline that the linewidth given by Eqs. (6.91) and (6.92) represents the best result that can be obtained from an optical resonator. In fact, the actual emission spectrum is affected by the mechanical and thermal stability of the cavity: vibrations and change of temperature may slightly modify the cavity length, thus "moving" the position of the resonator mode along the frequency axis, and therefore the average emission spectrum will be wider than the ideal linewidth of the cavity resonance. Temperature may also affect the refractive index of the gain medium, contributing to a shift of the mode frequency.

In any case, the emission spectrum of a laser source is very narrow with respect to that of a conventional white source like a common lamp; this means that, according to the expression (1.155), the electromagnetic waves coming out from a laser source have a coherence time much longer than that of a conventional white source (i.e. *high temporal coherence*). Such a property is evident at a glance by the appearance of the sharp color (small $\delta\nu$) of a typical laser beam (monochromatic).

Beam divergence

The spatial counterpart of the monochromatic laser beam is its low spatial divergence. This property is due to the combination of two effects: the process of stimulated emission leading to photons emitted with the same wave vector of the stimulating ones, and the characteristics of the open optical resonator allowing stable oscillations within a limited solid angle. Actually, we have worked out the oscillation conditions for a Fabry–Perot resonator by considering a high number of reflections on the cavity mirrors ($\to \infty$), thus considering waves traveling with the wave vector perpendicular to the mirror planes. Therefore, the oscillation conditions apply only to waves with a single value of the wave vector (that would mean a perfect plane wave). On the other hand, at the output mirror the wave front is limited by the finite mirror size, and thus diffraction occurs at the mirror edge. As a consequence, from a plane wave inside the cavity we should expect an output beam with a divergence angle due to diffraction that can be calculated using the expression (1.171). However, as outlined in Section 6.3, a more detailed description of the field is given by the Gaussian beam; in this case, we can use the expression (6.49) for the beam divergence outside the resonator, which actually gives a value very close to the one provided by Eq. (1.171). In this way, we usually approximate the divergence angle of a laser beam as

$$\alpha \approx \frac{\lambda}{\pi w_M}, \tag{6.93}$$

with w_M being the mirror transversal dimension (the radius in the common case of a circular mirror). A typical value for the common laser is $\alpha \approx 10^{-3}$ rad. When the laser divergence reaches the minimum value given by Eq. (6.93), we speak of a *diffraction-limited* beam. This is achieved when a Gaussian wave is oscillating in the resonator. Short laser cavities, scattering effects and mechanical instabilities may contribute to generating broader divergence of laser beams.

Such low divergence is expressed by a wave vector $\mathbf{k}$ pointing in a definite direction (high directionality) with a small transversal spread $\delta\mathbf{k}$, as sketched in Fig. 6.16. If we take the main propagation direction as the $\mathbf{z}$ axis, $\mathbf{k} = k_z\hat{\mathbf{z}} + \delta k_x\hat{\mathbf{x}} + \delta k_y\hat{\mathbf{y}}$, such that

$$\alpha = \frac{\delta k_{x,y}}{k_z}. \tag{6.94}$$

This low divergence is the evidence of the *high spatial coherence* typical of most laser beams, which is probably the most important property of the laser for applications. In fact, it not only makes available a reference line that can travel long distances and enable precise length measurements, but also allows focusing light on extremely small surfaces. By considering a Gaussian beam, the minimum radius w_f of the focal spot obtained by a lens with focal length f can be calculated to be

$$w_f \approx \frac{2\lambda}{\pi}\frac{f}{2w_a} = \frac{2\lambda}{\pi}f^\#, \tag{6.95}$$

where w_a is the radius of the aperture of the lens. The quantity $f^\#$ is called the *f number*, defined as the ratio between the focal length and the diameter of the aperture of the lens.

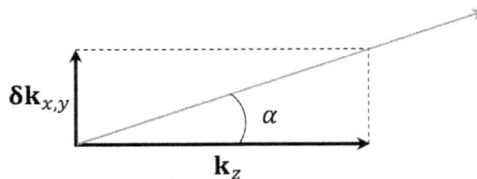

Fig. 6.16.   The relationship between the wave vector components and the beam divergence.

Therefore, the higher w_a is (a wider and uniform wave front means a quasi-plane wave), the smaller w_f will be. The ability to reach w_f values of the order of $1\,\mu$m is the basis of the present technology of optical storage on DVD.

Finally, a small focal spot means that high intensity is easily achievable. In fact, as pointed out in Section 6.3, for a given optical power P the light intensity on this spot is

$$I_0 = \frac{2P}{\pi w_f^2}. \tag{6.96}$$

Then the possibility of focusing on extremely small spots allows getting very high intensity. With a high-power laser it is possible to get intensities able to produce atomic melting!

6.6. Special resonators for additional laser features

Starting from the basic Fabry–Perot laser cavity, the resonator design can be modified to provide additional features to the emitted radiation. We briefly discuss here the principles and methods for obtaining (i) tunable emission frequency; (ii) Q-switching of the laser cavity to get high-power nanosecond pulses (10^{-9} s); (iii) mode-locking of oscillating modes to get repetitive emission of pulses in the picosecond (10^{-12} s) or femtosecond (10^{-15} s) regime.

Tunable lasers

Many applications may require different laser wavelengths to be used. Of course, this can be done by choosing the appropriate laser medium to get the right light emission, since we have several different materials able to produce laser radiation and for each part of the optical spectrum we can find a good one. However, it is obviously easier and cheaper to have a single laser source able to provide a wavelength to be selected in a quite broad range. In order to design a laser having this property, we need the following:

(a) An active medium with wide $\gamma(\nu)$ given by homogeneous broadening;
(b) The ability to modulate the threshold gain as a function of the emission frequency in order to have a low threshold value γ_{th} only in a narrow range of frequencies allowing laser emission only in this range;
(c) The ability to control the frequency position of γ_{th} in order to move the frequency range of the emission.

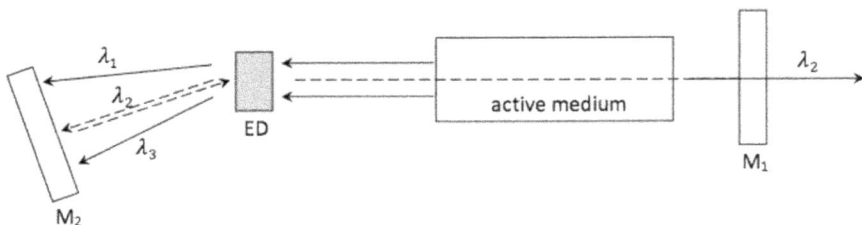

Fig. 6.17.   Scheme of laser-tuning using a dispersive device (ED) in the cavity.

Broad emission bandwidth is required in order to have a broad range of tunability, and homogeneous broadening is required in order to have efficient exploitation of pumping, thus channeling the excitation energy into the narrow range of emission.

The functionalities (b) and (c) can be obtained by including in the cavity a dispersive device able to have a cavity quality factor $Q(\nu)$ high only in a narrow range of frequencies, which can be shifted inside the emission bandwidth.

A simple scheme for laser-tuning is shown in Fig. 6.17, where the dispersive device ED acts just as a ray deflector with the deflection angle dependent on frequency. In this way, most of the waves transmitted by ED reach the back mirror at a slant angle and are not able to oscillate in the cavity; losses for these waves are very high and the corresponding quality factor Q is very low. Only waves within a narrow range of frequencies have a propagation direction perpendicular to the back mirror plane and can fulfill the oscillation condition.

Tuning is achieved by a rotation of the dispersive medium or of the back mirror realizing the oscillation condition for a different range of frequencies.

The tuning concept is sketched in Fig. 6.18. The narrow frequency range with low γ_{th} can be shifted within the emission line shape.

One of the most-used dispersive device is the diffraction grating, which is employed in the reflection mode, replacing the back mirror of the laser as shown in Fig. 6.19.

Q-*switching*

We have pointed out that in the steady state the oscillation gain is clamped at the threshold value γ_{th} and an increase in emission power is possible until all the available volume of the active medium is excited, remaining anyway related to the threshold inversion ΔN_{th}. In order to increase the emission

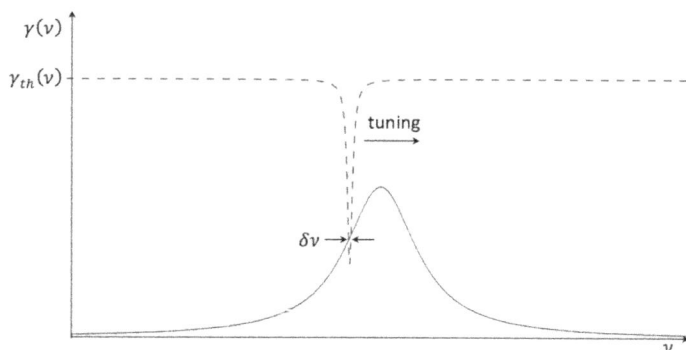

Fig. 6.18. Sketch of the concept of laser-tuning. In this figure, the cavity modes are supposed to be very close to each other and cannot be observed.

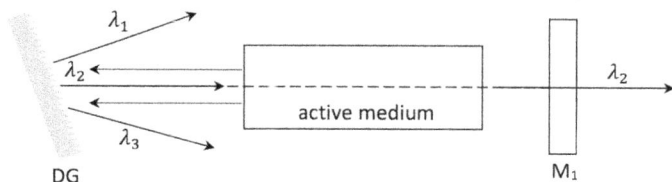

Fig. 6.19. Scheme of laser-tuning using a diffraction grating (DG) as the back mirror.

power, it would be necessary to increase the population inversion over this value, which is not possible under steady state conditions. To overcome this difficulty, the basic idea is to produce a sudden change of the cavity quality factor Q, called Q-switching. The working principle is based on the following concept. While the active medium is being pumped, the quality factor Q of the cavity is kept at a very low value; in this way, energy can be stored by the system and oscillations are prevented. When a large population inversion is achieved, the quality factor is suddenly increased, actually establishing a good optical resonator. For this cavity, the system is now much over the threshold condition ($\Delta N \gg \Delta N_{th}$); therefore, one obtains a fast buildup of laser oscillation. This excess of energy converted into photons comes out of the cavity with the characteristic photon lifetime t_c given by Eq. (6.90), which is typically in the range of 5–50 ns. Therefore, all the stored energy is emitted as an optical pulse of short duration with a typical power of the order of 10–100 MW per pulse. Of course, in order to realize efficient Q-switching it is necessary to have a switching time t_s shorter than the lifetime of the excited level τ_2 ($t_s \ll \tau_2$).

Details on this process (to be found in specialized books) allow writing the peak power of the optical pulse as

$$P_{\text{peak}} = \frac{h\nu}{2t_c} V_m \left[\Delta N_{\text{th}} \ln \left(\frac{\Delta N_{\text{th}}}{\Delta N_i} \right) - (\Delta N_{\text{th}} - \Delta N_i) \right], \qquad (6.97)$$

where ΔN_i is the initial inversion at the Q-switching. Since $\Delta N_i \gg \Delta N_{\text{th}}$, Eq. (6.97) can be approximated as

$$P_{\text{peak}} \approx \frac{h\nu}{2t_c} \Delta N_i V_m. \qquad (6.98)$$

The time dependence of the population inversion and photon density in the laser emission after Q-switching is shown in Fig. 6.20.

Different methods can be used to get Q-switching. The most practical one consists in including inside the cavity an electro-optical shutter to get a fast switching from a high-loss to a low-loss cavity. The shutter is made up of an electro-optical crystal coupled to a polarizing element; a possible configuration is shown in Fig. 6.21.

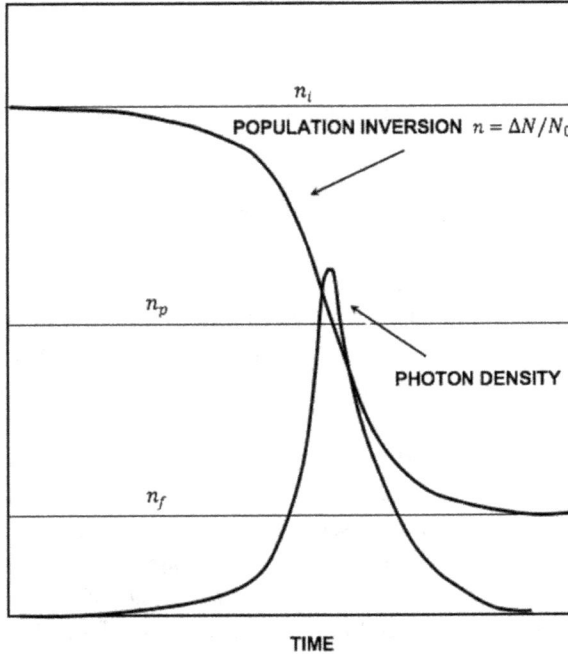

Fig. 6.20. Population inversion and photon density versus time for a Q-switched laser. Here $n_p = \Delta N_{\text{th}}/N_0$, the population inversion at threshold, while n_i and n_f are its initial and final values. [Reproduced and adapted from: W.G. Wagner *et al.*, *J. Appl. Phys.* **34**, 2040 (1963), with the permission of AIP Publishing.]

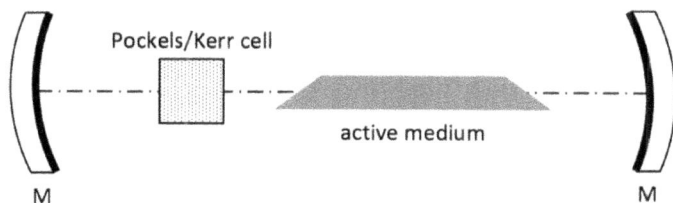

Fig. 6.21. Scheme of a Q-switched laser based on an electro-optical shutter.

During pumping, a voltage is applied to the electro-optical crystal in order to introduce a polarization rotation of $\pi/4$. After reflection on the back mirror, a second pass through the crystal induces an additional rotation of $\pi/4$, leading to a final polarization rotated by $\pi/2$ and then crossed with respect to the polarizer. In this way, radiation is blocked and cannot "see" the second mirror, and cavity losses are high (low Q). When population inversion reaches a large value, the system is synchronized to bring to zero the voltage applied to the crystal, thus reducing to zero the optical rotation. The quality factor Q becomes suddenly high and Q-switching is realized.

Mode-locking

The range of applications of very short laser pulses is very broad, as it includes communications, investigation of ultrafast phenomena (for example related to biomolecules), and ablation or induced structural changes of materials. The method of mode-locking allows achieving light pulses as short as a few femtoseconds (10^{-15} s); in this extreme case, optical waves have a length of a few micrometers and are therefore very far from the representation of an ideal plane wave, opening a completely new branch in the investigation of light–matter interaction, since some of the basic approximations used when one is dealing with longer wave packets are no longer valid. Nevertheless, in the following we still describe the field as a plane wave, neglecting the case of extremely short pulses.

The concept of mode-locking is easily understood if we consider multimode oscillation in a laser cavity. In this case, the total field due to superposition of N oscillating modes (considering linear polarization to simplify the mathematics) can be written as

$$E(t) = \sum_{-(N-1)/2}^{(N-1)/2} E_n e^{-i(\omega_n t + \phi_n)}, \tag{6.99}$$

where E_n, ω_n and ϕ_n are the amplitude, angular frequency and phase of the nth longitudinal mode. Taking $E_n = E_0$ for all the modes, when their phases are randomly distributed there is no interference among them and the total power is proportional to $|E(t)|^2$, being just the sum of the power of the N modes:

$$P \propto |E(t)|^2 = NE_0^2. \tag{6.100}$$

On the contrary, let us suppose the phases of all the modes to be fixed: $\phi_n = \phi_0$. According to Eq. (6.62), the frequency difference between the modes is

$$\delta\omega = \frac{\pi c}{n_0 l}. \tag{6.101}$$

We can define

$$\omega_n = \omega_0 + n\delta\omega, \tag{6.102}$$

where ω_0 is the frequency of the central mode; then Eq. (6.99) becomes

$$E(t) = E_0 e^{-i\phi_0} \sum_{-(N-1)/2}^{(N-1)/2} e^{-i(\omega_0+n\delta\omega)t}$$

$$= E_0 e^{-i\phi_0} e^{-i\omega_0 t} \sum_{-(N-1)/2}^{(N-1)/2} e^{-in\delta\omega t}. \tag{6.103}$$

By calculating the sum, we have

$$E(t) = E_0 e^{-i\phi_0} e^{-i\omega_0 t} \frac{\sin(N\delta\omega t/2)}{\sin(\delta\omega t/2)}. \tag{6.104}$$

The sum in Eq. (6.103) can be easily calculated by setting

$$x = e^{-i\delta\omega t}, \quad k = \frac{N-1}{2} = \frac{N}{2} - \frac{1}{2}.$$

In this way, we can write

$$S = \sum_{-(N-1)/2}^{(N-1)/2} e^{-in\delta\omega t} = x^{-k} + \cdots + x^k,$$

and multiplying by x,

$$xS = x^{-k+1} + \cdots + x^{k+1}.$$

By subtracting the last two expressions, we have

$$(1 - x)S = x^{-k} - x^{k+1}.$$

Then

$$S = \frac{x^{-k} - x^{k+1}}{1 - x} = \frac{x^{k+1/2} - x^{-(k+1/2)}}{x^{1/2} - x^{-1/2}} = \frac{x^{N/2} - x^{-N/2}}{x^{1/2} - x^{-1/2}},$$

and taking into account the definition of x,

$$S = \frac{e^{-iN\delta\omega t/2} - e^{iN\delta\omega t/2}}{e^{-i\delta\omega t/2} - e^{i\delta\omega t/2}} = \frac{\sin(N\delta\omega t/2)}{\sin(\delta\omega t/2)},$$

as shown in Eq. (6.104).

Finally, we get the total power as

$$P = K\left| E_0 e^{-i\phi_0} e^{-i\omega_0 t} \frac{\sin(N\delta\omega t/2)}{\sin(\delta\omega t/2)} \right|^2 = K E_0^2 \frac{\sin^2(N\delta\omega t/2)}{\sin^2(\delta\omega t/2)}, \qquad (6.105)$$

with K being a constant depending on the used units. This function is plotted in Fig. 6.22 versus $\delta\omega t/2$ for $N = 10$ and $N = 20$.[8] We can see that

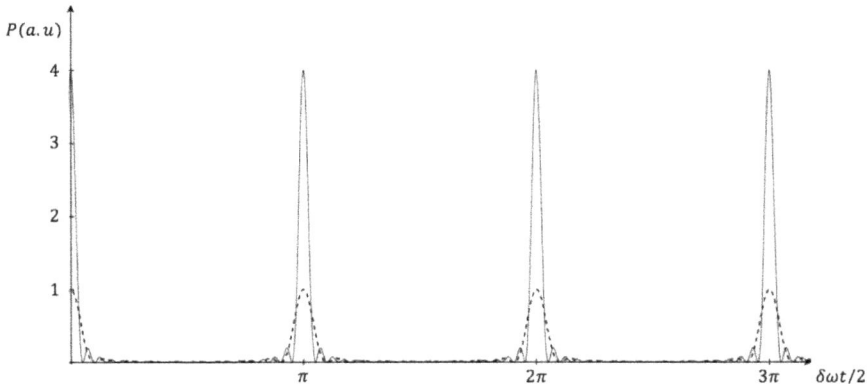

Fig. 6.22. The power of laser pulses originated by mode-locking versus the normalized time $\delta\omega t/2$. Full line — $N = 20$; dashed line — $N = 10$.

[8]Usually the number of locked modes is much higher, while their numbers for the plot of Fig. 6.22 have been chosen to make easily visible the shape of the function (6.105).

power is emitted as a train of pulses with a period,

$$T = \frac{2\pi}{\delta\omega} = \frac{2n_0 l}{c}, \tag{6.106}$$

which is the time necessary for a round trip of the laser cavity. From the reported example, we see that by increasing the number of locked modes the peak power increases and the pulse width decreases.

We can look at the normalized power,

$$\frac{P}{KE_0^2} = \frac{\sin^2(N\delta\omega t/2)}{\sin^2(\delta\omega t/2)}, \tag{6.107}$$

to point out the dependence of the peak power on the number N of locked modes.

By calculating Eq. (6.107) for the peak occurring at $\delta\omega t/2 = 0$, we get

$$\lim_{\delta\omega t/2 \to 0} \frac{\sin^2(N\delta\omega t/2)}{\sin^2(\delta\omega t/2)} = N^2. \tag{6.108}$$

Hence, the locking of N modes leads to pulses with the peak power dependent on N^2. This is clearly seen in Fig. 6.22, where doubling the number of locked modes leads to a four-times-higher peak. It can also be observed that the number of locked modes decreases the pulse duration; in fact, according to Eq. (6.107), the full width half maximum of each peak can be calculated as

$$\tau = \frac{T}{N} = \frac{2\pi}{N\delta\omega}. \tag{6.109}$$

Since the number of modes that can be locked can be estimated as the ratio between the transition line shape $\Delta\omega$ and the mode separation $\delta\omega$, $N = \Delta\omega/\delta\omega = \Delta\nu/\delta\nu$, the minimum pulse duration can be achieved when all the available modes are locked together,

$$\tau \approx \frac{2\pi}{N\delta\omega} = \frac{2\pi}{\Delta\omega} = \frac{1}{\Delta\nu}, \tag{6.110}$$

which is approximately the inverse of the gain linewidth.

The result of the mode-locking is the generation of a light pulse traveling in the cavity back and forth. A fraction of its power comes out of the output mirror after each round trip in such a way that the output light beam is made up of a periodic train of pulses, whose time distance corresponds to the time necessary for a round trip of the laser cavity.

The pulse period given by Eq. (6.106) is usually in the range of a few nanoseconds, while the minimum pulse duration given by (6.110) has a

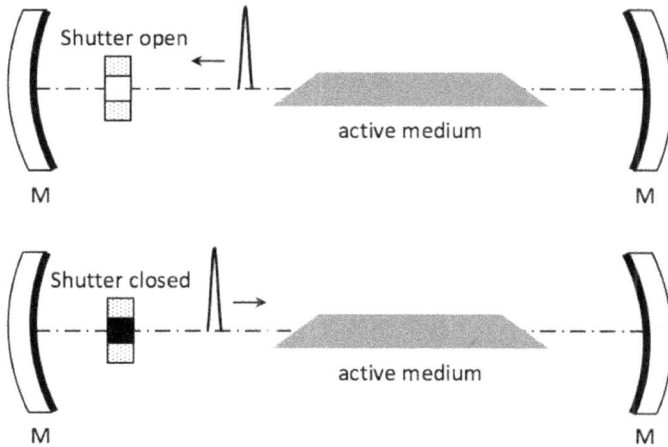

Fig. 6.23.   Scheme of a mode-locked laser with a longitudinal cavity and optical shutter.

broad range of values depending on the available gain bandwidth and goes from a few hundred picoseconds to a few femtoseconds (10^{-10}–10^{-14}s).

Mode-locking can be realized in a Fabry–Perot cavity using an electro-optical shutter close to one of the cavity mirrors in a configuration similar to that described above for Q-switching. When the shutter is open, all the oscillating modes have the maximum amplitude and are thus phase-locked. It should stay open for a very short time and switch on again after a time T corresponding to the pulse round trip in the cavity. The scheme for getting mode-locking is shown in Fig. 6.23.

Another scheme for producing ultrashort pulses is colliding pulse mode-locking (CPM), shown in Fig. 6.24. The basic configuration is that of a ring cavity where a saturable absorber is included and precisely located in a position that is reached at the same time by two opposite traveling pulses. The absorber works as a shutter, producing high losses when a single pulse is impinging on it (shutter closed) and saturating the absorption when two pulses get there at the same time (shutter open). The result can be the stable emission of a train of extremely short pulses.

6.7. Overview of laser systems

In this section, we will briefly present some of the most-used laser sources with the aim of showing how the concepts discussed in this chapter can find practical application in real devices. Nevertheless, we are not going to

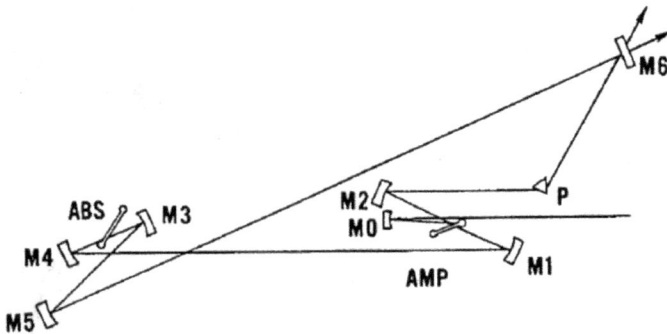

Fig. 6.24. Scheme of a ring cavity realizing mode-locking by the colliding pulse technique. M0 is the mirror focusing the pump beam on the dye jet acting as a gain medium (AMP); mirrors M1–M6 realize the ring laser cavity; ABS is the dye jet acting as an absorber, which induces the mode-locking; P is a prism used to control intracavity dispersion. [From J.-C.M. Diels *et al.*, *App. Opt.* **24**, 1270 (1985).]

discuss fabrication details, but only some main aspects of laser materials and cavities.

Solid state lasers

We start from the solid state laser, because the first laser was realized by Theodore Maiman using as active medium a crystal of ruby with emission at $\lambda = 694.3$ nm. The optical cavity was a simple Fabry–Perot resonator realized by coating reflecting mirrors on the base surface of a crystal having the shape of a cylinder and using a flash lamp for optical pumping of the system.

The medium consists of Al_2O_3 crystal with inclusion of Cr_3 impurities that give the characteristic red color of ruby. In this case, the energy level distribution corresponds to a three-level system where the final level involved in lasing is the ground state. For this reason, the pumping threshold is higher than in other solid state systems based on four levels; however, the extremely long lifetime of the excited state ($\sim$3 ms) makes the system interesting for Q-switching, which is the most-used configuration for a ruby laser.

The energy level diagram of interest for the ruby laser is shown in Fig. 6.25. In this diagram, as in the other shown in this section, the electronic levels are indicated with their typical spectroscopic denomination. Pumping occurs to the upper band $^4F_2$, then a fast relaxation from level 3 to level 2 allows getting population inversion between level ^2E and the

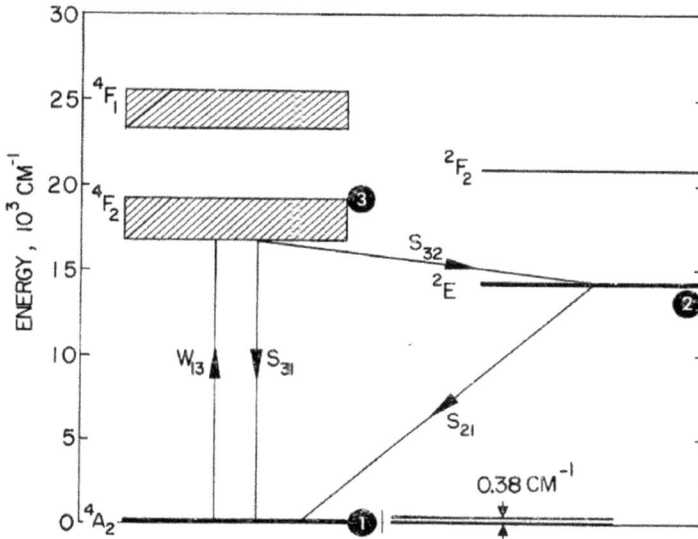

Fig. 6.25. Energy levels involved in the operation of a ruby laser. The numbers inside the black circles indicate the levels of interest for the laser process [Ref.: T.H. Maiman, *Phys. Rev. Lett.* **4**, 564 (1960); copyright (1960) by the American Physical Society.]

ground state, and after achieving the threshold condition a red beam at 694.3 nm comes out of the output mirror.

One of the most widely used solid state lasers exploits neodymium ions (Nd^{+3}) as impurities in different kinds of solid hosts: yttrium–aluminum garnet ($Y_3Al_5O_{12}$), yttrium vanadate (YVO_4), lithium–yttrium fluoride ($LiYF_4$) and glass. The laser materials are indicated as Nd:YAG, Nd:YVO_4, Nd:YLF and Nd:glass, respectively.

In this case, we have an efficient four-level system that is suitable for both continuous wave (CW) and pulsed operation. Optical pumping is realized either by flash lamp or by diode array, depending on the system and cavity, which can be a basic Fabry–Perot resonator or can include additional optical elements and devices. In this way, CW, Q-switched and mode-locked operation are possible. The emission wavelength depends on the host, but it is always in the near-IR region, being at $\lambda = 1064$ nm for the Nd:YAG laser. The scheme of energy levels for such a laser is depicted in Fig. 6.26.

The emission of ruby and neodymium lasers is characterized by a narrow $g(\nu)$, and thus single-mode oscillation is easily obtained, while laser tunability is not affordable. This can be achieved starting from a broad line shape, as in alexandrite (Cr^{+3} $BeAl_2O_4$) and titanium sapphire (Ti:Al_2O_3)

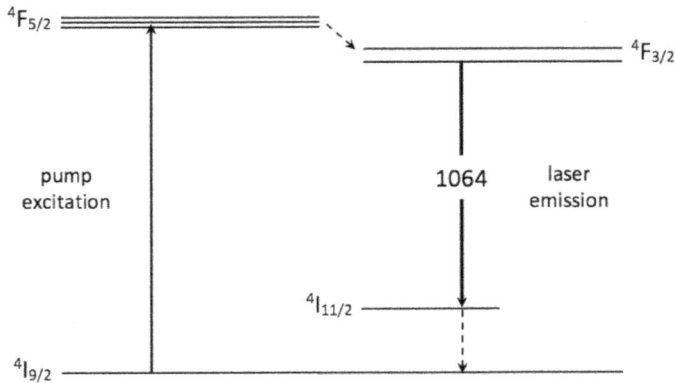

Fig. 6.26. Energy levels involved in the operation of a Nd:YAG laser. The wavelength (nm) of the laser emission is indicated.

Table 6.1. Basic characteristics and typical operation regimes of some common solid state lasers.

Active medium	pumping	Regime [CW/pulsed (s)]	λ (nm)
Ruby ($Cr^{+3}:Al_2O_3$)	Optical	10^{-8}	6943
Nd:YAG ($Nd:Y_3Al_5O_{12}$)	Optical	$CW/10^{-8}/10^{-12}$	1064
Nd:glass (silica glass)	Optical	$10^{-8}/10^{-12}$	1062
Nd:YVO$_4$	Optical	$CW/10^{-12}$	1064
Alexandrite ($Cr^{+3}:BeAl_2O_4$)	Optical	$CW/10^{-7}-10^{-10}$	700–820
Titanium sapphire ($Ti:Al_2O_3$)	Optical	$CW/10^{-14}$	660–1100

lasers. The former allows lasing in the range of 700–820 nm and the latter 660–1180 nm. Owing to the extremely broad emission bandwidth, the titanium sapphire laser is suitable for getting mode-locking with ultrashort pulse generation. In fact, the broad band allows phase-locking of a large number of longitudinal laser modes, leading to ultrashort laser pulses, in agreement with Eq. (6.110).

In Table 6.1, some characteristics of solid state lasers are summarized. Emission power or energy/pulse spans a broad range, depending on the specific laser configuration.

Gas lasers

The simple scheme of a gas laser is depicted in Fig. 6.27. In this case, pumping is realized by electric discharge in the gas tube, whose window are

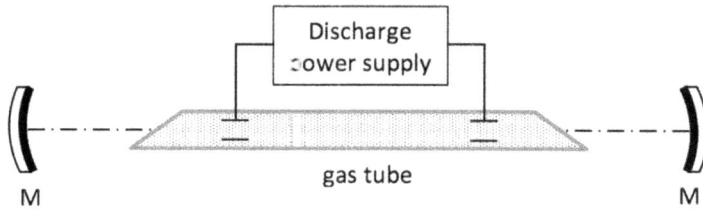

Fig. 6.27. Typical scheme of a gas laser.

usually mounted at a Brewster angle with respect to the optic axis in order
to get minimum reflection losses for one light polarization.

One of the most common gas lasers is the He–Ne laser. In this case,
the active medium is a mixture of helium and neon. The electrons of the
discharge current excite the helium atoms to several energy states, with
consequent decay to metastable states that nearly coincide with excited
states of the neon atoms. Therefore, owing to atomic collisions, energy can
be transferred between the two different atoms and population inversion
can be induced between couples of Ne levels, as shown in Fig. 6.28, which
presents the He–Ne energy level scheme. The most-exploited emission
occurs at $\lambda = 632.8$ nm, but other laser lines are also available.

Indeed, gas lasers are based on atomic or molecular sharp levels, and
therefore they are not suitable as tunable sources. However, the oscillation
condition may be reached at the same time by different couples of electronic
levels, and thus lasing occurs at different wavelengths.

This occurs, for instance, in the argon laser, where Ar^+ ions provide
the lasing levels. In this case, under CW operation, we have emission
at different wavelengths. They can be separated by a dispersive element
outside the cavity, but usually tuning among the different laser lines is
realized including in the cavity a dispersive device such as a simple glass
prism. The strongest emission of the argon laser is at 514.5 nm, but there
is a significant emission at least on the following wavelengths: 501.7 nm,
496.5 nm, 488.0 nm, 476.5 nm, 457.9 nm. It is quite interesting that the same
laser resonator can be used for a different noble gas as active medium, such
as krypton. In this case, the strongest line is at 647.1 nm, but significant
emission occurs also at 676.4 nm, 530.9 nm, 568.2 nm and 413.1 nm.

A number of different gas lasers have been developed for a wide
spectrum of applications, such as CO_2, copper vapor and excimer lasers.
These are based on "excited dimers," i.e. molecules made by two subunits
that can be linked only in the excited state when one unit is made by a

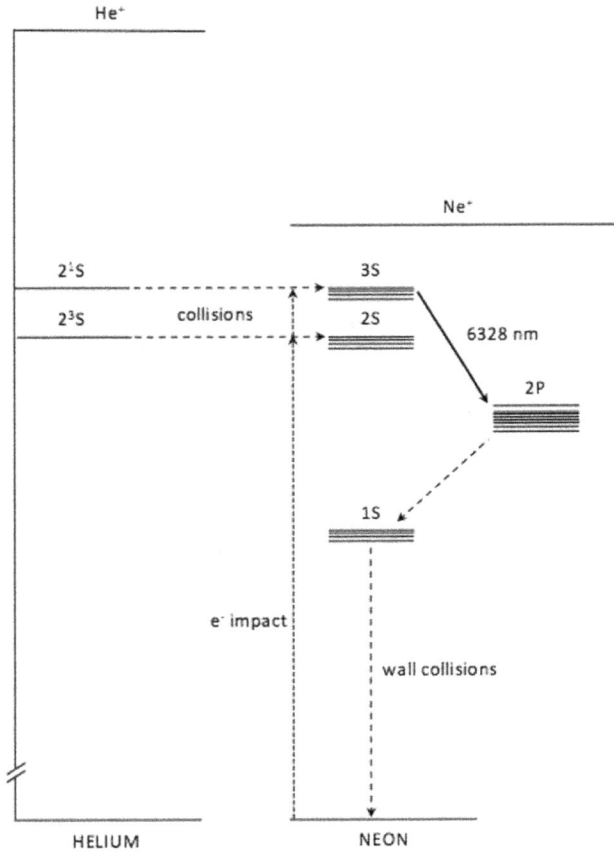

Fig. 6.28.   Energy levels involved in the operation of a He–Ne laser. Only the main laser line is indicated. [Adapted from: W.R. Bennet, *App. Opt.* **1(S1)**, 24 (1962).]

noble gas. Typical excimer molecules are ArF, KrF, XeCl, XeF and KrCl. The emission of the excimer laser is usually in the ultraviolet region of the electromagnetic spectrum, as shown in Table 6.2, where the characteristics of some common gas lasers are summarized.

Dye lasers

These lasers are based on solutions of organic dye molecules in different kinds of liquids, most frequently ethyl alcohol or ethylene glycol. These molecules have a large molecular weight ranging from about 200 to 1000 and show a broad emission spectrum which makes them attractive for the

Table 6.2. Basic characteristics and typical operation regimes of some common gas lasers.

Active medium	Pumping	Regime [CW/pulsed (s)]	λ (nm)
He–Ne	Electric discharge	CW	632.8
Ar$^+$	Electric discharge	CW/10^{-10}	514.5 (multilines)
Kr$^+$	Electric discharge	CW	647.1 (multilines)
CO$_2$	Electric discharge	CW/10^{-8}	10,600/9600
ArF	Electric discharge	10^{-8}	193
KrF	Electric discharge	10^{-8}	248
XeF	Electric discharge	10^{-8}	351

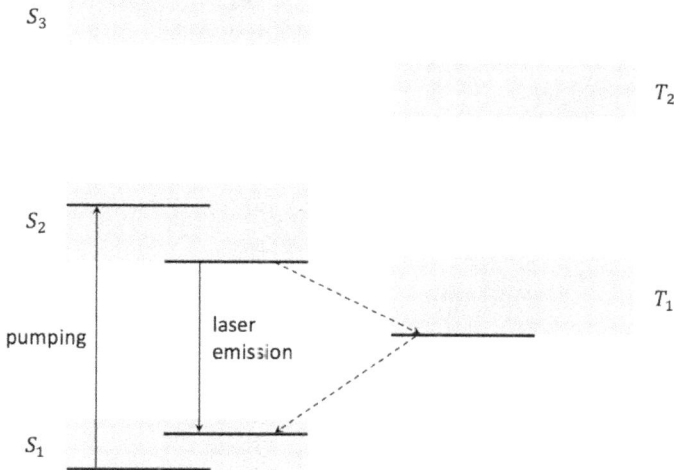

Fig. 6.29. Scheme of the energy levels of a dye molecule. The gray ranges are a simplified sketch of the existing rotational and vibrational energy levels of the molecule, as discussed in Chapter 3 (see Fig. 3.12). The four-level system involves the singlet states S_1 and S_2. The additional energy levels T_1, T_2 corresponding to triplet states are also shown. The dashed arrows indicate undesired transitions that reduce the laser efficiency.

tunable laser or for ultrashort pulse generation. Pumping is accomplished by optical excitation using either white lamps or other laser beams, depending on the cavity design and the working conditions.

A scheme of the energy level of this type of molecules is shown in Fig. 6.29. A ground electronic state S_1 includes a quasicontinuum of rotovibrational states. From there the electron can be excited to the singlet state S_2, where the spin of the excited electron is opposite to the spin of the remaining molecules. Triplet excited states T_1, T_2, where the spin of the excited electron is parallel to the spin of the remaining molecules, can play a role in the process. In fact, after absorption that occurs fulfilling the selection rule $\Delta l = \pm 1$ for the rotational quantum number, the electron relaxes to the lowest rotovibrational state in S_2, and then it can emit a photon and come back to the ground state or make a nonradiative transition to the triplet state T_1 and come back to S_1 slowly (on the microsecond time scale) without photon emission. Higher energy excited states like S_3 and T_2 can further subtract pumping energy by allowing additional electron excitation.[9] Of course, to make this system efficient, one should avoid all the ways of excitation/de-excitation except the one directly connecting S_1 to S_2.

In order to reduce the undesired effects of energy transfer to the triplet state or reabsorption from the excited state, optimization of the concentration of molecules in solution is required and recirculation or mixing is usually necessary.

This energy level distribution gives rise to a broad absorption and emission band. A typical example is shown in Fig. 6.30. Since the emission band has a strong homogeneous broadening character, such a

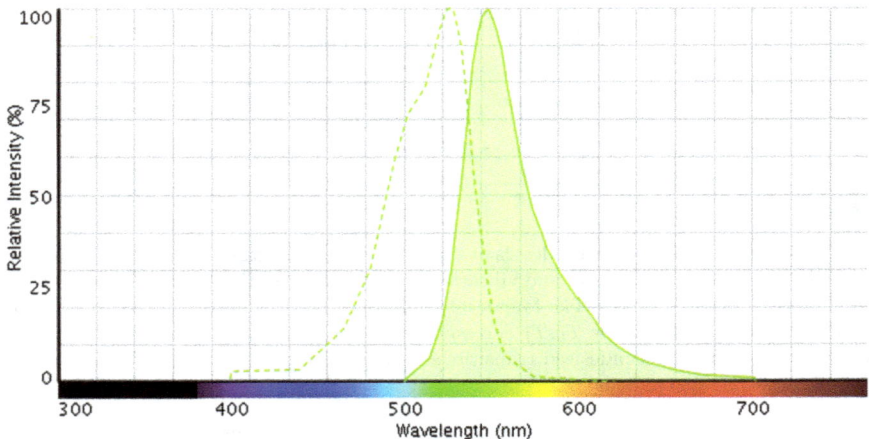

Fig. 6.30.   The absorption and emission band of a typical dye molecule (rhodamine 6G). The signal S is reported in arbitrary units versus the light wavelength. [Courtesy of ThermoFisher Scientific; www.thermofisher.com.]

---

[9]For a detailed description of the process, see also: B. Snavely, *Proc. IEEE*, **57**, 1374 (1969).

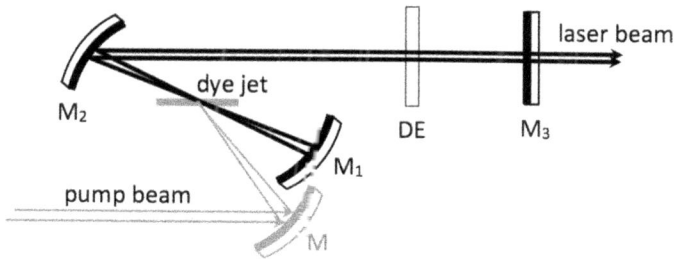

Fig. 6.31. Typical cavity design for a CW-laser-pumped dye laser. A laser beam is used to get optical pumping of the dye solution appearing as a fluid jet (normal to the scheme plane). This is a three-mirror, cavity (M_1, M_2, M_3) with the inclusion of a dispersive element (DE) necessary for tuning the laser emission.

large linewidth becomes the ideal condition for realizing tunable lasers. Additionally, owing to the liquid nature of the active medium, tunability can be extended to a spectrum larger than the one available from a single dye just by changing the dye solution in the same optical resonator.

A typical design for a CW dye laser is shown in Fig. 6.31. In this case, we have a three-mirror cavity with the inclusion of a dispersive device for wavelength tuning. The dye is flowing through a nozzle to have laminar flow producing a stable thin jet in the area of focusing of the laser pump beam.

By adjusting the mirrors' orientation it is possible to find the oscillation condition, and by acting on the dispersive device it is possible to tune the laser emission within the luminescence line shape. Of course, owing to the Stokes shift, laser emission always occurs at a wavelength longer than that of the pump laser. As already outlined, by changing the flowing solution it will be possible to change the range of tuning in such a way that using a limited number of different dyes it is possible to tune the emission over the whole visible spectrum, when pumping with a wavelength short enough. The emission spectra of different dyes are shown in Fig. 6.32.

Since short-pulse generation requires many phase-locked modes, a broad gain spectrum is the basic condition for obtaining a femtosecond pulse train. In Fig. 6.24, we have shown an example of a ring dye laser, which is able to generate ultrashort laser pulses of the order of 10^{-14}s. The cavity has two main blocks: one concerns the amplifier dye jet (mirrors M_0, M_1, M_2) and the other concerns the saturable absorber (mirrors M_3, M_4), used to realize the required mode locking. The role of the intracavity glass prism is to adjust the wavelength dispersion in order to minimize the pulse length.

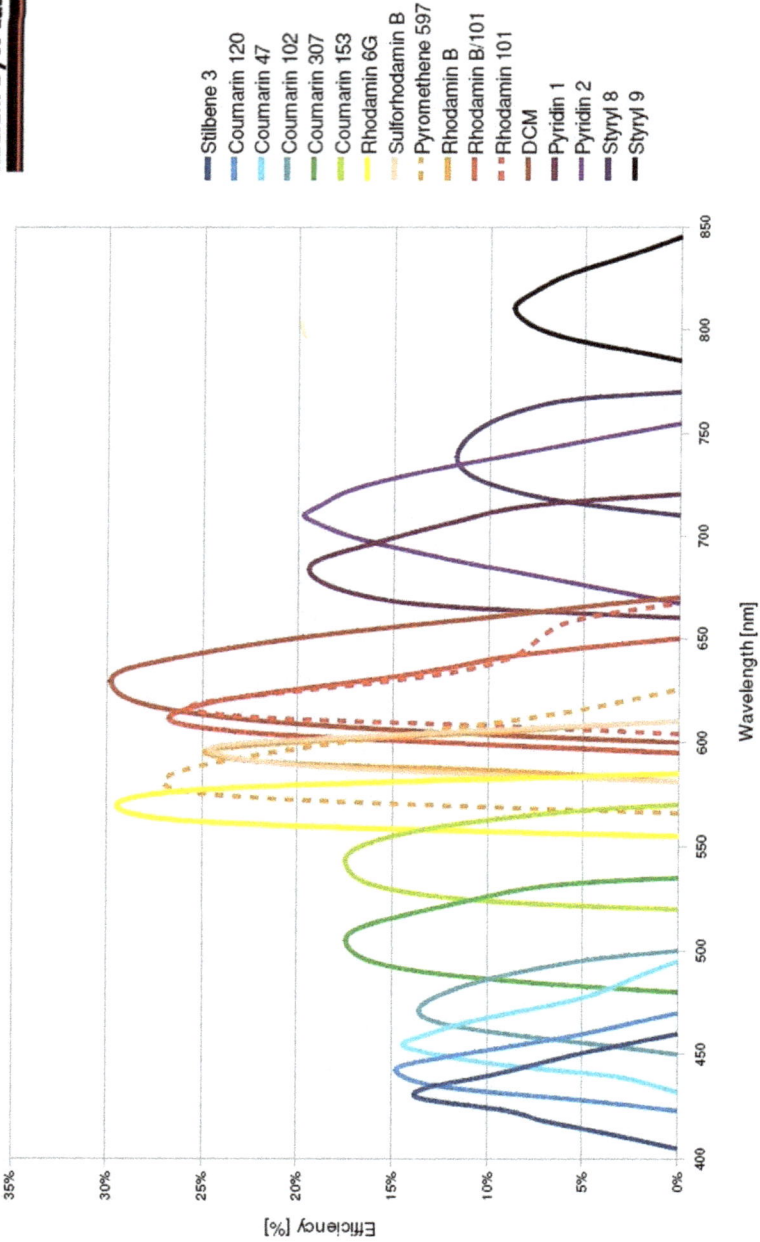

Fig. 6.32. Emission spectra of different dyes covering the whole visible range (courtesy of Radiant Dye Laser; http://www.radiant-dyes.com.)

A specific class of dye lasers concerns the ones realized under microfluidic conditions, i.e. when the active solution flows through microchannels (a cross section of the order of $10^2 \, \mu m^2$) and the optical resonator is embedded in a small chip with a cavity length in the range of 0.01–2 mm. These systems, called *optofluidic lasers*, are developed for the lab-on-a-chip (LOC) technology, with the aim of building a complete optical analysis laboratory on an $\sim 1 \, cm^2$ chip for application in biosensing and environment and security control.

An example of an optofluidic laser embedded in glass is shown in Fig. IN6.2.

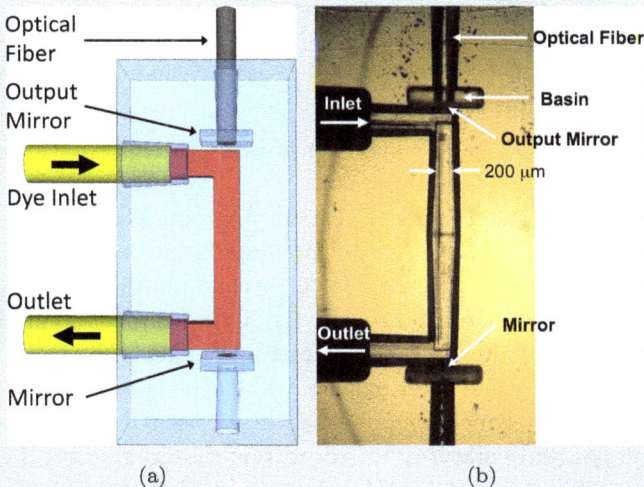

Fig. IN6.2. Optofluidic laser cavity embedded in a glass chip: (a) design of the device; (b) microscope picture of the fabricated laser. The laser dye flows in a channel of $200 \, \mu m$ diameter. The laser output is collected by an optical fiber and the indicated basins are realized to fabricate mirrors through inkjet printing [see also F. Simoni et al., *Opt. Exp.* **24**, 17416 (2016)].

Semiconductor lasers

In Section 6.2, we have described the specific mechanism that can lead to population inversion in semiconductors exploiting the properties of a p–n junction. This is a consequence of the peculiarity of these systems with respect to other laser media, due to the role of the energy band structure, whose occupation is determined by the Fermi–Dirac distribution function.

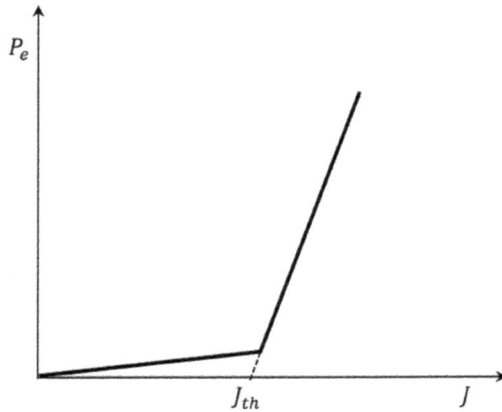

Fig. 6.33.   The output power of a semiconductor laser versus the input current density.

As a result, the optical transition occurs between two distributions of energy levels.

As already discussed, in these systems pumping is achieved by current injection able to produce the necessary population inversion. Hence, for semiconductor lasers a curve similar to the one reported in Fig. 6.12 can be drawn where the laser output power is plotted versus the input current density rather than the input pump power. A clear change in the slope above the threshold value J_{th} is expected when the laser action takes place.

Many different semiconductor lasers have been developed where the specific emission wavelength is determined by the existing band gap in the semiconducting material. Direct band gap materials are usually exploited for their higher emission efficiency, so that many semiconductor lasers are made of III–V compounds (combination of elements of the third group with elements of the fifth group of the periodic table). In this case, the emission wavelength is in the red–near-infrared region of the spectrum (600–1600 nm).

The wavelength emission at 1550 nm is the most interesting for optical communications and it is realized with InGaAsP or InGaAsNSb lasers. The traditional CD and DVD technologies were based on AlGaInP or GaInP lasers emitting, in the range of 635–660 nm, while the realization of GaN lasers emitting at the range 405 nm allowed developing the Blu-ray technology for increasing the storage capacity of optical disks. However, semiconductor lasers with the emission wavelength in the green range of the visible spectrum have also been developed.

Working operation can span a broad range of pulse durations from c.w. to picoseconds, tailored to the specific application of the device.

The actual cavity configurations of semiconductor lasers are also many in number. We recall here the traditional ones: the homojunction laser, where the effect is based on a single p–n junction, and the heterojunction laser, where two p–n junctions are exploited. In the latter case, the materials' refractive indices are chosen in order to realize a guiding condition in the active region so as to increase the efficiency of the emission process. Sketches of these configurations are shown in Figs. 6.34 and 6.35.

Optical cavities usually allow emission from the edge with the laser beam parallel to the surface of the junction; however, in another configuration, emission occurs from the top surface of the device, leading to the so-called vertical cavity surface-emitting laser (VCSEL).

In the conventional case, the resonator is realized either by coating reflectors on the edge of the semiconductors or by exploiting the reflection properties of a Bragg grating in the distributed feedback (DFB) or distributed reflector (DBR) configuration.

The DFB configuration is realized by recording a periodic modulation of the refractive index on one side of the active region. In this way, the wave traveling in the active region is affected by the Bragg grating acting in the reflecting mode. The wavelength fulfilling the Bragg condition is reflected

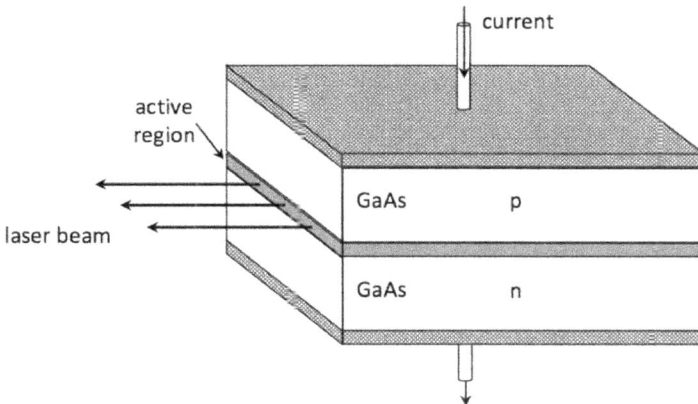

Fig. 6.34. Scheme of a semiconductor laser based on the homojunction, such as the GaAs laser: the active region is achieved at the junction between two semiconductor crystals of the same material with opposite doping (p–n). Pumping is realized by current injection.

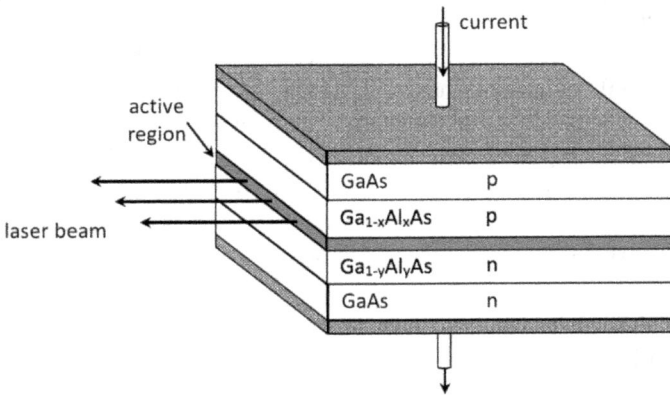

Fig. 6.35. Scheme of a semiconductor laser based on the heterojunction, such as the GaAs–GaAlAs laser. This structure allows higher charge confinement with improvement of the laser efficiency.

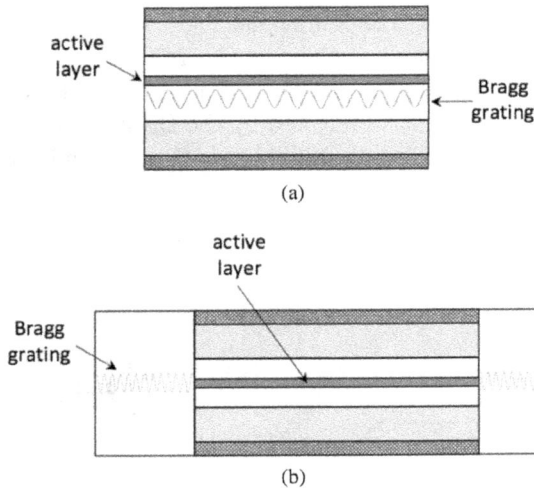

Fig. 6.36. Scheme of a semiconductor laser resonator realized by DFB (a) or DBR (b).

back and forth along the active layer and can reach the oscillation condition. In the DBR configuration, the Bragg reflecting gratings act as narrow bandwith mirrors, selecting the proper wavelength which will experience low cavity losses and will be able to achieve the threshold gain.

In Fig. 6.36, sketches of the DFB and DBR resonators are given.

Appendix

A. Physical Constants and Unit Systems

A.1. *Table of constants*

Mass of electron	m_e	$9.11 \cdot 10^{-31}$ kg
Mass of proton	m_p	$1.673 \cdot 10^{-27}$ kg
Mass of neutron	m_n	$1.675 \cdot 10^{-27}$ kg
Atomic mass unit	u	$1.66 \cdot 10^{-27}$ kg
Charge of electron	e	$1.602 \cdot 10^{-19}$ C $= 4.806 \cdot 10^{-10}$ esu
Velocity of light in vacuum	c	$2.998 \cdot 10^8$ m/s $= 2.998 \cdot 10^{10}$ cm/s
Bohr radius	a	$0.529 \cdot 10^{-10}$ m $= 0.529$ Å
Planck constant	h	$6.626 \cdot 10^{-34}$ J $\cdot$ s $= 4.136 \cdot 10^{-15}$ eV·s
	$\hbar = h/2\pi$	$1.055 \cdot 10^{-34}$ J $\cdot$ s $= 0.658 \cdot 10^{-15}$ eV·s
Avogadro number	N_A	$6.022 \cdot 10^{23}$ atoms/mol
Boltzmann constant	k_B	$1.381 \cdot 10^{-23}$ J/K $= 8.616 \cdot 10^{-5}$ eV/K
Gas constant	R	8.314 J/mol·K
Bohr magneton	μ_B	$9.274 \cdot 10^{-24}$ A·m^2
Dielectric permittivity of	ϵ_0	$8.854 \cdot 10^{-12}$ F/m
vacuum	$k = 1/4\pi\epsilon_0$	$8.988 \cdot 10^9$ N·m^2/C^2
Magnetic permeability of	μ_0	$1.257 \cdot 10^{-6}$ H/m
vacuum		
Universal gravitational constant	G	$6.674 \cdot 10^{-11}$ m^3 kg^{-1}· s^{-2}

A.2. *Unit systems*

The *International System of Units* (*SI*) is based on four base units related to the corresponding physical quantities: meter (*length*), kilogram (*mass*), second (*time*), ampere (*electrical current*). Additional associate units are: Kelvin (*temperature*), mole (*amount of substance*), candela (*luminous intensity*).

Other units systems uses three base units: centimeter (*length*), gram (*mass*), second (*time*), and fix different units for the electromagnetic physical quantities. The *electrostatic system of units* (*esu*) fix

$$\frac{1}{4\pi\epsilon_0} = 1 \quad \text{then } \mu_0 = \frac{4\pi}{c^2}$$

and the *electromagnetic system of units* (*emu*) fix

$$\frac{1}{4\pi\mu_0} = 1 \quad \text{then } \epsilon_0 = \frac{4\pi}{c^2}.$$

The *Gaussian system of units* uses the *esu* units for the electrostatic quantities and the *emu* units for the magnetostatic quantities. The connection between electric and magnetic quantities is given by the Faraday-Neumann law of induction, where it is necessary to take into account that if the electric field E is measured in *esu* the corresponding value in *emu* is cE; as a consequence several equations must be properly modified. In this system the dielectric and magnetic permittivities of vacuum do not appear any more, so that the dielectric and magnetic permittivity of a material correspond to the quantities ϵ_r and μ_r.

Some useful formulae in the two systems

	SI	Gauss
Coulomb force	$F = \dfrac{1}{4\pi\epsilon_0}\dfrac{q_1 q_2}{r^2}$	$F = \dfrac{q_1 q_2}{r^2}$
Coulomb potential	$\mathcal{V} = \dfrac{1}{4\pi\epsilon_0}\dfrac{q_1 q_2}{r}$	$\mathcal{V} = \dfrac{q_1 q_2}{r}$
e.m energy density	$u = \epsilon E^2$	$u = \dfrac{\epsilon E^2}{4\pi}$
Polarization vector	$\mathbf{P} = \epsilon_0 \chi \mathbf{E}$	$\mathbf{P} = \chi \mathbf{E}$
Dielectric permittivity	$\epsilon = \epsilon_0 \epsilon_r$	$\epsilon = \epsilon_r$
Phase velocity	$\dfrac{c}{\sqrt{\epsilon_r \mu_r}}$	$\dfrac{c}{\sqrt{\epsilon \mu}}$
Force on charge	$\mathbf{F} = q(\mathbf{E} + \mathbf{v} \times \mathbf{B})$	$\mathbf{F} = q\left(\mathbf{E} + \dfrac{\mathbf{v}}{c} \times \mathbf{B}\right)$
Field relationship	$\lvert\mathbf{E}\rvert = \dfrac{c}{\sqrt{\epsilon_r \mu_r}}\lvert\mathbf{B}\rvert$	$\lvert\mathbf{E}\rvert = \lvert\mathbf{B}\rvert$
Poynting vector	$\mathbf{S} = \mathbf{E} \times \mathbf{B}$	$\mathbf{S} = \dfrac{c}{4\pi}\mathbf{E} \times \mathbf{B}$

A.3. *Useful unit conversions*

Time	Microsecond	$1\ \mu s = 10^{-6}$ m	Second
Time	Nanosecond	$1\ ns = 10^{-9}$ m	Second
Time	Picosecond	$1\ ps = 10^{-12}$ m	Second
Time	Femtosecond	$1\ fs = 10^{-15}$ m	Second
Length	Micrometer	$1\ \mu m = 10^{-6}$ m	Meter
Length	Nanometer	$1\ nm = 10^{-9}$ m	Meter
Length	Ångstrom	$1\ \text{Å} = 10^{-10}$ m	Meter
Mass	Gram	$1\ g = 10^{-3}$ kg	Kilogram
Force	Dyne	$1\ dyn = 10^{-5}$ N	Newton
Pressure	Pascal	$1\ \text{Pa} = 1\ N/m^2 = 10\ dyn/cm^2$	Newton/(meter)^2
Pressure	Torr	$1\ torr = 133.3$ Pa	Pascal
Pressure	Atmosphere	$1\ atm = 1.013 \cdot 10^5$ Pa	Pascal
Energy	Erg	$1\ erg = 10^{-7}$ J	Joule
Energy	Electron volt	$1\ eV = 1.602 \cdot 10^{-19}$ J	Joule
Energy	Calorie	$1\ cal = 4.186$ J	Joule
Power	Watt	$1\ W = 1\ J/s = 10^7\ erg/s$	Joule/second
Charge	esu	$1\ esu = 0.333 \cdot 10^{-9}$ C	Coulomb
Electric potential	esu	$1\ esu = 3 \cdot 10^2$ V	Volt
Electric field	esu	$1\ esu = 3 \cdot 10^4$ V/m	Volt/meter
Dipole moment	Debye	$1\ D = 3.336 \cdot 10^{-30}\ C \cdot m$	Coulomb $\cdot$ meter
Electric current	Ampere	$1\ A = 1\ C/s = 3 \cdot 10^9$ esu	Coulomb/second
Electric resistance	Ohm	$1\ \Omega = 1\ V/A = 0.111 \cdot 10^{-11}$ esu	Volt/ampere
Magnetic induction	Gauss	$1\ G = 10^{-4}$ T	Tesla
Temperature	Degree Celsius	$0°C = 273.15$ K	Degree Kelvin

A.4. *Wave parameters*

Period:

$$T = \frac{1}{\nu} = \frac{2\pi}{\omega} = \frac{\lambda}{c};$$

Frequency:

$$\nu = \frac{1}{T} = \frac{\omega}{2\pi} = \frac{c}{\lambda} = c\tilde{\nu} \quad \text{or} \quad \omega = \frac{2\pi}{T} = 2\pi\nu = 2\pi\frac{c}{\lambda};$$

Wavelength:

$$\lambda = cT = \frac{c}{\nu} = \frac{1}{\tilde{\nu}} = 2\pi\frac{c}{\omega};$$

Wave vector:

$$k = \frac{2\pi}{\lambda} = \frac{\omega}{c} = \frac{2\pi\nu}{c} = 2\pi\tilde{\nu};$$

Wave number:

$$\tilde{\nu} = \frac{1}{\lambda} = \frac{\nu}{c} = \frac{k}{2\pi} = \frac{\omega}{2\pi c}.$$

These expressions concern wave propagation in vacuum. For propagation in a medium having a refractive index n, the following substitutions must be done:

$$c \to \frac{c}{n}, \quad \lambda \to \frac{\lambda}{n}.$$

A.5. *Photon parameters*

$Energy: \quad h\nu = \dfrac{hc}{\lambda}; \quad momentum: p = \hbar k; \quad angular\ momentum: \boldsymbol{\ell} = \pm\hbar\hat{\mathbf{k}}.$

Useful expression to evaluate the photon energy from wavelength:

$$h\nu = \frac{hc}{\lambda} = \frac{1.986 \cdot 10^{-26}(J \cdot m)}{\lambda(m)} = \frac{1.24 \cdot 10^4(eV \cdot \text{Å})}{\lambda(\text{Å})}$$

A.6. *Conversion from mass to number of molecules*

Given:

M — *mass of substance* (g)

A — *atomic/molecular weight of substance* (given in the unit of g)

$\dfrac{M}{A}$ *is the number of moles*

the number N of atoms/molecules included in the mass M is obtained multiplying the number of moles by the Avogadro number N_A, that is the number of atoms/molecules per mole:

$$N = N_A \frac{M}{A}.$$

B. Vector Fields and Operators

B.1. *Products of vectors*

Given two vectors: $\mathbf{a} = a_x\hat{\mathbf{x}} + a_y\hat{\mathbf{y}} + a_z\hat{\mathbf{z}}$; $\mathbf{b} = b_x\hat{\mathbf{x}} + b_y\hat{\mathbf{y}} + b_z\hat{\mathbf{z}}$, the following definitions hold:

Scalar product:

$$\mathbf{a} \cdot \mathbf{b} = a_x b_x + a_y b_y + a_z b_z$$

Vector product:

$$\mathbf{a} \times \mathbf{b} = \begin{bmatrix} \hat{\mathbf{x}} & \hat{\mathbf{y}} & \hat{\mathbf{z}} \\ a_x & a_y & a_z \\ b_x & b_y & b_z \end{bmatrix}$$

$$= (a_y b_z - a_z b_y)\hat{\mathbf{x}} + (a_z b_x - a_x b_z)\hat{\mathbf{y}} + (a_x b_y - a_y b_x)\hat{\mathbf{z}}$$

According to it, the triple products with a third vector $\mathbf{c} = c_x\hat{\mathbf{x}} + c_y\hat{\mathbf{y}} + c_z\hat{\mathbf{z}}$ fulfill the following properties:

$$\mathbf{a} \times \mathbf{b} \cdot \mathbf{c} = \begin{bmatrix} a_x & a_y & a_z \\ b_x & b_y & b_z \\ c_x & c_y & c_z \end{bmatrix}$$

$$= (b_y c_z - b_z c_y)a_x + (b_z c_x - b_x c_z)a_y + (b_x c_y - b_y c_x)a_z$$

$$\mathbf{a} \times (\mathbf{b} \times \mathbf{c}) = (\mathbf{a} \cdot \mathbf{c})\mathbf{b} - (\mathbf{a} \cdot \mathbf{b})\mathbf{c}$$

B.2. *Differential operators*

$$nabla \quad \nabla = \hat{\mathbf{x}}\frac{\partial}{\partial x} + \hat{\mathbf{y}}\frac{\partial}{\partial y} + \hat{\mathbf{z}}\frac{\partial}{\partial z}$$

$$Laplacian \quad \nabla^2 = \frac{\partial^2}{\partial x^2} + \frac{\partial^2}{\partial y^2} + \frac{\partial^2}{\partial z^2}$$

Given a scalar function $f(x, y, z)$, the following identity holds:

$$\nabla f = \hat{\mathbf{x}}\frac{\partial f}{\partial x} + \hat{\mathbf{y}}\frac{\partial f}{\partial y} + \hat{\mathbf{z}}\frac{\partial f}{\partial z} \quad (grad\ f)$$

Given a vector field $\mathbf{v} = v_x\hat{\mathbf{x}} + v_y\hat{\mathbf{y}} + v_z\hat{\mathbf{z}}$, the following identities hold:

$$\nabla \cdot \mathbf{v} = \frac{\partial v_x}{\partial x} + \frac{\partial v_y}{\partial y} + \frac{\partial v_z}{\partial z} \quad (div\ \mathbf{v})$$

$$\nabla \times \mathbf{v} = \begin{bmatrix} \hat{\mathbf{x}} & \hat{\mathbf{y}} & \hat{\mathbf{z}} \\ \frac{\partial}{\partial x} & \frac{\partial}{\partial y} & \frac{\partial}{\partial z} \\ v_x & v_y & v_z \end{bmatrix}$$

$$= \left(\frac{\partial v_z}{\partial y} - \frac{\partial v_y}{\partial z}\right)\hat{\mathbf{x}} + \left(\frac{\partial v_x}{\partial z} - \frac{\partial v_z}{\partial x}\right)\hat{\mathbf{y}} + \left(\frac{\partial v_y}{\partial x} - \frac{\partial v_x}{\partial y}\right)\hat{\mathbf{z}} \quad (rot\ \mathbf{v})$$

Considering scalar functions $f(x, y, z)$ and $g(x, y, z)$ and vector fields $\mathbf{a} = a_x\hat{\mathbf{x}} + a_y\hat{\mathbf{y}} + a_z\hat{\mathbf{z}}$ and $\mathbf{b} = b_x\hat{\mathbf{x}} + b_y\hat{\mathbf{y}} + b_z\hat{\mathbf{z}}$, the following properties hold:

$$\nabla \cdot (f\mathbf{a}) = f(\nabla \cdot \mathbf{a}) + \mathbf{a} \cdot (\nabla f)$$

$$\nabla \cdot (\mathbf{a} \times \mathbf{b}) = \mathbf{b} \cdot (\nabla \times \mathbf{a}) - \mathbf{a} \cdot (\nabla \times \mathbf{b})$$

$$\nabla \times (\nabla \times \mathbf{a}) = \nabla(\nabla \cdot \mathbf{a}) - \nabla^2\mathbf{a}$$

$$\nabla \times (\nabla f) \equiv 0$$

$$\nabla \cdot (\nabla \times \mathbf{a}) \equiv 0$$

$$\nabla^2(fg) = f\nabla^2(g) + 2(\nabla f) \cdot (\nabla g) + g\nabla^2(f)$$

Bibliography

Books

M. Alonso and E.J. Finn, *Fundamental University Physics*, Vol. III, Quantum and Statistical Physics (Addison–Wesley, 1968).

N.W. Ashcroft and N.D. Mermin. *Solid State Physics* (Harcourt, 1976).

N. Bloembergen, *Nonlinear Optics*, 4th ed. (World Scientific, 1996).

M.L. Boas, *Mathematical Methods in the Physical Sciences*, 2nd ed. (Wiley & Sons, 1983).

M. Born and E. Wolf, *Principles of Optics*, 6th ed. (Pergamon, 1980).

B.H. Bransden and C.J. Joachin, *Physics of Atoms and Molecules* (Longman, 1983).

L. Brillouin, *Wave Propagation and Group Velocity* (Academic, 1960).

R.H. Bube, *Electrons in Solids: An Introductory Survey* (Academic, 1981).

P.A.M. Dirac, *The Principles of Quantum Mechanics* (Clarendon, 1958).

M.S. Dresselhous, *Solid State Physics* MIT Lecture Notes (2001).

F. Duan and J. Guojun, *Introduction to Condensed Matter Physics* (World Scientific, 2005).

P.A. Egelstaff, *An Introduction to the Liquid State*, 2nd ed. (Oxford University Press, 1950).

R. Eisberg and R. Resnik, *Quantum Physics of Atoms, Molecules, Solids, Nuclei and Particles*, 2nd ed. (Wiley & Sons, 1985).

I.S. Gradshteyn and I.M. Ryzhik, *Table of Integrals, Series, and Products* (Academic, 1980).

H.T. Grahn, *Introduction to Semiconductor Physics* (World Scientific, 1999).

D.J. Griffiths, *The Principles of Quantum Mechanics*, 2nd ed. (Prentice Hall, 2005).

P. Hariharan, *Optical Interferometry*, 2nd ed. (Elsevier, 2003).

G. Herzberg, *Molecular Spectra and Molecular Structure* I. Spectra of Diatomic Molecules (D. Van Nostrand, 1950).

J.R. Hook and H.E. Hall, *Solid State Physics*, 2nd ed. (Wiley & Sons, 1995).

R.E. Hummel, *Electronic Properties of Materials*, 4th ed. (Springer, 2010).

M. Karplus and R.N. Porter, *Atoms & Molecules* (Benjamin, 1970).

I.C. Khoo and F. Simoni (eds.), *Physics of Liquid Crystalline Materials* (Gordon & Breach, 1991).

C. Kittel, *Introduction to Solid State Physics*, 8th ed. (Wiley & Sons, 2005).

S.A. Main, *Plasmonics: Fundamentals and Applications* (Springer, 2007).

A.C. Phillips, *Introduction to Quantum Mechanics* (Wiley & Sons, 2003).

G.A. Reider, *Photonics* (Springer, 2016).

G.O. Reynolds, J.B. DeVelis, G.B. Parrent, Jr. and B.J. Thompson, *The New Physical Optics Notebook: Tutorials in Fourier Optics* (SPIE Optical Engineering Press, 1989).

D. Röss, *Laser Light Amplifiers and Oscillators* (Academic, 1969).

B.E.A. Saleh and M.C. Teich, *Fundamental of Photonics* (Wiley & Sons, 1991).

F.P. Schäfer, *Dye Lasers*, 3rd ed. (Springer, 1990).

L.I. Schiff, *Quantum Mechanics* (McGraw–Hill, 1955).

E.F. Schubert, *Light-Emitting Diodes*, 2nd ed. (Cambridge University Press, 2006).

A.E. Siegman, *Lasers* (University Science, 1986).

W.T. Silfvast, *Laser Fundamentals*, 2nd ed. (Cambridge University Press, 2004).

J.H. Simmons and K.S. Potter, *Optical Materials* (Academic, 2000).

F. Simoni, *Nonlinear Optical Properties of Liquid Crystals and PDLC* (World Scientific, 1997).

A. Sommerfeld, *Thermodynamics and Statistical Mechanics* (Academic, 1964).

O. Svelto, *Principles of Lasers*, 5th ed. (Springer, 2010).

P.A. Tipler and G. Mosca, *Physics for Scientists and Engineers* Vol. 3, Modern Physics (W.H. Freeman, 2008).

W.J. Tropf, M.E. Thomas and T.J. Harris, *Handbook of Optics*, Vol. IV, Chap. 33, Optical and Physical Properties of Materials, 3rd ed. (McGraw–Hill, 2009).

J.T. Verdeyen, *Laser Electronics* (Prentice Hall, 1981).

H.E. White, *Introduction to Atomic Spectra* (McGraw–Hill, 1934).

F. Wooten, *Optical Properties of Solids* (Academic, 1972).

A. Yariv, *Optical Electronics*, 4th ed. (Saunders College Pub., 1991).

A. Yariv, *Quantum Electronics*, 3rd ed. (Wiley & Sons, 1989).

J.M. Ziman, *Principles of the Theory of Solids* (Cambridege University Press, 1964).

Complete citation of papers referred to in the book

W.R. Bennet, Gaseous optical masers, *Appl. Opt.* **1**(S1), 24 (1962).

J.F.H. Custers and F.A. Raal, Fundamental absorption edge of diamond, *Nature* **179**, 268 (1957).

C. Davisson and L.H. Germer, Diffraction of electrons by a crystal of nickel, *Phys. Rev.* **30**, 705 (1927).

J.C. Diels, J.J. Fontaine, I.C. McMichael and F. Simoni, Control and measurement of ultrashort pulse shapes (in amplitude and phase) with femtosecond accuracy, *Appl. Opt.* **24**, 1270 (1985).

A. Einstein, Über die von der molekularkinetischen Theorie der Wärme geforderte Bewegung von in ruhenden Flüssigkeiten suspendierten Teilchen [Investigations on the Theory of the Brownian Movement (Dover, (1956)], *Annalen der Physik* **17**, 549 (1905).

R.H. French, Electronic Band Structure of Al_2O_3, with comparison to Alon and AlN, *J. Am. Ceram. Soc.* **73**, 477 (1990).

W. Gerlach and O. Stern, *Der experimentelle Nachweis der Richtungsquantelung im Magnetfeld*, Z. Phys. **9**, 349 (1922).

R.A. Heaton and C.C. Lin, Electronic energy-band structure of the calcium fluoride crystal, *Phys. Rev. B* **22**, 3629 (1980).

H. Kogelnik and T. Li, Laser Beams and Resonators, *Appl. Opt.* **5**, 1550 (1966).

J. Li, Doctoral thesis, University of Central Florida (2005); http://stars.library.ucf.edu/etd.

D.K.C. Mac Donald and K. Mendelssohn, Resistivity of pure metals at low temperatures, I. The alkali metals, *Proc. R. Soc.* **A202**, 103 (1950).

W. Maier and A. Saupe, Eine einfache molekulare Theorie desnematischen kristallinflüssigen Zustandes, *Z. Naturforsch.* **A13**, 564 (1958).

W. Maier and A. Saupe, Eine einfache molekular-statistische Theorie der nematischen kristallinflüssigen Phase. Teil I, *Z. Naturforsch.* **A14**, 882 (1959).

W. Maier and A. Saupe, Eine einfache molekular-statistische Theorie der nematischen kristallinflüssigen Phase. Teil II, *Z. Naturforsch.* **A15**, 287 (1960).

T.H. Maiman, Optical and microwave optical experiments in ruby, *Phys. Rev. Lett.* **4**, 564 (1960).

B.T. Matthias, T.H. Geballe and V.B. Compton, Superconductivity, *Rev. Mod. Phys.* **35**, 1 (1963).

H.B. Michaelson, The work function of the elements and its periodicity, *J. Appl. Phys.* **48**, 4729 (1977).

S.S. Nekrashevich and V.A. Gritsenko, Electronic structure of silicon dioxide (a review), *Phys. Solid State* **56**, 207 (2014).

W. Saslow, T.K. Bergstresser and M.L. Cohen, Band structure and optical properties of diamond, *Phys. Rev. Lett.* **16**, 354 (1966).

P. Schiebener and J. Straub, Refractive index of water and steam as function of wavelength, temperature and density, *J. Phys. Chem. Ref. Data* **19**, 677 (1990).

F. Simoni, S. Bonfadini, P. Spegni, S. Lo Turco, D.E. Lucchetta and L. Criante, Low threshold Fabry–Perot optofluidic resonator fabricated by femtosecond laser micromachining, *Opt. Exp.* **24**, 17416 (2016).

B. Snavely, Flashlamp-excited organic dye lasers, *Proc. IEEE* **57**, 1374 (1969).

M.D. Sturge, Optical absorption of gallium arsenide between 0.6 and 2.75 eV, *Phys. Rev.* **127**, 768 (1962).

E.A. Taft and H.R. Phillip, Optical constants of silver, *Phys. Rev.* **121**, 1100 (1961).

G.P. Thomson and A. Reid, Diffraction of cathode rays by a thin film, *Nature* **119**, 890 (1927).

Y.P. Varshni, Temperature dependence of the energy gap in semiconductors, *Physica* **34**, 149 (1967).

W.G. Wagner and B.A. Lengyel, Evolution of the giant pulse in a laser, *J. Appl. Phys.* **34**, 2040 (1963).

Index*

---

*Numbers in bold indicate this page onwards.